Springer Tracts in Modern Physics 87

W0246052

Springer Tracts in Modern Physics

66* **Quantum Statistics in Optics and Solid-State Physics**
With contributions by R. Graham, F. Haake

67* **Conformal Algebra in Space-Time and Operator Product Expansion**
By S. Ferrara, R. Gatto, A. F. Grillo

68* **Solid-State Physics** With contributions by D. Bäuerle, J. Behringer, D. Schmid

69* **Astrophysics** With contributions by G. Börner, J. Stewart, M. Walker

70* **Quantum Statistical Theories of Spontaneous Emission and their Relation to Other Approaches** By G. S. Agarwal

71 **Nuclear Physics** With contributions by J. S. Levinger, P. Singer, H. Überall

72 **Van der Waals Attraction:** Theory of Van der Waals Attraction By D. Langbein

73 **Excitons at High Density** Edited by H. Haken, S. Nikitine. With contributions by V. S. Bagaev, J. Biellmann, A. Bivas, J. Goll, M. Grosmann, J. B. Grun, H. Haken, E. Hanamura, R. Levy, H. Mahr, S. Nikitine, B. V. Novikov, E. I. Rashba, T. M. Rice, A. A. Rogachev, A. Schenzle, K. L. Shaklee

74 **Solid-State Physics** With contributions by G. Bauer, G. Borstel, H. J. Falge, A. Otto

75 **Light Scattering by Phonon-Polaritons** By R. Claus, L. Merten, J. Brandmüller

76 **Irreversible Properties of Type II Superconductors** By. H. Ullmaier

77 **Surface Physics** With contributions by K. Müller, P. Wißmann

78 **Solid-State Physics** With contributions by R. Dornhaus, G. Nimtz, W. Richter

79 **Elementary Particle Physics** With contributions by E. Paul, H. Rollnick, P. Stichel

80* **Neutron Physics** With contributions by L. Koester, A. Steyerl

81 **Point Defects in Metals I:** Introduction to the Theory 2nd Printing
By G. Leibfried, N. Breuer

82 **Electronic Structure of Noble Metals, and Polariton-Mediated Light Scattering**
With contributions by B. Bendow, B. Lengeler

83 **Electroproduction at Low Energy and Hadron Form Factors**
By E. Amaldi, S. P. Fubini, G. Furlan

84 **Collective Ion Acceleration** With contributions by C. L. Olson, U. Schumacher

85 **Solid Surface Physics** With contributions by J. Hölzl, F. K. Schulte, H. Wagner

86 **Electron-Positron Interactions** By B. H. Wiik, G. Wolf

87 **Point Defects in Metals II:** Dynamical Properties and Diffusion Controlled Reactions
With contributions by P. H. Dederichs, K. Schroeder, R. Zeller

88 **Excitation of Plasmons and Interband Transitions by Electrons**
By H. Raether

* denotes a volume which contains a Classified Index starting from Volume 36.

Point Defects in Metals II

Dynamical Properties and Diffusion Controlled Reactions

Contributions by
P. H. Dederichs K. Schroeder R. Zeller

With 91 Figures

Springer-Verlag
Berlin Heidelberg GmbH 1980

Professor Dr. Peter H. Dederichs
Dr. Rudolf Zeller
Dr. Kurt Schroeder

Institut für Festkörperforschung der Kernforschungsanlage Jülich GmbH,
Postfach 1913, D-5170 Jülich 1, Fed. Rep. of Germany

Manuscripts for publication should be addressed to:

Gerhard Höhler
Institut für Theoretische Kernphysik der Universität Karlsruhe
Postfach 6380, D-7500 Karlsruhe 1, Fed. Rep. of Germany

Proofs and all correspondence concerning papers in the process of publication should be addressed to:

Ernst A. Niekisch
Haubourdinstrasse 6, D-5170 Jülich 1, Fed. Rep. of Germany

ISBN 978-3-662-15396-3 ISBN 978-3-540-34769-9 (eBook)
DOI 10.1007/978-3-540-34769-9

Originally published by Springer-Verlag Berlin Heidelberg New York in 1980
Softcover reprint of the hardcover 1st edition 1980

2153/3130 — 5 4 3 2 1 0

To the late

Günther Leibfried and *Klaus Dettmann*

whose friendship and guidance will always be remembered

Contents

Dynamical Properties of Point Defects in Metals

By *P.H. Dederichs* and *R. Zeller*. With 66 Figures

1. Introduction and Survey 1

2. Dynamical Green's Functions of Ideal and Defect Lattices 3
   - 2.1 Equation of Motion and Eigenfunctions 3
   - 2.2 Real and Imaginary Part of Green's Function and its Connection with Correlation Functions 6
   - 2.3 The Local Frequency Spectrum 10
   - 2.4 Behaviour of $G(\omega)$ for High and Low Frequencies 13
   - 2.5 Asymptotic Expansion for Large Distances 16
   - 2.6 Singularities at Critical Points 18
   - 2.7 Analytical and Numerical Solutions for Cubic Ideal Lattices 22

3. Lattice with an Isolated Point Defect 28
   - 3.1 Scattering States and Localized States 28
   - 3.2 The Isotopic Defect 33
   - 3.3 Variational Method for Localized States 40

4. Description of Resonant and Localized Defect Vibrations 41
   - 4.1 Method of Krumhansl and Matthew 42
   - 4.2 Resonant Modes 45
   - 4.3 Effective Force Constant and Effective Mass 49
   - 4.4 Damping of Resonant Modes 54
   - 4.5 Localized Modes 57
   - 4.6 Resonance and Localized Modes near the Band Edge ω_{max} 61

5. Dynamics of Substitutional Defects 65
   - 5.1 Nearest Neighbour Model for Substitutional Defects 65
   - 5.2 Mössbauer Studies of Fe, Sn and Au Impurities 74
   - 5.3 Dynamical Behaviour of Vacancies 81

6. Vibrational Properties of Interstitials 83
   - 6.1 Vibrations of H in Metals 83
   - 6.2 Qualitative Explanation of the Dynamics of Self-interstitials 89

6.3 Octahedral Interstitial 91
6.4 100-Dumbbell Interstitial 97

7. Effects on Phonon Dispersion Curves 104
7.1 The Average Green's Function 105
7.2 Theory of Thermal Neutron Scattering 112
7.3 Change of Phonon Dispersion Curves 116
7.4 Effects of Different Scattering Lengths and Static Displacements .. 121
7.4.1 Effects of Different Scattering Lengths 121
7.4.2 Incoherent Scattering 123
7.4.3 Effects of Static Displacements 125
7.5 Results of Neutron Scattering Experiments 128
7.5.1 Effects due to Resonance Modes 128
7.5.2 Observation of Localized Modes 134
7.5.3 Incoherent Scattering Experiments 136
7.6 Change of Elastic Constants 138

8. Thermodynamic Properties 146
8.1 Free Energy of Defect Crystal 146
8.2 Properties of the Changed Density of States $\Delta z(\omega)$ 149
8.3 Change of the Specific Heat 157
8.4 Formation Entropy of Point Defects 160

References 165

Theory of Diffusion Controlled Reactions of Point Defects in Metals

By *K. Schroeder*. With 25 Figures

1. Introduction 171

2. Diffusion in Ideal Crystals 173
2.1 Lattice of Equilibrium Sites ("Diffusion Lattice") 176
2.2 Jump Frequency 179
2.3 Diffusion Equation and Green's Function 183
2.3.1 The Stationary Green's Function 186
2.3.2 Asymptotic Behaviour 188
2.3.3 Spectral Representation 188
2.3.4 Analytical Behaviour 189
2.3.5 Numerical Results 190
2.4 Experimental Methods 192

3. Interaction of Defects in Metals 194
3.1 Static Displacements Around Defects 194
3.2 Point Defect Interaction 196
3.3 Interaction of Point Defects with Dislocations 199
3.4 Short-Range Interaction 202

4. Diffusion in Force Fields 203
4.1 Derivation of Continuum Theory from Lattice Theory 205
4.1.1 Cubic Defects 205
4.1.2 Noncubic Defects 209
4.2 Diffusion in a Homogeneously Deformed Crystal 212
4.3 Discussion 214

5. Phenomenological Theory for Reactions of Point Defects 214
5.1 Stationary Diffusion and Boundary Condition 215
5.2 Reaction Probability of a Single Defect with a Single Sink 218
5.3 Independent Sink Approximation 219
5.4 Reaction of Point Defects with Straight Dislocations 220
5.5 Finite Sink Densities 222
5.5.1 Spherical Cell Approximation 222
5.5.2 Random Distribution of Sinks 223

6. Lattice Theory for the Reaction Probability 224
6.1 General Expression for the Reaction Probability 226
6.2 Asymptotic Form 229
6.3 Variational Principle for Calculating R_a 229
6.4 Results 231
6.4.1 Exact Results for the bcc Lattice 231
6.4.2 Asymptotic Results 232
6.4.3 Other Cubic Lattices 233
6.4.4 Noncompact Reaction Regions 234
6.5 Discussion and Conclusion 236

7. Influence of Long-Range Potentials on the Rate Constant 237
7.1 Variational Principles 239
7.1.1 Upper Bound for the Rate Constant 239
7.1.2 Lower Bound for the Rate Constant 241
7.1.3 Discussion 242
7.1.4 Variational Principle with Production 242
7.2 Effective Rate Constant for Spherical Sinks 243
7.2.1 Temperature Dependence 244
7.2.2 Spherically Symmetric Potentials 244
7.2.3 Test of Variational Principles 246
7.2.4 Nonspherical Potentials 247

7.3 Effective Rate Constant for Straight Dislocations 250
7.3.1 Cylindrically Symmetric Potentials 251
7.3.2 Edge Dislocations 252
7.4 Discussion and Applications 253

8. Summary and Outlook 256

References 258

Dynamical Properties of Point Defects in Metals

P. H. Dederichs and R. Zeller

1. Introduction and Survey

The theory of lattice dynamics for crystals with defects goes back to the work of LIFSHITS /1.1/ in the years 1943 and 1944. Due to the development of new experimental methods such as the Mössbauer effect, neutron scattering, and infrared absorption this field expanded very quickly from about 1960 on and the theory has been worked out in detail, especially by Maradudin, Ludwig and others. All essential parts of the theory were developed more than ten years ago. Several review articles by LUDWIG /1.2,3/, MARADUDIN /1.4/, MARADUDIN et al. /1.5/, ELLIOTT /1.6/, KLEIN /1.7/, and LIFSHITS and KOSEVICH /1.8/ appeared at that time as well as a conference report /1.9/.

In the last ten to fifteen years our knowledge of point defects has increased considerably. Compared to the situation about ten years ago when one understood the principal effects of point defects on the dynamics of lattices, such as the occurrence of localized and resonant modes, we have now a much better knowledge about the different types of defects and their vibrational characteristics. Not only have a very large number of substitutional defects been studied, but we begin also to learn about such important defects like vacancies or self-interstitials. Also theoretically we can now make quite sophisticated model calculations which is mainly due to the development of numerical programs for the lattice Green's function.

The aim of this article is to review our present knowledge about point defects. Since this field has expanded so tremendously, several restrictions have to be made:

i) We will confine ourselves to the harmonic theory. Experimental studies of anharmonic effects on the dynamics of point defects are very rare and the same is true for theoretical studies. Certainly the most important features can be studied within the harmonic approximation.

ii) We will only consider very small concentrations of point defects. This is quite natural, when one wants to study the physical behaviour of the single point defect. A discussion of higher concentration effects and such theories as the coherent potential approximation (CPA) are beyond the scope of this article. They have been reviewed recently by ELLIOTT et al. /1.10/.

iii) Finally we restrict ourselves to a discussion of point defects in metals only, since our experience is restricted to these materials.

The article was intended to be self-contained. The heart of the theory is a discussion of the lattice Green's function which we feel is most essential for the understanding of the point defect dynamics.

Chapter 2 represents an introduction to the frequency-dependent Green's function. Special emphasis is put on the local frequency spectrum of a specific atom being a natural generalization of the frequency spectrum of the ideal lattice. The behaviour of the Green's functions for small and large frequencies and near critical points is discussed as well as the asymptotic expansion for large distances. Finally we discuss analytical and numerical solutions for the three cubic lattices.

Chapter 3 treats an isolated point defect in an otherwise ideal crystal, i.e. the occurrence of localized and resonant modes. The isotopic defect is discussed as the simplest example.

In Chapter 4 we give a description of the local vibrational behaviour of the point defect by employing a method of Krumhansl and Matthew. The exact defect Green's function can be written in the structural form of a single oscillator Green's function. For resonant and localized modes it can be expressed in terms of effective force constants and effective masses which include the effect of the coupling to the lattice. A special chapter is devoted to the damping of resonant modes.

Chapter 5 deals with substitutional defects. First we discuss a model where only the nearest neighbour force constants are changed which allows an analytical solution. The experimental results of Mössbauer measurement of Fe^{57}, Au^{191} and Sn^{117} impurities in metals are reviewed. Finally we discuss the vibrational behaviour of vacancies.

Chapter 6 is devoted to interstitial defects. Gases like H, O, N and C go into solution as interstitials. Due to their small masses they are characterized by localized vibrations with very high frequencies. Selfinterstitials can only be produced by irradiation. They show an especially interesting vibrational behaviour, since they vibrate with localized vibrations as well as with resonant vibrations. Two models for such interstitials in fcc are presented: the octahedral interstitial and the 100-dumbbell which has been found to be the stable configuration in Cu and Al.

In Chapters 7 and 8 we will discuss the effects of many randomly distributed defects on the dynamical properties of the crystal. Here one has to introduce the average Green's function (Sect. 7.1) which is obtained by a configurational average over all possible microscopic configurations. After a short introduction to thermal neutron scattering (Sect. 7.2) we discuss the possible effects of point defects on the phonon dispersion curves. The relevant results of neutron-scattering measurements are reviewed. We treat also the change of elastic constants due to point defects and their connection with the phonon shifts. In Chapter 8 we then discuss how point defects affect the thermal properties of a crystal. The change of the free energy can be expressed in terms of the change of the total frequency spectrum which is directly related to the single defect. In detail we discuss the change of the specific heat including some experimental measurements. Finally, we calculate the formation and solution entropy of point defects in metals. Especially we discuss the solution entropy for hydrogen in metals and the formation entropy of vacancies in fcc crystals.

Unfortunately we cannot claim completeness for the list of references, despite the fact that we tried hard. At this point already we therefore ask the reader to consult also other reviews /1.2-9/, especially the more recent ones by TAYLOR /1.11/, BARKER and SIEVERS /1.12/, WOOD /1.13/, and NICKLOW /1.14/.

2. Dynamical Green's Functions of Ideal and Defect Lattices

The frequency-dependent Green's function plays a central role in the theory of lattice vibrations especially for crystals with defects. In this chapter we will therefore summarize the most important properties of the dynamical Green's function both for the general case, i.e. a defect lattice, as well as for the ideal lattice. We will further discuss analytical and numerical results for the fcc and bcc lattices.

2.1 Equation of Motion and Eigenfunctions

The classical equations of motions for the displacement $s_i^{\underline{m}}(t)$ of atom No. $\underline{m}$ are in the harmonic approximation, (see e.g. /2.1 - 4/)

$$M^{\underline{m}}\, \ddot{s}_i^{\underline{m}}(t) + \sum_{\underline{n}j} \phi_{ij}^{\underline{mn}}\, s_j^{\underline{n}}(t) = F_i^{\underline{m}}(t) \tag{2.1}$$

where the $F_i^{\underline{m}}(t)$ represent external forces. The $\phi_{ij}^{\underline{mn}}$ are the well known coupling parameters, the properties of which are discussed in /2.1-4/.

A partial solution may be written by means of the Green's function $G_{ij}^{\underline{mn}}(t)$ as

$$s_i^{\underline{m}}(t) = \sum_{\underline{n}j} \int_{-\infty}^{+\infty} dt'\, G_{ij}^{\underline{mn}}(t-t')\, F_j^{\underline{n}}(t') \tag{2.2}$$

where this so-called retarded Green's function is determined by

$$M^{m} \ddot{G}^{mn}_{ij}(t) + \sum_{pk} \phi^{mp}_{ik} G^{pn}_{kj}(t) = \delta^{mn}_{ij}\delta(t) \tag{2.3}$$

with

$$G^{mn}_{ij}(t) = 0 \quad \text{for } t < 0.$$

In the following we will, whenever possible, use a matrix notation so that the equations (2.2) and (2.3) read

$$\underline{s}(t) = \int_{-\infty}^{+\infty} dt'\; G(t-t')\, \underline{F}(t') \quad \text{and} \quad M\,\ddot{G}(t) + \phi\, G(t) = \delta(t)\ . \tag{2.4}$$

Note that M is a diagonal matrix: $M^{m}\delta^{mn}$. For the case of equal masses we have $M = M\underline{1}$ where $\underline{1}$ is the unit matrix which will not be written explicitly in the following.

By Fourier transformation

$$\underline{s}(t) = \int_{-\infty}^{+\infty} \frac{d\omega}{2\pi}\, e^{-i\omega t}\, \underline{s}(\omega) \tag{2.5}$$

the equation of motion becomes

$$(\phi - M\omega^2)\, \underline{s}(\omega) = \underline{F}(\omega)\ . \tag{2.6}$$

The partial solution is therefore

$$\underline{s}(\omega) = G(\omega)\, \underline{F}(\omega) \tag{2.7}$$

where $G(\omega)$ is the Fourier transform of the retarded Green's function (2.3) [1]

$$G(\omega) = \frac{1}{\phi - M(\omega + i\eta)^2} \quad , \quad \eta \to +0\ . \tag{2.8}$$

Here the infinitesimal positive quantity η guarantees the retardation: $G(t) = 0$ for $t < 0$. For $\omega = 0$ $G^{mn}_{ij}(0)$ is identical with the static Green's funtion /2.4/.

The homogeneous solutions of (2.6) with frequencies ω_α

$$(\phi - M\omega_\alpha^2)\, \underline{s}(\alpha) = 0 \tag{2.9}$$

are found by the $\sqrt{M}$ -transformation. By setting

$$\underline{s}(\alpha) = \frac{1}{\sqrt{M}}\, \underline{\xi}(\alpha) \quad , \quad s^{m}_{i}(\alpha) = \frac{1}{\sqrt{M^{m}}}\, \xi^{m}_{i}(\alpha) \tag{2.10}$$

[1] The corresponding advanced Green's function ($G(t) = 0$ for $t > 0$) is obtained by replacing $\omega + i\eta$ by $\omega - i\eta$.

(2.9) is transformed into an eigenvalue form

$$(D - \omega_\alpha^2)\, \underline{\xi}(\alpha) = 0 \qquad (2.11)$$

so that the $\underline{\xi}(\alpha)$'s are the eigenvectors and the ω_α^2's are the eigenvalues of the "dynamical matrix" D:

$$D = \frac{1}{\sqrt{M}}\, \phi \,\frac{1}{\sqrt{M}} \quad \text{or} \quad D_{ij}^{\underline{mn}} = \frac{1}{\sqrt{M^{\underline{m}}}}\, \phi_{ij}^{\underline{mn}} \,\frac{1}{\sqrt{M^{\underline{n}}}} = D_{ji}^{\underline{nm}} \,. \qquad (2.12)$$

Since D is symmetrical and real, the eigenvectors $\underline{\xi}(\alpha)$ form a complete and orthonormal system and can be chosen real:

$$\sum_{\underline{m}i} \xi_i^{\underline{m}}(\alpha)\, \xi_i^{\underline{m}}(\beta) = \delta_{\alpha\beta} \,, \qquad \sum_\alpha \xi_i^{\underline{m}}(\alpha)\, \xi_j^{\underline{n}}(\alpha) = \delta_{ij}^{\underline{mn}} \,. \qquad (2.13)$$

With these eigenfunctions, the general solution $\underline{s}(t)$ in the absence of external forces may be written as

$$\underline{s}(t) = \frac{1}{\sqrt{M}} \sum_\alpha A_\alpha(t)\, \underline{\xi}(\alpha) \qquad (2.14)$$

where the "normal coordinates" $A_\alpha(t)$ describe the independent normal modes of the system, since

$$\ddot{A}_\alpha(t) + \omega_\alpha^2\, A_\alpha(t) = 0 \quad \text{or} \quad A_\alpha(t) = A_\alpha(0)\, e^{\pm i\omega_\alpha t} \,. \qquad (2.15)$$

Also the Green's function $G(\omega)$ can be expanded in terms of the $\underline{\xi}(\alpha)$'s. First we have

$$G(\omega) = \frac{1}{\sqrt{M}}\, \mathcal{G}(\omega) \,\frac{1}{\sqrt{M}} \quad \text{with} \quad \mathcal{G}(\omega) = \frac{1}{D - (\omega + i\eta)^2} \,. \qquad (2.16)$$

Expanding now the modified Green's function $\mathcal{G}(\omega)$ into the eigenfunctions of D we obtain

$$\mathcal{G}(\omega) = \sum_\alpha \frac{|\underline{\xi}(\alpha)\rangle \langle \underline{\xi}(\alpha)|}{\omega_\alpha^2 - (\omega + i\eta)^2} \quad \text{or} \quad G_{ij}^{\underline{mn}}(\omega) = \frac{1}{\sqrt{M^{\underline{m}} M^{\underline{n}}}} \sum_\alpha \frac{\xi_i^{\underline{m}}(\alpha)\, \xi_j^{\underline{n}}(\alpha)}{\omega_\alpha^2 - (\omega + i\eta)^2} \,. \qquad (2.17)$$

In the special case of an ideal, infinite, and primitive lattice, all masses are equal ($M^{\underline{m}} = \overset{\circ}{M}$) and the coupling parameters have translation symmetry: $\phi_{ij}^{\underline{mn}} = \overset{\circ}{\phi}{}_{ij}^{(\underline{m}-\underline{n})}$ [2]. Thus the eigenfunctions $\overset{\circ}{\underline{\xi}}(\alpha)$'s are plane waves labelled by a wave vector $\underline{k}$ confined to the first Brillouin zone V_B and by a polarization index σ

[2] In the following all quantities referring to an ideal lattice are denoted by an upper index $^\circ$, e.g., $\overset{\circ}{G}$, $\overset{\circ}{\phi}$, etc.

$$\overset{\circ}{\xi}{}^{\underline{m}}_{i}(\underline{k}\sigma) = \frac{1}{\sqrt{V_B}} e_i(\underline{k}\sigma) \exp(i\underline{k}\underline{R}^{\underline{m}}) \ . \tag{2.18}$$

Instead of these complex eigenvectors also real ones can be chosen which are obtained by replacing $\exp(i\underline{k}\underline{R}^{\underline{m}})$ by $\sqrt{2}\cos \underline{k}\underline{R}^{\underline{m}}$ and $\sqrt{2}\sin \underline{k}\underline{R}^{\underline{m}}$. In an infinite lattice, all $\underline{k}$-vectors are allowed. Thus the ideal lattice Green's function is given by an integral over the first Brillouin zone:

$$\overset{\circ}{G}{}^{\underline{mn}}_{ij}(\omega) = \sum_{\sigma} \int_{V_B} \frac{d\underline{k}}{V_B} \frac{e_i(\underline{k}\sigma)\, e_j^*(\underline{k}\sigma) \exp[i\underline{k}(\underline{R}^{\underline{m}} - \underline{R}^{\underline{n}})]}{\overset{\circ}{M}\left(\omega_\sigma^2(\underline{k}) - (\omega + i\eta)^2\right)} \tag{2.19}$$

where the $\omega_\sigma(\underline{k})$ are the eigenfrequencies.

Since the Green's function is essentially the inverse of D, it has all the symmetries of the dynamical matrix, so that for the ideal Green's function we have, e.g.,

$$\overset{\circ}{G}{}^{\underline{mn}}_{ij}(\omega) = \overset{\circ}{G}{}^{(\underline{m}-\underline{n})}_{ij}(\omega) = \overset{\circ}{G}{}^{-(\underline{m}-\underline{n})}_{ij}(\omega) = \overset{\circ}{G}{}^{(\underline{m}-\underline{n})}_{ji}(\omega) \ . \tag{2.20}$$

2.2 Real and Imaginary Parts of Green's Function and its Connection with Correlation Functions

By using the standard relation

$$\frac{1}{x - i\eta} = \frac{x}{x^2 + \eta^2} + i \frac{\eta}{x^2 + \eta^2} = P\left(\frac{1}{x}\right) + i\pi\delta(x) \tag{2.21}$$

the Green's functions $G(\omega)$ and $\mathscr{G}(\omega)$ can be split into their real and imaginary parts

$$\mathrm{Re}\{G(\omega)\} = \frac{1}{\sqrt{M}} P \frac{1}{D - \omega^2} \frac{1}{\sqrt{M}} = \frac{1}{\sqrt{M}} \sum_\alpha P \frac{|\xi(\alpha)\rangle \langle\xi(\alpha)|}{\omega_\alpha^2 - \omega^2} \frac{1}{\sqrt{M}} \tag{2.22}$$

$$\mathrm{Im}\{G(\omega)\} = (\mathrm{sgn}\ \omega) \frac{1}{\sqrt{M}} \pi\delta(\omega^2 - D) \frac{1}{\sqrt{M}} = (\mathrm{sgn}\ \omega) \frac{1}{\sqrt{M}} \pi \sum_\alpha \delta(\omega^2 - \omega_\alpha^2)\, |\xi(\alpha)\rangle\langle\xi(\alpha)| \frac{1}{\sqrt{M}} \tag{2.23}$$

$\mathrm{Re}\{G(\omega)\}$ is an even function in ω, whereas $\mathrm{Im}\{G(\omega)\}$ is odd

$$\left.\begin{aligned} \mathrm{Re}\{G(\omega)\} &= \mathrm{Re}\{G(-\omega)\} \\ \mathrm{Im}\{G(\omega)\} &= -\mathrm{Im}\{G(\omega)\} \end{aligned}\right\} \quad \text{or} \quad G(\omega) = G^*(-\omega) \ . \tag{2.24}$$

Further the real and imaginary parts are not independent from each other, but connected by the dispersion relation (Kramers-Kronig relation)

$$\mathrm{Re}\{G(\omega)\} = \frac{1}{\pi} \int_0^\infty d\omega'^2\, P \frac{1}{\omega'^2 - \omega^2} \mathrm{Im}\{G(\omega')\} \tag{2.25}$$

from which we obtain the identity

$$G(\omega) = \frac{1}{\pi} \int_{Q}^{\infty} d\omega'^2 \frac{1}{\omega'^2 - (\omega + i\eta)^2} \operatorname{Im}\{G(\omega')\} \tag{2.26}$$

expressing $G(\omega)$ in terms of $\operatorname{Im}\{G(\omega)\}$ alone.

From this or from (2.17) one can see, that $G(\omega^2)$ is - as a function of the complex variable ω^2 - analytic in the whole complex ω^2 plane with the exception of the points $\omega^2 = \omega_\alpha^2$ on the positive real axis, where $G(\omega^2)$ has simple poles which may line up to form branch cuts as, e.g., in the ideal lattice.

For the sake of completeness we also give the back transformation of (2.25)

$$\operatorname{Im}\{G(\omega^2)\} = \frac{-1}{\pi} \int_{-\infty}^{\infty} d\omega'^2 \, P \frac{1}{\omega'^2 - \omega^2} \operatorname{Re}\{G(\omega')\} \, . \tag{2.25a}$$

In quantum theory the classical displacements $s_i^m(t)$ are replaced by hermitian operators which together with their conjugate impulses p_i^m satisfy the commutation relations

$$[p_i^m, s_i^n] = \frac{\hbar}{i} \delta_{ij}^{mn} \, . \tag{2.27}$$

In the Heisenberg picture the time-dependent operators obey the same equations (2.1) as the classical coordinates and therefore the same applies for the time dependent *correlation function*

$$\langle s_i^m(t) \, s_j^n(0) \rangle = \frac{\operatorname{Tr}\{e^{-\beta H} \, s_i^m(t) \, s_j^n(0)\}}{\operatorname{Tr}\{e^{-\beta H}\}} \tag{2.28}$$

where

$$H = \sum_{mi} \frac{1}{2M^m} (p_i^m)^2 + \frac{1}{2} \sum_{\substack{mn \\ ij}} s_i^m \, \phi_{ij}^{mn} \, s_j^n \, .$$

This correlation function allows one to describe a wide range of physical processes such as thermodynamic properties (like specific heat, etc.), X-ray and neutron-scattering in very simple terms. Since it obeys the same equation of motion as the Green's function $G_{ij}^{mn}(t)$, there should be a simple relation between both functions which will be derived in the following (see, e.g., /2.3,4/).

First we expand $s_i^m(t)$ into the real eigenfunctions $\xi_i^m(\alpha)$

$$s_i^m(t) = \sum_\alpha \frac{1}{\sqrt{M^m}} A_\alpha(t) \, \xi_i^m(\alpha) \, . \tag{2.29}$$

The normal coordinate $A_\alpha(t)$ is now an operator which satisfies with its conjugate impulse $\dot{A}_\alpha(t)$ the commutation relation

$$[A_\alpha(t),\dot{A}_\beta(t)] = \frac{\hbar}{i}\,\delta_{\alpha\beta} \; . \tag{2.30}$$

It obeys the classical oscillator equation (2.15). Thus as for the one-dimensional linear oscillator, we can introduce creation and annihilation operators $a_\alpha^+(t)$ and $a_\alpha(t)$ for the mode α, i.e.

$$\begin{aligned} A_\alpha(t) &= \sqrt{\frac{\hbar}{2\omega_\alpha}}\left(a_\alpha(t) + a_\alpha^+(t)\right) \\ \dot{A}_\alpha(t) &= \frac{1}{i}\sqrt{\frac{\hbar\omega_\alpha}{2}}\left(a_\alpha(t) - a_\alpha^+(t)\right) \end{aligned} \qquad \text{with } [a_\alpha, a_\beta^+] = \delta_{\alpha\beta} \; . \tag{2.31}$$

The energy is then a sum of the energies for all modes α:

$$H = \sum_\alpha \frac{1}{2}\left\{\left(\dot{A}_\alpha(t)\right)^2 + \omega_\alpha^2\left(A_\alpha(t)\right)^2\right\} = \sum_\alpha \hbar\omega_\alpha\left\{a_\alpha^+ a_\alpha + \frac{1}{2}\right\} . \tag{2.32}$$

By inserting (2.29) into the correlation function, one obtains

$$<s_i^m(t)\; s_j^n(0)> \;=\; \sum_{\alpha\beta} \xi_i^m(\alpha)\; \xi_j^n(\beta)\; \frac{\hbar}{2\sqrt{M^m M^n \omega_\alpha \omega_\beta}} <[a_\alpha(t) + a_\alpha^+(t)]\,[a_\beta(0) + a_\beta^+(0)]> . \tag{2.33}$$

Since the different states α and β are independent, only $\alpha = \beta$ gives a contribution. The time dependence of $a_\alpha(t)$ is

$$a_\alpha(t) = \exp(-i\omega_\alpha t)\; a_\alpha(0) \; , \quad a_\alpha^+(t) = \exp(i\omega_\alpha t)\; a_\alpha(0) \; . \tag{2.34}$$

For the expectation values we have

$$<a_\alpha(0)\; a_\alpha(0)> \;=\; 0 \;=\; <a_\alpha^+(0)\; a_\alpha^+(0)> \tag{2.35}$$

whereas

$$<a_\alpha^+(0)\; a_\alpha(0)> \;=\; <a_\alpha(0)\; a_\alpha^+(0)> - 1 = n(\omega_\alpha) = \frac{1}{\exp(\hbar\omega_\alpha/kT) - 1} = -\left(n(-\omega_\alpha) + 1\right) \tag{2.36}$$

where $n(\omega)$ is the Bose distribution. Thus

$$\begin{aligned} <s_i^m(t)\; s_j^n(0)> \;=\; &\sum_\alpha \xi_i^m(\alpha)\; \xi_j^n(\alpha)\; \frac{\hbar}{2\omega_\alpha\sqrt{M^m M^n}} \; \cdot \\ &\left\{\exp(-i\omega_\alpha t)\left(n(\omega_\alpha) + 1\right) - \exp(i\omega_\alpha t)\left(n(-\omega_\alpha) + 1\right)\right\} . \end{aligned} \tag{2.37}$$

Finally we obtain by taking the Fourier transform and by comparing with (2.23)

$$\int_{-\infty}^{\infty} dt\; \exp(i\omega t) <s_i^m(t)\; s_j^n(0)> \;=\; 2\hbar\left(n(\omega) + 1\right)\; \mathrm{Im}\{G_{ij}^{mn}(\omega)\} \tag{2.38}$$

and by transforming back

$$<s_i^m(t)\ s_j^n(0)> = \int_{-\infty}^{\infty} \frac{d\omega}{2\pi}\, e^{-i\omega t}\ 2\hbar\Big(n(\omega)+1\Big)\ \mathrm{Im}\{G_{ij}^{mn}(\omega)\}\ . \qquad (2.39)$$

By using the relation $n(-\omega) = -n(\omega) - 1$ and $\mathrm{Im}\{G(\omega)\} = -\mathrm{Im}\{G(-\omega)\}$ the integral can be written as an integral over positive frequencies only:

$$<s_i^m(t)\ s_j^n(0)> = \frac{\hbar}{\pi}\int_0^{\infty} d\omega\ \Big\{(\cos\omega t)\Big(2n(\omega)+1\Big) - i\sin\omega t\Big\}\ \mathrm{Im}\{G_{ij}^{mn}(\omega)\}\ . \qquad (2.40)$$

In the same way the velocity or momentum correlation functions can be calculated. Since $p_i^m(t) = M^m \dot{s}_i^m(t)$, we insert $\dot{s}_i^m(t)$ of (2.29) and $\dot{A}_\alpha(t)$ of (2.31) into the momentum correlation function and obtain the same expression (2.40) but multiplied by $M^m M^n \omega^2$:

$$<p_i^m(t)\ p_j^n(0)> = \frac{\hbar}{\pi}\int_0^{\infty} d\omega\ \Big\{(\cos\omega t)\Big(2n(\omega)+1\Big) - i\sin\omega t\Big\}\ M^m M^n \omega^2\ \mathrm{Im}\{G_{ij}^{mn}(\omega)\}\ . \qquad (2.41)$$

Of special interest, e.g., for X-ray diffraction, are the equal time correlation functions ($t = 0$). They are always real and especially simple in the high-temperature limit. First by using the representation (2.23) for $\mathrm{Im}\{G(\omega)\}$, the integral over ω can be done leading to

$$<s_i^m(0)\ s_j^n(0)> = \frac{\hbar}{\sqrt{M^m M^n}}\left\{\Big(n(\sqrt{D}) + \frac{1}{2}\Big)\frac{1}{\sqrt{D}}\right\}_{ij}^{mn} \qquad (2.42)$$

and

$$<p_i^m(0)\ p_j^n(0)> = \hbar\sqrt{M^m M^n}\left\{\Big(n(\sqrt{D}) + \frac{1}{2}\Big)\sqrt{D}\right\}_{ij}^{mn}\ . \qquad (2.43)$$

For high temperatures $n(\omega) + \frac{1}{2} \cong kT/\hbar\omega$ we obtain therefore

$$<s_i^m(0)\ s_j^n(0)> \cong \frac{kT}{\sqrt{M^m M^n}}\left\{D^{-1}\right\}_{ij}^{mn} = kT\left\{\phi^{-1}\right\}_{ij}^{mn} = kT\ G_{ij}^{mn}(\omega = 0) \qquad (2.44)$$

whereas the momentum correlation follows as

$$<p_i^m(0)\ p_j^n(0)> \cong \frac{kT}{M^m}\,\delta_{ij}^{mn}\ . \qquad (2.45)$$

Thus in the classical limit the displacement correlations are independent of the masses and given by the static Green's function, whereas the momentum correlations are independent of the coupling parameters and diagonal. This is, of course, clear since in classical physics the distributions of displacements and momenta are independent, the first one depending on the coupling parameters only and the second one only on the masses.

2.3 The Local Frequency Spectrum

The vibrational properties of a single atom, e.g. of a substitutional point defect, can be described conveniently by the so-called local frequency spectrum: $z_i^{\underline{m}}(\omega)$ is the spectrum of atom $\underline{m}$ for vibrations in direction i and is defined by

$$z_i^{\underline{m}}(\omega) = \frac{2\omega M^{\underline{m}}}{\pi} \mathrm{Im}\{G_{ii}^{\underline{mm}}(\omega)\} = \frac{2\omega}{\pi} \mathrm{Im}\{\mathscr{G}_{ii}^{\underline{mm}}(\omega)\} = \sum_{\alpha} |\xi_i^{\underline{m}}(\alpha)|^2 \, \delta(\omega - \omega_\alpha) \quad \text{for } \omega > 0 \,. \tag{2.46}$$

As a proper distribution, the spectrum $z_i^{\underline{m}}(\omega)$ is always positive and normalized with respect to ω. By using the completeness relation (2.13) we have

$$\int_0^\infty d\omega \; z_i^{\underline{m}}(\omega) = \sum_{\alpha} |\xi_i^{\underline{m}}(\alpha)|^2 = 1 \,. \tag{2.47}$$

$z_i^{\underline{m}}(\omega)$ counts the number of eigenfrequencies in the interval $(\omega, d\omega)$ multiplied by the square of the amplitude of atom $\underline{m}$ in the direction i. We can interpret $z_i^{\underline{m}}(\omega)$ as follows: The single degree of freedom of atom $\underline{m}$ for vibration in i direction is a weighted superposition of different eigenmodes $\underline{\xi}(\alpha)$, and $z_i^{\underline{m}}(\omega)$ gives a quantitative measure of how much the modes of frequency ω contribute to the vibrational behaviour of atom $\underline{m}$. In general we will obtain a different local spectrum for each atom and each direction. Only for an ideal primitive lattice we obtain the same local spectrum for each atom due to the plane wave character of the eigenmodes $|\overset{\circ}{\xi}{}_i^{\underline{m}}|^2 = \frac{1}{V_B} |e_i(\underline{k}\sigma)|^2$. By summing over all directions, we obtain the usual spectrum of the ideal lattice, i.e. the number of frequencies in the interval $(\omega, d\omega)$:

$$\overset{\circ}{z}(\omega) = \frac{1}{3} \sum_i \overset{\circ}{z}_i(\omega) = \sum_\sigma \int \frac{d\underline{k}}{3V_B} \, \delta\big(\omega - \omega_\alpha(\underline{k})\big) \,. \tag{2.48}$$

Without explicitly calculating the Green's function, we can obtain very valuable information about the local spectrum by means of the follwing sum rule for $\mathrm{Im}\{\mathscr{G}(\omega)\}$:

$$\frac{1}{\pi} \int d\omega^2 \, \omega^{2n} \, \mathrm{Im}\{\mathscr{G}(\omega)\} = \int d\omega^2 \, \omega^{2n} \, \delta(\omega^2 - D) = (D)^n \,. \tag{2.49}$$

If we take the diagonal $\underline{m}$i-matrix element of this equation, the normalization condition is recovered for n = 0. For n = 1 we obtain the second moment of the $\underline{m}$i spectrum

$$\langle\omega^2\rangle = \int_0^\infty d\omega \, \omega^2 \, z_i^{\underline{m}}(\omega) = D_{ii}^{\underline{mm}} = \phi_{ii}^{\underline{mm}} / M^{\underline{m}} \tag{2.50}$$

which is given by the Einstein-frequency ω_E^2, i.e. the frequency which we obtain by fixing all neighbours and allowing the atom $\underline{m}$ to vibrate in i direction. Analogously the higher moments can be calculated, e.g.,

$$<\omega^4> = \int d\omega\, \omega^4\, z_i^{\underline{m}}(\omega) = \left\{D^2\right\}_{ii}^{\underline{mm}} = \sum_{\underline{n}j} \left(\phi_{ij}^{\underline{mn}}\right)^2 / M^{\underline{m}} M^{\underline{n}} \ . \tag{2.51}$$

From the second moment $<\omega^2>$ we can immediately conclude that atoms with a large mass or/and with a weak coupling to their neighbours will vibrate mostly with low-frequency modes, whereas atoms with light masses or atoms which are strongly bound vibrate mostly with high-frequency modes.

Since these moments are weighting the spectrum $z_i^{\underline{m}}(\omega)$ by powers of ω^2, they are very sensitive to the behaviour of the spectrum for the higher frequencies, but tell us essentially nothing about the behaviour for very low frequencies. However this information can be obtained by the inverse moment $<1/\omega^2>$ of the spectrum

$$<1/\omega^2> = \int_0^\infty d\omega\, \frac{1}{\omega^2}\, z_i^{\underline{m}}(\omega) = \left\{D^{-1}\right\}_{ii}^{\underline{mm}} = M^{\underline{m}} \left\{\phi^{-1}\right\}_{ii}^{\underline{mm}} = M^{\underline{m}} G_{ii}^{\underline{mm}}(\omega = 0) \ . \tag{2.52}$$

Thus $<1/\omega^2>$ is essentially given by the static Green's function.

Higher inverse moments (e.g. $<1/\omega^4>$) can also be calculated. However special care has to be taken to avoid divergencies (see next section).

The local frequency spectrum $z_i^{\underline{m}}(\omega)$ allows the calculation of all vibrational properties, which do not depend on correlations $\underline{m} \neq \underline{n}$ of different atoms. Examples are the thermal displacements-squared and the momentum-squared, which follow from (2.40) and (2.41) for $t = 0$, $\underline{m} = \underline{n}$ and $i = j$

$$<(s_i^{\underline{m}})^2> = \int_0^\infty d\omega\, \frac{\hbar}{M^{\underline{m}}\omega} \left(n(\omega) + \frac{1}{2}\right) z_i^{\underline{m}}(\omega) \tag{2.53}$$

$$<(s_i^{\underline{m}})^2> = \int_0^\infty d\omega\, \frac{\varepsilon(\omega,T)}{M^{\underline{m}}\omega^2}\, z_i^{\underline{m}}(\omega) = \begin{cases} \dfrac{kT}{M^{\underline{m}}} \displaystyle\int_0^\infty d\omega\, \frac{z_i^{\underline{m}}(\omega)}{\omega^2} & \text{for high T} \\ \dfrac{\hbar}{2M^{\underline{m}}} \displaystyle\int_0^\infty d\omega\, \frac{z_i^{\underline{m}}(\omega)}{\omega} & \text{for low T} \end{cases} \tag{2.54}$$

Here

$$\varepsilon(\omega,T) = \hbar\omega\left(n(\omega) + \frac{1}{2}\right) = \begin{cases} kT & \text{for } kT \gg \hbar\omega \\ \dfrac{\hbar\omega}{2} & \text{for } kT \ll \hbar\omega \end{cases}$$

is the average energy of an oscillator of frequency ω. Eq. (2.54) is a simple generalization of the corresponding equation of a single oscillator $<s^2> = \varepsilon(\omega,T)/M\omega^2$ which follows from the virial theorem $2E_{kin} = E_{total}$. For the square of the momentum we have

$$<(p_i^{\underline{m}})^2> = \int_0^\infty d\omega\, M^{\underline{m}}\, \varepsilon(\omega,T)\, z_i^{\underline{m}}(\omega) \cong \begin{cases} M^{\underline{m}} kT & \text{for high T} \\ \dfrac{M^{\underline{m}}\hbar}{2} \displaystyle\int_0^\infty d\omega\, \omega\, z_i^{\underline{m}}(\omega) & \text{for low T} \end{cases} \tag{2.55}$$

Both $\langle(s_i^{\underline{m}})^2\rangle$ and $\langle(p_i^{\underline{m}})^2\rangle$ can be measured by the Mössbauer effect for Mössbauer active nuclei such as Fe, Au and Sn (see Chapter 5). Thus one can measure three different nontrivial moments of the local frequency spectrum: From $\langle \underline{s}^2\rangle$ the moments $\langle 1/\omega^2\rangle$ for high T and $\langle 1/\omega\rangle$ for low T and from $\langle \underline{p}^2\rangle$ the moment $\langle\omega\rangle$ for low T.

Averaging the local frequency spectra of all N atoms of an arbitrarily disordered crystal, we obtain the total spectrum, which due to the normalization (2.13) of $\xi_i^{\underline{m}}(\alpha)$ is given by

$$Z(\omega) = \sum_{\underline{m}i} z_i^{\underline{m}}(\omega) = \sum_{\underline{m}i} \sum_{\alpha=1}^{3N} |\xi_i^{\underline{m}}(\alpha)|^2 \, \delta(\omega - \omega_\alpha)$$
$$Z(\omega) = \sum_{\alpha=1}^{3N} \delta(\omega - \omega_\alpha) \quad \text{with} \int d\omega \, Z(\omega) = 3N \, . \tag{2.56}$$

As we will discuss in Chapter 8, all thermodynamic properties of the crystal can be expressed in the harmonic approximation by the total spectrum $Z(\omega)$ and hence by the local frequency spectra of all atoms.

Another very intuitive meaning of $\mathrm{Im}\{G(\omega)\}$ and of the spectrum $z_i^{\underline{m}}(\omega)$ can be obtained as follows: Let us apply a force $F_i^{\underline{m}}(t) = f_i^{\underline{m}} \cos \omega t$ to atom $\underline{m}$ in direction i. The work done per second is then

$$W(t) = \dot{s}_i^{\underline{m}}(t) \, F_i^{\underline{m}}(t) \, . \tag{2.57}$$

By means of the Green's function, $s_i^{\underline{m}}(t)$ can be expressed by $F_i^{\underline{m}}(t)$ (2.2). Since $\cos \omega t = \frac{1}{2}[\exp(-i\omega t) + \exp(i\omega t)]$ we obtain using (2.7)

$$s_i^{\underline{m}}(t) = \left(f_i^{\underline{m}}/2\right)\left(G_{ii}^{\underline{mm}}(\omega)\exp(-i\omega t) + G_{ii}^{\underline{mm}}(-\omega)\exp(i\omega t)\right)$$
$$= f_i^{\underline{m}} \left(\mathrm{Re}\{G_{ii}^{\underline{mm}}(\omega)\} \cos \omega t + \mathrm{Im}\{G_{ii}^{\underline{mm}}(\omega)\} \sin \omega t\right) . \tag{2.58}$$

The first contribution with $\mathrm{Re}\{G(\omega)\}$ is in phase with the applied force, the second one with $\mathrm{Im}\{G(\omega)\}$ is out of phase by $\pi/2$. Inserting that into (2.57) and averaging over one period we obtain for the average work transferred to the system only a contribution from $\mathrm{Im}\{G(\omega)\}$:

$$\overline{W(t)} = (f_i^{\underline{m}})^2 \, \frac{\omega}{2} \, \mathrm{Im}\{G_{ii}^{\underline{mm}}(\omega)\} = \frac{\pi}{4M^{\underline{m}}} (f_i^{\underline{m}})^2 \, z_i^{\underline{m}}(\omega) \, . \tag{2.59}$$

This means, that we can transfer energy only for ω equal to one of the eigenfrequencies ω_α, since $z_i^{\underline{m}}(\omega)$ vanishes otherwise. For other frequencies we can only temporarily transfer energy which we get back in the same period.

2.4 Behaviour of $G(\omega)$ for High and Low Frequencies

As can be seen from (2.23), $\mathrm{Im}\{G(\omega)\}$ vanishes for frequencies ω larger than the maximum, ω_{max}, of all the eigenfrequencies ω_α. Therefore the Green's function is real in that range and can be expanded into powers of $1/\omega^2$:

$$G(\omega) = \mathrm{Re}\{G(\omega)\} = \frac{1}{\phi - M\omega^2} = -\frac{1}{M\omega^2} - \frac{1}{M\omega^2}\phi\frac{1}{M\omega^2} - \frac{1}{M\omega^2}\left(\phi\frac{1}{M\omega^2}\right)^2 - \dots \quad . \tag{2.60}$$

The first term $-1/M\omega^2$ represents the Green's function of a free atom. For high frequencies or for very short time the coupling to the neighbouring atoms can be neglected and the atom behaves as a free atom. As a function of the separation $\underline{m}-\underline{n}$, the Green's function $G_{ij}^{\underline{mn}}(\omega)$ decreases with increasing powers of $1/\omega^2$. E.g., if ϕ couples only nearest neighbours, the Green's function vanishes for a pair of nearest neighbours as $1/\omega^4$, for second nearest neighbours as $1/\omega^6$ and so on. Thus for large distances the Green's function decreases for $\omega \to \infty$ faster than any power of $1/\omega^2$ which makes the result of the next section, namely an exponential decrease for large distances, plausible.

We can obtain the same result (2.60) also from the dispersion relation (2.25) if we expand $1/(\omega'^2 - \omega^2)$:

$$G(\omega) = \frac{1}{\pi}\int_0^\infty d\omega'^2 \left\{-\frac{1}{\omega^2} - \frac{1}{\omega^4}\omega'^2 - \dots\right\} \mathrm{Im}\{G(\omega')\} \; . \tag{2.61}$$

Thus $G(\omega)$ is determined for large ω by the moments $\langle\omega^2\rangle$, $\langle\omega^4\rangle$ of $\mathrm{Im}\{G(\omega)\}$ which can be calculated by the sum rule (2.49) to yield the previous result.

This procedures give us a hint how to calculate the behaviour of $\mathrm{Re}\{G(\omega)\}$ for very low frequencies. Namely using the dispersion relation (2.25) we make a Taylor expansion of $\mathrm{Re}\{G(\omega)\}$

$$\begin{aligned}\mathrm{Re}\{G(\omega)\} &= \mathrm{Re}\{G(0)\} + \omega^2 \left\{\partial_{\omega^2} \mathrm{Re}\{G(\omega)\}\right\}_{\omega\to+0} + \dots \\ &= \frac{1}{\pi}\int d\omega'^2\, P\left(\frac{1}{\omega'^2}\right) \mathrm{Im}\{G(\omega')\} + \omega^2\left\{\frac{1}{\pi}\int d\omega'^2\, \tilde{P}\left(\frac{1}{(\omega'^2-\omega^2)^2}\right)\mathrm{Im}\{G(\omega')\}\right\}_{\omega\to+0} + \dots\end{aligned} \tag{2.62}$$

The coefficients are essentially the reciprocal moments $\langle 1/\omega^{2n}\rangle$ of $\mathrm{Im}\{G(\omega)\}$ which, however, have to be defined by the principal value and its derivative. For instance

$$\tilde{P}\left(\frac{1}{x^2}\right) = -\frac{d}{dx}P\left(\frac{1}{x}\right) = -\frac{d}{dx}\frac{x}{x^2+\eta^2} = \frac{x^2-\eta^2}{(x^2+\eta^2)^2} \; , \quad \eta\to+0 \; . \tag{2.63}$$

The functions $P\left(\frac{1}{x}\right)$ and $\tilde{P}\left(\frac{1}{x^2}\right)$ are shown in Fig. 1 for a finite value of η. From its definition as a derivative of $P\left(\frac{1}{x}\right)$ it follows that all integrals over $\tilde{P}\left(\frac{1}{x^2}\right)$ exist. The same applies to the higher derivatives. The second coefficient in (2.62) is

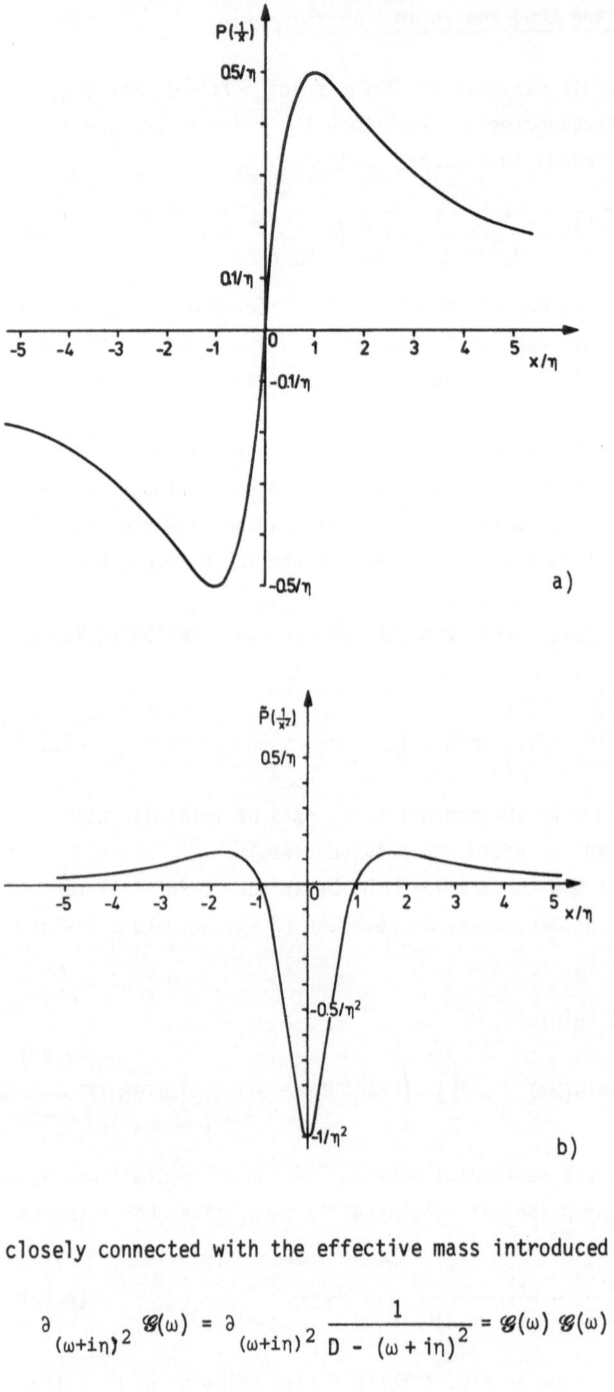

<u>Fig. 1.</u> The function $P\left(\frac{1}{x}\right) = \frac{x}{x^2+\eta^2}$, the principal value of $\frac{1}{x}$, (Fig. 1a), and its derivative $\tilde{P}\left(\frac{1}{x^2}\right) = -\frac{d}{dx} P\left(\frac{1}{x}\right) = \frac{x^2-\eta^2}{(x^2+\eta^2)^2}$ (Fig. 1b)

closely connected with the effective mass introduced in Chapter 4. Since

$$\partial_{(\omega+i\eta)^2} \mathscr{G}(\omega) = \partial_{(\omega+i\eta)^2} \frac{1}{D - (\omega + i\eta)^2} = \mathscr{G}(\omega)\, \mathscr{G}(\omega) \tag{2.64}$$

it can also be written as

$$\left\{\partial_{\omega^2}\, \mathrm{Re}\{G^{\underline{m}\underline{n}}_{ij}(\omega)\}\right\}_{\omega\to+0} = \mathrm{Re}\left\{\sum_{\underline{p}k} G^{\underline{m}\underline{p}}_{ik}(\omega)\; M^{\underline{p}}\; G^{\underline{p}\underline{n}}_{kj}(\omega)\right\}_{\omega\to+0}$$

$$= \left\{\frac{1}{\pi}\int d\omega'^2\; \widetilde{P}\left(\frac{1}{(\omega'^2-\omega^2)^2}\right) \mathrm{Im}\{G^{\underline{m}\underline{n}}_{ij}(\omega')\}\right\}_{\omega\to+0} . \tag{2.65}$$

In the last integral the limit $\eta\to 0$ has to be done before ω^2 goes to +0. Note that ∂_{ω^2} Re{G(ω)} does not exist for $\omega^2\to -0$ (see Sect. 2.6).

Whereas the imaginary part of G(ω) vanishes for $\omega > \omega_{max}$, its behaviour for very small frequencies is more complicated and will be derived in the following only for the ideal crystal. For small $\underline{k}$ we have

$$\omega_\sigma(\underline{k}) \cong c_\sigma(\underline{\kappa})k \quad \text{with } \underline{k} = k\underline{\kappa} \;, \quad \underline{\kappa} = \frac{\underline{k}}{k}$$

where $c_\sigma(\underline{\kappa})$ is the velocity of sound for the direction $\underline{\kappa}$. Thus Im{G(ω)} is given by

$$\mathrm{Im}\{\overset{o}{G}{}^{(\underline{h})}_{ij}(\omega)\} = \frac{\pi}{\overset{o}{M}V_B}(\mathrm{sgn}\,\omega)\sum_\sigma \int_{V_B} d\underline{k}\; e_i(\underline{k}\sigma)\; e_j^*(\underline{k}\sigma)\; \exp(i\underline{k}\underline{R}^{\underline{h}})\; \delta\left(\omega^2-\omega_\sigma^2(\underline{k})\right)$$

$$\cong \frac{\pi}{2\overset{o}{M}V_B\omega}\sum_\sigma \int k^2\, dk\, d\Omega\; e_i(\underline{\kappa}\sigma)\; e_j^*(\underline{\kappa}\sigma)\; \cos \underline{k}\underline{R}^{\underline{h}}\; \delta\left(\omega - c_\sigma(\underline{\kappa})k\right) , \quad \omega>0 \tag{2.66}$$

since for $k\to 0$ the polarization vectors depend only on the direction of $\underline{k}$. By performing the k-integration, the cosine may be expanded for $R^{\underline{h}} \ll c_\sigma(\underline{\kappa})/\omega$.

$$\mathrm{Im}\{\overset{o}{G}{}^{(\underline{h})}_{ij}(\omega)\} \cong \omega\, \frac{\pi}{2\overset{o}{M}V_B}\sum_\sigma \int d\Omega\, \frac{e_i(\underline{\kappa}\sigma)e_j^*(\underline{\kappa}\sigma)}{c_\sigma^3(\underline{\kappa})} - \omega^3\, \frac{\pi}{4\overset{o}{M}V_B}\sum_\sigma \int d\Omega\, \frac{e_i\; e_j^*}{c_\sigma^5(\underline{\kappa})}\,(\underline{\kappa}\underline{R}^{\underline{h}})^2 - \dots \tag{2.67}$$

The first term is independent of $\underline{h}$ and for cubic crystals proportional to δ_{ij}. In addition to these two terms we get a second term $\sim\omega^3$ which is independent of $\underline{h}$ and results from the dispersion for larger $\underline{k}$-values, i.e. the deviation of $\omega_\sigma(\underline{k})$ from $c_\sigma\cdot k$. Thus we obtain for cubic crystals

$$\mathrm{Im}\{\overset{o}{G}{}^{(\underline{h})}_{ij}(\omega)\} \cong \alpha\omega(1+\alpha'\omega^2)\,\delta_{ij} - \omega^3\sum_{k\ell} R^{\underline{h}}_k\, T_{ij\,k\ell}\, R^{\underline{h}}_\ell \,. \tag{2.68}$$

This means that the spectrum $\overset{o}{z}(\omega)$ of the ideal crystal is proportional to ω^2 for small ω and can be represented by a Debye-spectrum

$$\overset{o}{z}(\omega) \cong z_{Debye}(\omega) = \begin{cases} 3\dfrac{\omega^2}{\omega_D^3} & \text{for } \omega<\omega_D \\ 0 & \text{for } \omega>\omega_D \end{cases} . \tag{2.69}$$

The constants α, ω_D and $T_{ij\,k\ell}$ are given from (2.67) as angular integrals over the sound velocities. For instance,

$$\frac{3}{\omega_D^3} = \frac{2\overset{\circ}{M}}{\pi}\alpha = \frac{V_c}{6\pi^2}\int\frac{d\Omega}{4\pi}\sum_\sigma\frac{1}{\{c_\sigma(\underline{\kappa})\}^3}\,, \tag{2.70}$$

where V_c is the elementary volume. They are rather complicated functions of the elastic constants and can only be given explicitly for isotropic crystals ($2c_{44} = c_{11} - c_{12}$):

$$\frac{3}{\omega_D^3} = \frac{2\overset{\circ}{M}\alpha}{\pi} = \frac{V_c}{6\pi^2}\left(\frac{2}{c_t^3} + \frac{1}{c_\ell^3}\right) \quad \text{with} \quad c_t = \sqrt{\frac{c_{44}}{\rho}}\,, \quad c_\ell = \sqrt{\frac{c_{11}}{\rho}}\,, \quad \rho = \frac{\overset{\circ}{M}}{V_c}$$

$$T_{ij\,k\ell} = \frac{V_c}{60\pi\overset{\circ}{M}}\left\{\left(\frac{4}{c_t^5} + \frac{1}{c_\ell^5}\right)\delta_{ij}\,\delta_{k\ell} + \left(\frac{1}{c_\ell^5} - \frac{1}{c_t^5}\right)(\delta_{ij}\,\delta_{k\ell} + \delta_{i\ell}\,\delta_{jk})\right\}. \tag{2.71}$$

As we will show in Chapter 4,the result derived here for the spectrum of the ideal crystal is valid more generally. For instance, the local spectrum $z_i^{\underline{m}}(\omega)$ for the case of a single, arbitrary point defect agrees with the ideal spectrum $\overset{\circ}{z}(\omega)$ for sufficiently small ω except for a trivial factor.

2.5 Asymptotic Expansion for Large Distances

For distances $|\underline{R}^{\underline{m}} - \underline{R}^{\underline{n}}| \gg a$, the lattice constant, the ideal lattice Green's function can be expanded asymptotically by using the method of stationary phases /1.2,3/. For frequencies within the spectrum $0 \leqslant \omega \leqslant \omega_{max}$ we transform the denominator in (2.19) into an integral

$$\frac{1}{\omega_\sigma^2(\underline{k}) - (\omega + i\eta)^2} = i\int_0^\infty \exp\left[-i\left(\omega_\sigma^2(\underline{k}) - \omega^2\right)t\right]\exp(-2\eta\omega t)\,dt$$

leading to

$$\overset{\circ}{G}{}_{ij}^{(\underline{h})}(\omega) = \frac{1}{MV_B}\sum_\sigma\int_{V_B}d\underline{k}\int dt\; e_i(\underline{k}\sigma)\,e_j^*(\underline{k}\sigma)\exp(i\varphi - 2\eta\omega t)$$
$$\text{with} \quad \varphi(\underline{k},t) = \left(\omega^2 - \omega_\sigma^2(\underline{k})\right)t + \underline{k}\underline{R}^{\underline{h}}\,. \tag{2.72}$$

For large $\underline{R}^{\underline{h}}$ the exponential function oscillates very rapidly so that we obtain essentially only contributions from those $\underline{k}$,t values, for which the phase $\varphi(\underline{k},t)$ is stationary

$$\partial_{\underline{k}}\varphi = 0 = \underline{R}^{\underline{h}} - t\,\partial_{\underline{k}}\,\omega_\sigma(\underline{k})$$
$$\partial_t\varphi = 0 = \omega^2 - \omega_\sigma^2(\underline{k})\,. \tag{2.73}$$

If there is more than one point $\underline{k}_\nu$, t_ν where for given $\underline{R}^{\underline{h}}$ the phase is stationary, we have to sum over these. By expanding around a stationary point, we obtain

$$\varphi(\underline{k}_\nu + \delta\underline{k}\,,\, t_\nu + \delta t) = \underline{k}_\nu \underline{R}^{\underline{h}} - \frac{\partial\omega_\sigma^2(\underline{k}_\nu)}{\partial\underline{k}_\nu}\,\delta\underline{k}\;\delta t - \frac{1}{2}\sum_{ij}\frac{\partial^2\omega_\sigma^2}{\partial k_{\nu i}\,\partial k_{\nu j}}\,\delta k_i\,\delta k_j\, t - \dots \quad (2.74)$$

The integrals $d\underline{k}$, dt are extended to integrals from $-\infty$ to $+\infty$. The polarization vectors are replaced by their values at $\underline{k}_\nu$. By choosing a new $\delta\underline{k}$-coordinate system with the x axis along $\partial_{\underline{k}_\nu}\omega_\sigma^2(\underline{k}_\nu) = \underline{R}^{\underline{h}}/t$ and the y and z axis perpendicular to $\underline{R}^{\underline{h}}$ so that the mixed derivative $\partial_y\partial_z\omega_\sigma^2(\underline{k}_\nu)$ vanishes, the t integration yields a δ function $\delta(\delta k_x)$, whereas the subsequent δk_y and δk_z integrals are Fresnel integrals. Thus we obtain finally

$$\overset{o}{G}{}_{ij}^{(\underline{h})}(\omega) \cong \frac{1}{MV_B}\sum_{\sigma,\nu} e_i(\underline{k}_\nu\sigma)\, e_j(\underline{k}_\nu\sigma)\,\frac{\exp\!\left(-i\frac{\pi}{2}\,(\mathrm{sgn}\,\varepsilon_y + \mathrm{sgn}\,\varepsilon_z)\right)}{\sqrt{(|\varepsilon_y\cdot\varepsilon_z|)}}\,\frac{\exp(i\underline{k}_\nu\underline{R}^{\underline{h}})}{|\underline{R}^{\underline{h}}|} \quad (2.75)$$

$$\text{with}\quad \varepsilon_y = \frac{\partial^2\omega_\sigma^2(\underline{k}_\nu)}{\partial k_{\nu y}^2}\;,\qquad \varepsilon_z = \frac{\partial^2\omega_\sigma^2(\underline{k}_\nu)}{\partial k_{\nu z}^2}\;.$$

This solution holds, if $R^{\underline{h}} \gg a$ and in addition the phase φ is large, i.e., $\underline{k}\underline{R}^{\underline{h}} \cong \frac{\omega}{c}R^{\underline{h}} \gg 1$. The latter might not be the case for small frequencies, even if $R^{\underline{h}} \gg a$. In this limit, however the lattice Green's function approaches the continuum Green's function for which we have with $\omega_\sigma(\underline{k}) = c_\sigma(\underline{\kappa})k$ and $\rho = M/V_c$

$$G_{ij}(\underline{R}^{\underline{h}},\omega) = \frac{1}{\rho(2\pi)^3}\sum_\sigma\int_{-\infty}^{\infty} d\underline{k}\;\frac{e_i(\underline{\kappa}\sigma)\; e_j(\underline{\kappa}\sigma)}{c_\sigma^2(\underline{\kappa})k^2 - (\omega + i\eta)^2}\,\exp(i\underline{k}\underline{R}^{\underline{h}})\;. \quad (2.76)$$

Unfortunately the elastic Green's function cannot be calculated for the general anisotropic case. For isotropy [two constant sound velocities c_ℓ and c_t of (2.71)] the polarization vectors are

$$e_i(\underline{\kappa}1) = e_i^\ell(\underline{\kappa}) = \kappa_i \quad\text{and}\quad \underline{e}(\underline{\kappa}2)\cdot\underline{\kappa} = 0 = \underline{e}(\underline{\kappa}3)\underline{\kappa}$$
$$\text{so that}\quad \sum_{\sigma=2}^{3} e_i(\underline{\kappa}\sigma)\; e_j(\underline{\kappa}\sigma) = \delta_{ij} - \kappa_i\kappa_j\;. \quad (2.77)$$

Then we obtain by performing the integration

$$G_{ij}(\underline{R},\omega) = \frac{\delta_{ij}}{4\pi\rho c_t^2}\,\frac{\exp(i\omega R/c_t)}{R} + \frac{1}{4\pi\rho\omega^2}\,\frac{\partial^2}{\partial R_i\,\partial R_j}\left\{\frac{1}{R}\left(\exp(i\omega R/c_t) - \exp(i\omega R/c_\ell)\right)\right\}. \quad (2.78)$$

For the asymptotic limit $\omega R/c \gg 1$ we get the same result as we would obtain directly from (2.75), namely $1/R$ times an oscillatory function. Of special interest is the static limit for $\omega R \ll 1$:

$$G_{ij}(\underline{R},\omega) \cong \frac{\delta_{ij}}{8\pi\rho R}\left(\frac{1}{c_t^2}+\frac{1}{c_\ell^2}\right)+\frac{R_iR_j}{8\pi\rho R^3}\left(\frac{1}{c_t^2}-\frac{1}{c_\ell^2}\right)+i\,\frac{\delta_{ij}}{12\pi\rho}\left(\frac{2}{c_t^3}+\frac{1}{c_\ell^3}\right). \qquad (2.79)$$

The real part is the static Green's function /2.4/, whereas the imaginary part leads to the Debye spectrum (2.68,70).

Outside the allowed frequency band, i.e., for $\omega > \omega_{max}$, the Green's function shows a completely different behaviour: it decreases exponentially for $R^{\underline{h}} \gg a$. Here the method of stationary phases has to be modified, since the condition (2.73) $\omega^2 = \omega_\sigma^2(\underline{k})$ can no longer be satisfied for real $\underline{k}$ values, but only for complex $\underline{k} = \underline{k}' + i\underline{k}''$. Since for large $\underline{R}^{\underline{h}}$ we must have an outgoing plane wave, $\underline{k}''$ has to be parallel to $\underline{R}^{\underline{h}}$ and $\underline{k}''\underline{R}^{\underline{h}} > 0$. Thus by putting the x axis in the direction of $\underline{R}^{\underline{h}} = (R^{\underline{h}},0,0)$ we have

$$\underline{k} = (k_x' + ik_x'' , k_y , k_z) \quad \text{with} \quad k_y , k_z \text{ real and } k_x'' > 0.$$

Then all real ω compatible with $\omega^2 = \omega_\sigma^2(k_x' + ik_x'' , k_y , k_z)$ define a line in the complex k_x plane, along which the gradient $\partial\omega^2/\partial\underline{k}$ of (2.74) has to be calculated, since $R^{\underline{h}}$ and t are real. Because the phase φ is only stationary for complex k_x, the k_x integration has to be shifted into the complex plane which can be shown to be possible for simple examples.

The result is essentially the same as (2.75), but with $\exp(i\underline{k}_\nu\underline{R}^{\underline{h}})$ replaced by $\exp(i\underline{k}_\nu'\underline{R}^{\underline{h}})\exp(-\underline{k}_{\nu x}''\underline{R}^{\underline{h}})$. The imaginary part $k_{\nu x}''$ and therefore the exponential decrease is larger the larger ω^2.

2.6 Singularities at Critical Points

The most prominent feature of the frequency spectra discussed in the next section is the appearance of sharp peaks and kinks which give rise to singularities in the derivatives of the spectrum. These singularities have their origin in the periodicity of the lattice and in the behaviour of $\omega_\sigma(\underline{k})$ at certain points in the Brillouin zone.

The appearance of such singularities can best be seen by writing the frequency distribution $\mathring{z}(\omega^2)$ as a surface integral over the constant frequency surface $\omega^2 = \omega_\sigma^2(\underline{k})$:

$$\mathring{z}(\omega^2) = \frac{1}{3V_B}\sum_\sigma \int_{V_B} d\underline{k}\;\delta\left(\omega^2 - \omega_\sigma^2(\underline{k})\right) = \frac{1}{3V_B}\sum_\sigma \int \frac{dS}{|\mathrm{grad}\;\omega_\sigma^2(\underline{k})|} \quad \text{with } \mathring{z}(\omega^2) = \frac{\mathring{z}(\omega)}{2\omega} \qquad (2.80)$$

since $d\underline{k} = dS\; d\omega_\sigma^2(\underline{k}) \Big/ \left|\frac{\partial\omega_\sigma^2(\underline{k})}{\partial\underline{k}}\right|$.

Thus we expect a singular behaviour of $\mathring{z}(\omega^2)$ to result from those "critical points", for which grad $\omega_\sigma^2(\underline{k}) = 0$. E.g., $\underline{k} = 0$ is always such a point. Further there are always critical points at some places at the boundary of the Brillouin zone. E.g., if a part of the boundary coincides with a mirror plane, then along the boundary $\underline{n}\cdot\text{grad}\,\omega_\sigma^2(\underline{k}) = 0$, where $\underline{n}$ is the normal. Since in symmetry directions (e.g., 100, 110) $\underline{n} \parallel \text{grad}\,\omega_\sigma(\underline{k})$, grad $\omega_\sigma(\underline{k})$ has to vanish at these points on the Brillouin zone boundary.

By expanding $\omega_\sigma^2(\underline{k})$ at a critical point $\underline{k}^c$

$$\omega_\sigma^2(\underline{k}) = \omega_\sigma^2(k^c) + \frac{1}{2}\sum_{ij=1}^{3} \frac{\partial^2\omega_\sigma^2}{\partial k_i^c\,\partial k_j^c}(k_i - k_i^c)(k_j - k_j^c) + \dots \tag{2.81}$$

we can choose the coordinate system so that it coincides with the main axes of $\partial^2\omega_\sigma^2/\partial k_i\partial k_j$ at $\underline{k}^c$. Then we have

$$\omega^2(\underline{k}) = \omega_c^2 + \sum_{\alpha=1}^{3} \lambda_\alpha(k_\alpha - k_\alpha^c)^2 \tag{2.82}$$

and we can classify the critical points as follows

a)	$\lambda_1, \lambda_2, \lambda_3 > 0$	minimum of $\omega_\sigma^2(\underline{k})$ at $\underline{k}^c$
b)	$\lambda_1, \lambda_2 > 0$, $\lambda_3 < 0$	saddle point of type 1
c)	$\lambda_1 > 0$, $\lambda_2, \lambda_3 < 0$	saddle point of type 2
d)	$\lambda_1, \lambda_2, \lambda_3 < 0$	maximum.

The contribution of the region around such a critical point to $\overset{\circ}{z}(\omega^2)$ can be calculated as is illustrated in the following for the minimum. We get a contribution

$$\delta\overset{\circ}{z}(\omega^2) = \frac{1}{3V_B}\int d(\delta\underline{k})\;\delta(\omega^2 - \omega_c^2 - \sum_{\alpha=1}^{3}\lambda_\alpha^2\,\delta k_\alpha^2)\;. \tag{2.83}$$

By setting $\sqrt{\lambda_1}\,\delta k_1 = x$, $\sqrt{\lambda_2}\,\delta k_2 = y$, $\sqrt{\lambda_3}\,\delta k_3 = z$ and $x^2 + y^2 + z^2 = r^2$ we obtain for the minimum

$$\delta\overset{\circ}{z}(\omega^2) = \frac{2\pi}{3V_B\sqrt{\lambda_1\lambda_2\lambda_3}}\sqrt{\omega^2 - \omega_c^2}\,\theta(\omega^2 - \omega_c^2) \quad \text{with} \quad \theta(x) = \begin{cases} 1 & \text{for } x > 0 \\ 0 & \text{for } x < 0 \end{cases}\;. \tag{2.84}$$

Thus not $\overset{\circ}{z}(\omega^2)$, but its derivative becomes singular $\sim 1/\sqrt{\omega^2 - \omega_c^2}$ as ω approaches the critical frequency ω_c from above. (In 1 and 2 dimensions, also $\overset{\circ}{z}(\omega^2)$ is singular).

Analogously, the behaviour in the vicinity of the other critical points can be calculated and the results are illustrated in Fig. 2a-d. For a saddle point of type 1, $\overset{\circ}{z}(\omega^2)$ rises to $\overset{\circ}{z}(\omega_c^2)$ as $-\sqrt{\omega_c^2 - \omega^2}$ as ω approaches ω_c from below. For the saddle point of type 2, we have a $\sqrt{\omega^2 - \omega_c^2}$-decrease on the high-frequency side, whereas the maximum behaves analogously to the minimum, but with reversed frequency scale.

We will not further discuss the theory of these critical points which has been developed by van HOVE /2.5/, PHILLIPS /2.6/, ROSENSTOCK /2.7/ and others, and which can be found in detail in /1.5/. We want to point out that this singular be-

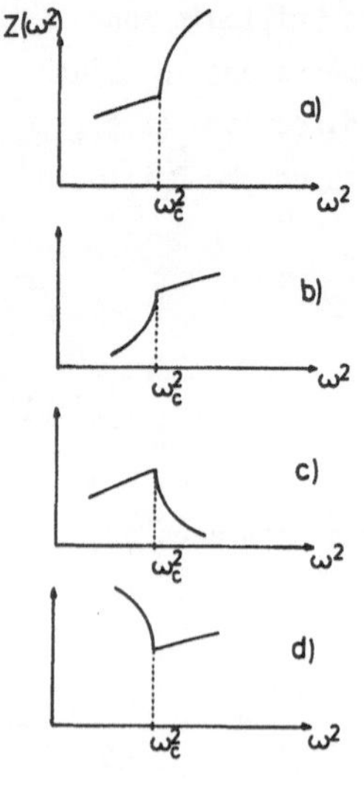

Fig. 2a-d. Behaviour of the spectrum $z(\omega^2)$ near critical points

a) minimum ($\lambda_1 , \lambda_2 , \lambda_3 > 0$) eq. (2.84)

b) saddle point of type 1 ($\lambda_1 , \lambda_2 > 0$; $\lambda_3 < 0$)

c) saddle point of type 2 ($\lambda_1 > 0$; $\lambda_2 , \lambda_3 < 0$)

d) maximum ($\lambda_1 , \lambda_2 , \lambda_3 < 0$)

haviour does show up not only in the spectrum but also in all matrix elements $G_{ij}^{\underline{m}\underline{n}}(\omega)$ of the ideal and defect lattice Green's functions.

This is made clear by writing $\mathrm{Im}\{\overset{\circ}{G}(\omega)\}$ as a surface integral over the constant energy surface $\omega = \omega_\sigma(\underline{k})$:

$$\mathrm{Im}\{\overset{\circ}{G}{}_{ij}^{(\underline{m}-\underline{n})}(\omega)\} = \frac{\pi}{MV_B} \sum_\sigma \int \frac{dS}{|\mathrm{grad}\ \omega_\sigma^2(\underline{k})|} e_i(\underline{k}\sigma)\ e_j^*(\underline{k}\sigma) \cdot \exp\left(i\underline{k}(\underline{R}^{\underline{m}} - \underline{R}^{\underline{n}})\right) . \qquad (2.85)$$

Thus we obtain for $\mathrm{Im}\{\overset{\circ}{G}{}_{ij}^{(\underline{m}-\underline{n})}(\omega)\}$ essentially the same square root singularities at the critical points. Further this behaviour of $\mathrm{Im}\{\overset{\circ}{G}(\omega)\}$ induces a similar behaviour of $\mathrm{Re}\{\overset{\circ}{G}(\omega)\}$. For if

$$\mathrm{Im}\{\overset{\circ}{G}(\omega)\} = \alpha \sqrt{\omega^2 - \omega_c^2}\, \theta(\omega^2 - \omega_c^2) \quad , \quad \partial_{\omega^2} \mathrm{Im}\{G(\omega)\} = \frac{\alpha}{2\sqrt{\omega^2 - \omega_c^2}} \theta(\omega^2 - \omega_c^2) \ , \quad \alpha \neq 0 \qquad (2.86)$$

we obtain from the dispersion relation (2.26) for $\partial_{\omega^2}\overset{\circ}{G}(\omega)$ by partial integration

$$\begin{aligned} \partial_{\omega^2}\overset{\circ}{G}(\omega) &= \frac{1}{\pi} \int_0^\infty d\omega'^2 \frac{1}{\omega'^2 - (\omega + i\eta)^2} \partial_{\omega'^2} \mathrm{Im}\{\overset{\circ}{G}(\omega')\} \\ &\cong \frac{1}{\pi} \int_0^\infty d\omega'^2 \frac{1}{\omega'^2 - (\omega + i\eta)^2} \frac{\alpha}{2\sqrt{\omega'^2 - \omega_c^2}} = i \frac{\alpha}{2\sqrt{\omega^2 - \omega_c^2}} . \end{aligned} \qquad (2.87)$$

The last integral can be evaluated by complex integration, i.e., making a branch-cut on the real axis from ω_c^2 to ∞ and writing the integral as a line integral along both sides of the cuts which can be calculated by residuum technique. Thus $\overset{o}{G}(\omega)$ and $\mathrm{Re}\{\overset{o}{G}(\omega)\}$ behave at ω_c as

$$\overset{o}{G}(\omega) = i\alpha \sqrt{\omega^2 - \omega_c^2} \;, \quad \mathrm{Re}\{\overset{o}{G}(\omega)\} = -\alpha \sqrt{\omega_c^2 - \omega^2}\, \theta(\omega_c^2 - \omega^2) \;. \tag{2.88}$$

On the other hand, if

$$\mathrm{Im}\{\overset{o}{G}(\omega)\} = \beta \sqrt{\omega_c^2 - \omega^2}\, \theta(\omega_c^2 - \omega^2) \quad \text{with } \beta \neq 0 \tag{2.89}$$

we obtain by the same arguments as before

$$\overset{o}{G}(\omega) = \beta \sqrt{\omega^2 - \omega_c^2} \;, \quad \mathrm{Re}\{\overset{o}{G}(\omega)\} = +\beta \sqrt{\omega^2 - \omega_c^2}\, \theta(\omega_c^2 - \omega^2) \;. \tag{2.90}$$

Thus $\overset{o}{G}(\omega)$ has a square root singularity at $\omega^2 = \omega_c^2$ and therefore $\mathrm{Re}\{\overset{o}{G}(\omega)\}$ shows such a singularity on the opposite side of ω_c as $\mathrm{Im}\{\overset{o}{G}(\omega)\}$. This behaviour of $\mathrm{Im}\{\overset{o}{G}(\omega)\}$ and $\mathrm{Re}\{\overset{o}{G}(\omega)\}$ is illustrated in Fig. 3a,b.

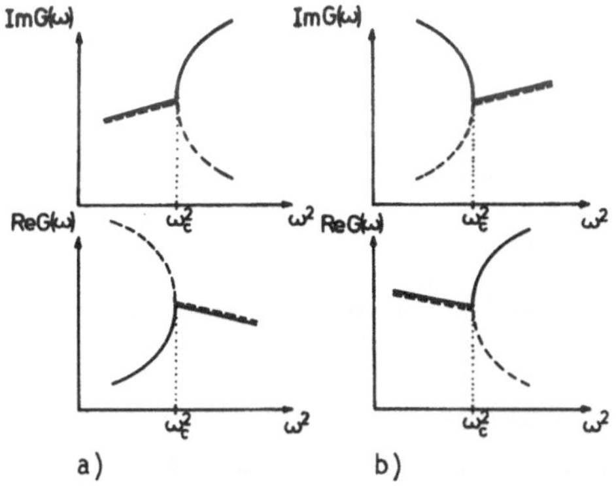

Fig. 3a,b. Behaviour of the real part $\mathrm{Re}\{\overset{o}{G}(\omega)\}$ and imaginary part $\mathrm{Im}\{\overset{o}{G}(\omega)\}$ near a critical point according to eq. (2.86,88) and (2.89,90).

a) $\mathrm{Im}\{\overset{o}{G}(\omega)\} = \alpha \sqrt{\omega^2 - \omega_c^2}\, \theta(\omega^2 - \omega_c^2) \leftrightarrow$
$\mathrm{Re}\{\overset{o}{G}(\omega)\} = -\alpha \sqrt{\omega_c^2 - \omega^2}\, \theta(\omega_c^2 - \omega^2)$

b) $\mathrm{Im}\{\overset{o}{G}(\omega)\} = \beta \sqrt{\omega_c^2 - \omega^2}\, \theta(\omega_c^2 - \omega^2) \leftrightarrow$
$\mathrm{Re}\{\overset{o}{G}(\omega)\} = \beta \sqrt{\omega^2 - \omega_c^2}\, \theta(\omega^2 - \omega_c^2)$.

The full lines refer to $\alpha > 0$ and $\beta > 0$, the dashed lines to $\alpha < 0$ and $\beta < 0$

Finally we note that these singularities also appear for some classes of defect lattice Green's functions. If we have an isolated point defect or in general a bounded three-dimensional defect lattice region embedded in an infinite ideal lattice, then the eigenfunctions for $\omega < \overset{o}{\omega}_{max}$ are scattering states $\underline{\xi}(\underline{k}\sigma)$ which can be labelled by the incident plane wave $|\underline{k}\sigma\rangle$. Most important is that the eigenfrequencies $\omega_\sigma(\underline{k})$ are the same as in the ideal lattice. Thus the critical points are the same and the corresponding Green's functions have the same kind of square root singularities at the critical frequencies.

2.7 Analytical and Numerical Solutions for Cubic Ideal Lattices

Despite the importance of the Green's function in lattice dynamics there exist practically no analytical solutions except for rather trivial one-dimensional problems as is discussed in /1.3,5/.

The only nontrivial analytical solutions in three dimensions are the ones for the "scalar model". This model is also known as Potts-Montroll model. If the coupling matrix is diagonal in i and j, the Green's function is diagonal, too

$$\overset{\circ}{\phi}{}^{\underline{m}\underline{n}}_{ij} = \delta_{ij}\,\varphi^{(\underline{m}-\underline{n})} \quad , \quad \overset{\circ}{G}{}^{(\underline{m}-\underline{n})}_{ij}(\omega) = \delta_{ij}\, g^{(\underline{m}-\underline{n})}(\omega) \tag{2.91}$$

and reduces to a *scalar Green's function* $g^{(\underline{m}-\underline{n})}$. Due to the scalar coupling, the eigenfrequencies $\omega_\sigma(\underline{k}) = \omega(\underline{k})$ for the different branches are degenerate, so that the polarization vector $e_i(\underline{k}\sigma)$ can be chosen arbitrarily. Thus $g(\omega)$ is given by

$$g^{(\underline{m}-\underline{n})}(\omega) = \frac{1}{MV_B}\int_{V_B} d\underline{k}\,\frac{\exp\left(i\underline{k}(\underline{R}^{\underline{m}}-\underline{R}^{\underline{n}})\right)}{\omega^2(\underline{k}) - (\omega + i\eta)^2} \; . \tag{2.92}$$

For the special case of nearest neighbour interaction with force constant f, $\omega^2(\underline{k})$ can directly be given

$$\omega^2(\underline{k}) = \begin{cases} 6\,\frac{f}{M}\left\{1 - \frac{1}{3}\left(\cos k_x a + \cos k_y a + \cos k_z a\right)\right\} & \text{for scc} \\ 8\,\frac{f}{M}\left\{1 - \left(\cos k_x\frac{a}{2}\cdot\cos k_y\frac{a}{2}\cdot\cos k_z\frac{a}{2}\right)\right\} & \text{for bcc} \\ 12\,\frac{f}{M}\left\{1 - \frac{1}{3}\left(\cos k_x\frac{a}{2}\cos k_y\frac{a}{2} + \cos k_y\frac{a}{2}\cos k_z\frac{a}{2} + \cos k_z\frac{a}{2}\cos k_x\frac{a}{2}\right)\right\} & \text{for fcc} \end{cases} \tag{2.93}$$

This scalar Green's function is widely used for problems of spinwave theory of magnetism (see references given in /2.8/) and for random walk processes in lattices ("lattice diffusion") /2.8-10/. In the latter case the square of the frequency ω^2 has to be replaced by $i\omega$ due to the irreversible time behaviour in diffusion caused by the first derivative d/dt in the equation of motion.

In the last years a considerable amount of work has been invested to obtain analytical solutions of the scalar Green's function for nearest neighbour interaction. As a result, all matrix elements $g^{(\underline{m}-\underline{n})}(\omega)$ can in principle be expressed in terms of complete elliptic integrals of the first and second kind [K(k) and E(k)]. The results for the bcc lattice are given in /2.8,11/, for fcc in /2.12,13/ and for scc in /2.14,15/. For instance, for bcc $g^{(0)}(\omega)$ is given by

$$\begin{aligned} \mathrm{Re}\{g^{(0)}(\omega)\} &= \frac{2}{f\pi^2}\,K(k_+)\,K(k_-) & &\text{with } k_\pm = \frac{1}{2} \pm \sqrt{\frac{\omega^2}{\omega_{max}^2}\left(1 - \frac{\omega^2}{\omega_{max}^2}\right)} \\ \mathrm{Im}\{g^{(0)}(\omega)\} &= \frac{1}{f\pi^2}\left(K(k_+)^2 - K(k_-)^2\right) & &\text{and } \omega_{max}^2 = 16\,\frac{f}{M} \end{aligned} \tag{2.94}$$

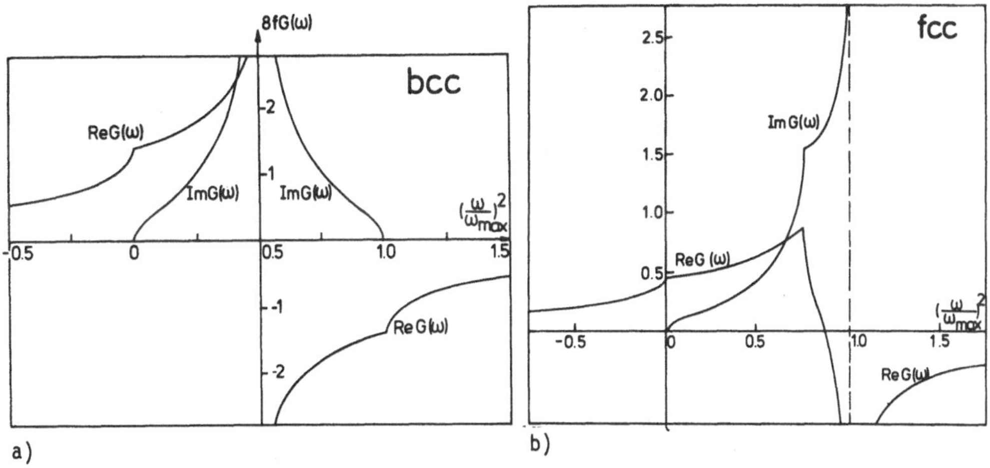

Fig. 4a,b. Real and imaginary part of the scalar Green's function $G(\omega) = g^{(0)}(\omega)$ as a function of $(\omega/\omega_{max})^2$; a) for the bcc lattice; b) for the fcc lattice

The real and imaginary parts are plotted in Fig. 4a. As a function of ω^2, $\text{Im}\{g^{(0)}(\omega)\}$ is symmetrical around $\omega^2 = \omega^2_{max}/2 = 8\frac{f}{M}$. There are ordinary critical points at $\omega^2 = 0$, arising from the minimum at $\underline{k} = 0$ and at $\omega^2 = \omega^2_{max}$ due to the maximum at $\underline{k} = \frac{2\pi}{a}$ (111). At $\omega^2 = \omega^2_{max}/2$, there is a log singularity arising, e.g., from the critical region $k = \frac{\pi}{a}$ (111), where in addition to grad $\omega^2(\underline{k}) = 0$ also the determinant of the second derivatives $\partial^2\omega^2/\partial k_i \partial k_j$ vanishes, leading to a log divergence of $g^{(0)}(\omega)$ /1.5/.

Fig. 4b shows the real and imaginary parts of the Green's function for the scalar fcc lattice model for nearest neighbours. Here the Green's function diverges logarithmically for $\omega^2 = \omega^2_{max}$. From both figures the reversed behaviour of Re{G} and Im{G} at the critical points can be seen.

The scalar model has some serious drawbacks. First the divergences of the Green's functions for the fcc and bcc nearest neighbour model as shown in Fig. 4a,b signalize a dynamic instability of the lattice. Such a dynamic instability would be removed by anharmonic effects which broaden the phonon dispersion curves. Secondly, and more important, the model is also elastically instable.

Since all velocities of sound are independent of the direction of $\underline{k}$ in this model, the lattice is isotropic ($c_{11} - c_{12} = 2c_{44}$). Further, the longitudinal and the transversal velocities are equal ($c_{11} = c_{44}$). Thus, the compression modulus K, $K = \frac{1}{3}(c_{11} + 2c_{12}) = -c_{44}/3$, is negative and the crystal is instable against compression. This is not a contradiction to $\omega^2(\underline{k}) > 0$ for all $\underline{k}$ since a compression cannot be described by plane waves.

Since other analytical methods are not available, one has to *calculate the Green's function* and also the spectrum *numerically*. The simplest and most straightforward method for the spectrum is the root sampling method: One determines the frequencies $\omega_\sigma(\underline{k})$ at a large number of uniformly distributed points $\underline{k}_\nu$ and then plots a histogramme of the frequencies. To calculate the spectrum in this way one needs an extremely large number of points in the first Brillouin zone since otherwise the fine features of the spectrum, expecially near the critical points, cannot be described.

A much more efficient way of calculating the spectrum and also the Green's functions was given by GILAT and RAUBENHEIMER /2.16/. In this method the irreducible $1/48^{th}$ of the first Brillouin zone is covered by a dense simple cubic mesh of points $\underline{k}_\nu$ which form the centers of mini-cubes of edge 2b. The spectrum is then the sum of all contributions from the single cubes. But instead of representing the contribution of a single cube merely by it values $\omega_\sigma(\underline{k}_\nu)$, the contribution of a single cube is calculated exactly under the assumption that $\omega_\sigma(\underline{k})$ is slowly varying in the cube, so that a linear expansion

$$\omega_\sigma(\underline{k}) = \omega_\sigma(\underline{k}\ ) + \text{grad}\ \omega_\sigma(\underline{k}_\nu)\cdot(\underline{k} - \underline{k}_\nu)$$

is sufficient. The gradient at each mesh point $\underline{k}_\nu$ is calculated by using the deviation of the dynamical matrix from its value at the mesh point as a perturbation. The advantage of this method is that one indeed calculates the contribution of "all" frequencies to the spectrum and not merely of individual points in the first Brillouin zone. Thus without increasing the numerical effort the resolution is improved enormously.

The imaginary parts of the Green's function $\text{Im}\{\overset{o}{G}{}^{\underline{m}\underline{n}}_{ij}(\omega)\}$ are calculated in essentially the same way by multiplying the contribution to the spectrum due to the minicube $\underline{k}_\nu$ by $e_i(\underline{k}_\nu\sigma)e_j^*(\underline{k}_\nu\sigma)\cdot\exp\left(i\underline{k}_\nu(\underline{R}^{\underline{m}} - \underline{R}^{\underline{n}})\right)$and by summing over all cubes. From the imaginary parts then the real parts $\text{Re}\{\overset{o}{G}{}^{(\underline{m}-\underline{n})}_{ij}(\omega)\}$are calculated by evaluating the finite one-dimensional integration $d\omega'^2$ in the Kramers-Kronig relation (2.25) numerically.

To increase the accuracy of $z(\omega)$ and $\text{Im}\{G(\omega)\}$ for very small ω, the cube mesh near the origin is subdivided twice to meshes three times finer making a 9^3 times higher cube density at the origin. Since only the behaviour for very small ω, $\underline{k}$ determines the elastic limit of the Green's function (2.75), this procedure allows a very accurate calculation of the asymptotic behaviour for large R and small ω. For instance, it could be shown for the static case /2.17/ that the lattice Green's function agrees well with its elastic counterpart, varying $\sim 1/R$ outside the effective range of the coupling parameters, which means, e.g., for Cu and Al, from the first neighbours on.

Another efficient method of calculating densities of states is the tetrahedron method of JEPSEN and ANDERSEN /2.18/, and LEHMANN and TAUT /2.19/ which will, however, not be discussed here.

The simplest stable lattice model is the nearest neighbour model for the fcc lattice with one longitudinal force constant f (spiral spring) which for a central-

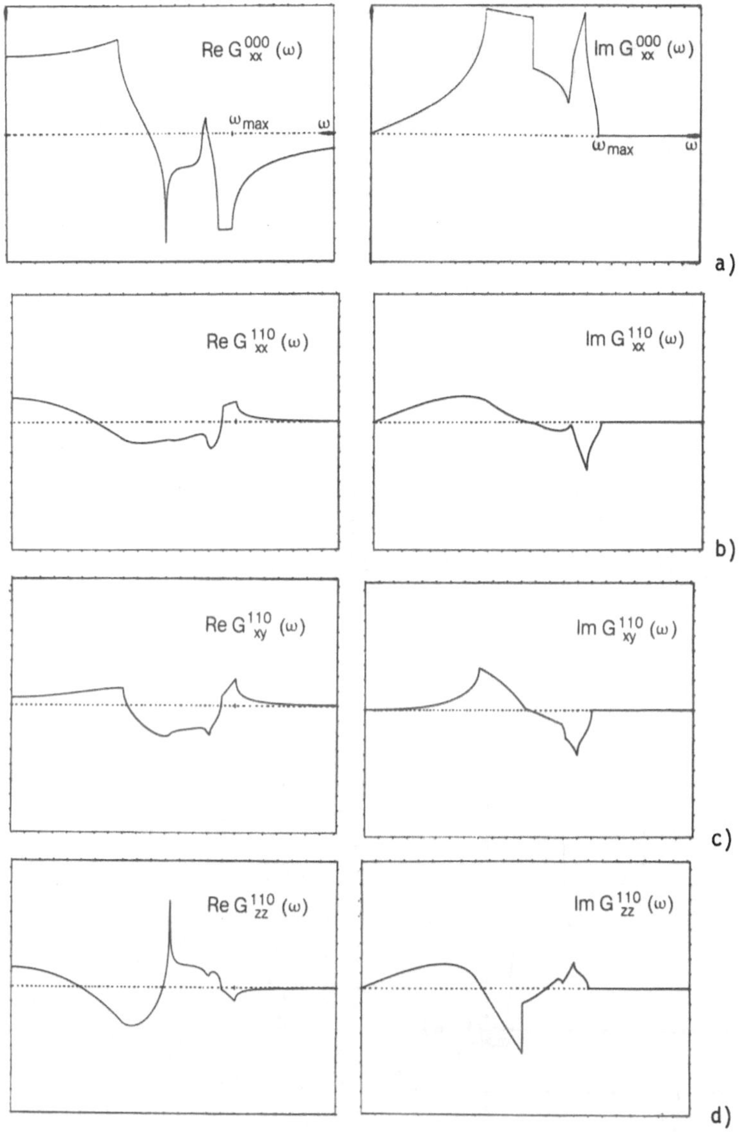

Fig. 5a-d. Real and imaginary part of the Green's functions for the central-force nearest neighbour model (fcc).

a) Green's function at the origin $G_{xx}^{(000)}(\omega)$

b) nearest neighbour Green's function $G_{xx}^{(110)}(\omega)$

c) $G_{xy}^{(110)}(\omega)$ d) $G_{zz}^{(110)}(\omega)$

force interaction is given by $f = V''(R)$ where $V(R)$ is the potential at the equilibrium position R. This model is sometimes also referred to as Leighton's model or central-force nearest neighbour model. It gives the following connection between the elastic data $c_{44} = c_{12} = c_{11}/2$. For this model the real and imaginary part of the Green's function at the origin $G_{xx}^{(0)}(\omega)$ and at the nearest neighbour $\left(G_{xx}^{(110)}(\omega), G_{xy}^{(110)}(\omega), G_{zz}^{(110)}(\omega)\right)$ are plotted in Fig. 5a-d. All other elements are equal to the ones given here, e.g., $G_{yy}^{(110)} = G_{xx}^{(110)}$ or vanish identically, e.g., $G_{xy}^{(110)} = 0$.

Whereas the imaginary part of all diagonal elements $G_{ii}^{\underline{mm}}(\omega)$ is positive [e.g., $G_{xx}^{(0)}(\omega)$], all nondiagonal ones [e.g. $G_{ii}^{(110)}(\omega)$] show an oscillatory behaviour which can be regarded as a consequence of the sum rule (2.49)

$$\int d\omega'^2 \, \mathrm{Im}\{G_{ij}^{\underline{mn}}(\omega')\} = 0 \qquad \text{for } \underline{m} \neq \underline{n} \text{ or } i \neq j \; .$$

For large ω, the elements for the nearest neighbour decrease much faster [$\sim 1/\omega^4$ (2.60)] than $G_{xx}^{(0)}(\omega)$ ($\sim 1/\omega^2$). This simple model should be approximately valid for noble gas solids. In Fig. 6a we have therefore plotted the spectrum $\overset{\circ}{z}(\omega)$ for this

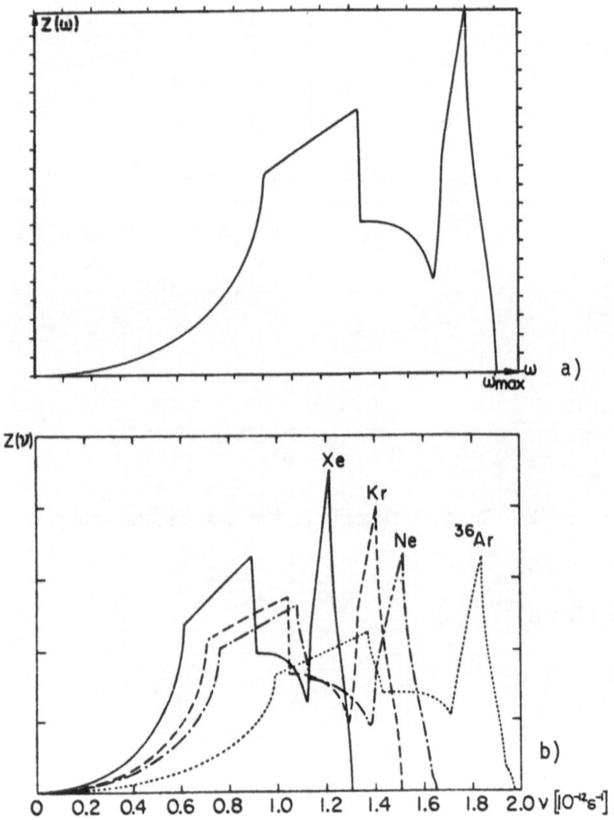

Fig. 6. a) Frequency spectrum $\overset{\circ}{z}(\omega)$ for the central force nearest neighbour model in an fcc lattice b) Frequency spectra for noble gas solids according to the results of neutron scattering experiments /2.20/

model and compared it with the spectra of the various noble gases (Fig. 6b) /2.20/. The general structure of the spectra agrees quite well.

Finally, we give in Fig. 7 the spectra for three typical fcc metals: Al, Cu, Au. The corresponding coupling parameters used in the calculation were obtained from Born-von Karman fits of the dispersion curves of Al /2.21/, Cu /2.22/ and Au /2.23/.

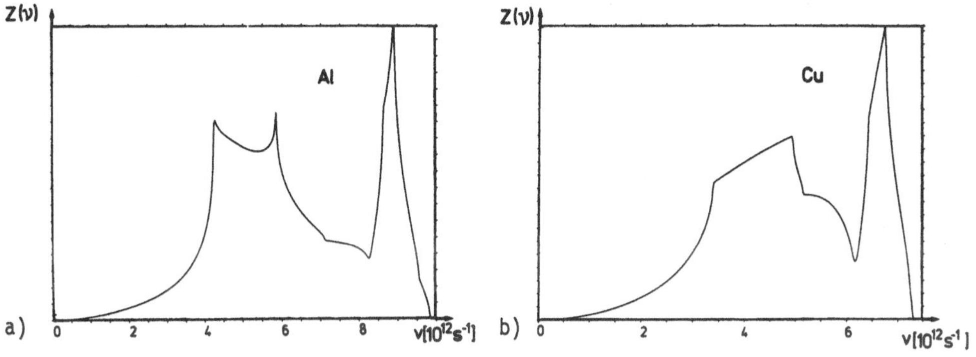

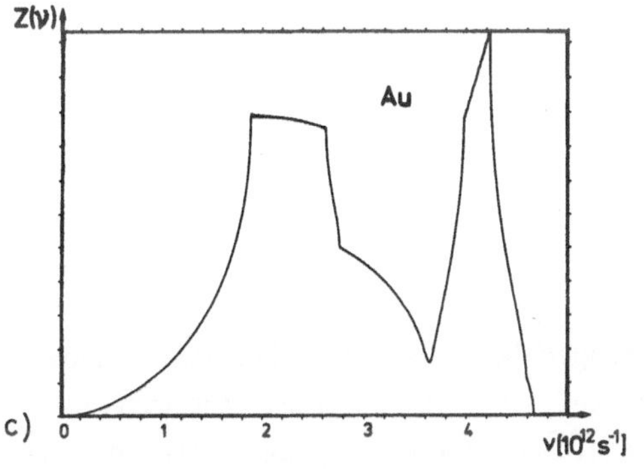

Fig. 7a-c. Spectra for three typical fcc metals: Al a), Cu b), Au c)

Whereas the Al and Cu spectra agree reasonably well with the nearest neighbour spectrum of Fig. 6b, the Au spectrum shows larger deviations. Fig. 8 shows the spectra of three bcc crystals: Na, Fe, Nb, the Born-von Karman fits of which were obtained from /2.24/ for Na, /2.25/ for Fe and /2.26/ for Nb. Whereas the Na and Fe spectra are "normal", the Nb spectrum shows distinct differences arising from anomalies in the phonon dispersion curves.

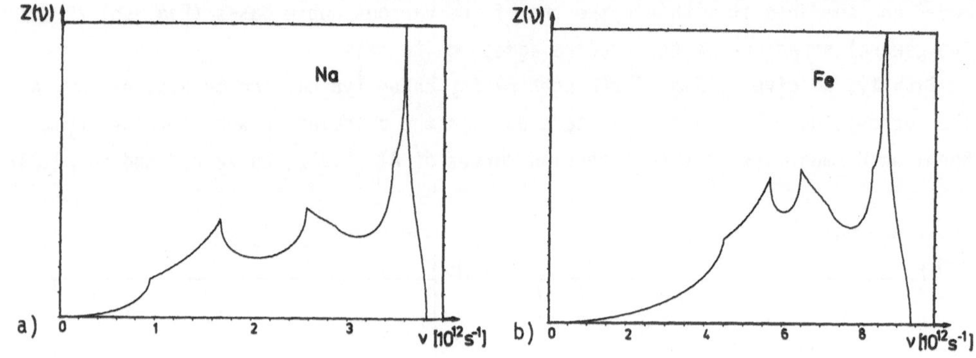

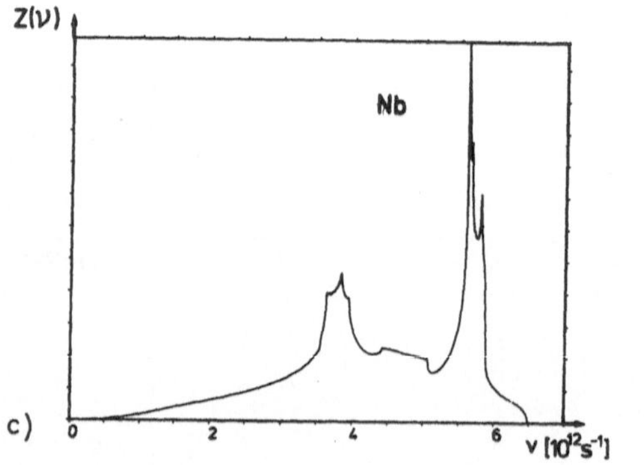

Fig. 8a-c. Spectra for the bcc metals Na a), Fe b) and Nb c)

3. Lattice with an Isolated Point Defect

In this chapter we discuss the vibrational properties, i.e., the vibrational eigenmodes and eigenfrequencies, of a single isolated point defect in an otherwise ideal lattice. We will see that the eigenmodes are either scattering states with frequencies inside the ideal spectrum $0 \leq \omega \leq \overset{\circ}{\omega}_{max}$ or localized states with $\omega > \overset{\circ}{\omega}_{max}$. As the simplest example we treat in detail the isotopic defect. Further we describe a variational method for localized modes.

3.1 Scattering States and Localized States

A point defect, e.g., a substitutional defect, disturbs the translational symmetry of the lattice. First, the mass M of the defect differs in general from the masses $\overset{\circ}{M}$ of the host atoms. Secondly, the coupling parameters are changed in the vicinity of the defect partly due to the changed defect-host interactions and partly due to the changed host-host interactions induced by the static displacements of the defect. This situation is sketched in Fig. 9 for a substitutional point defect coupled to

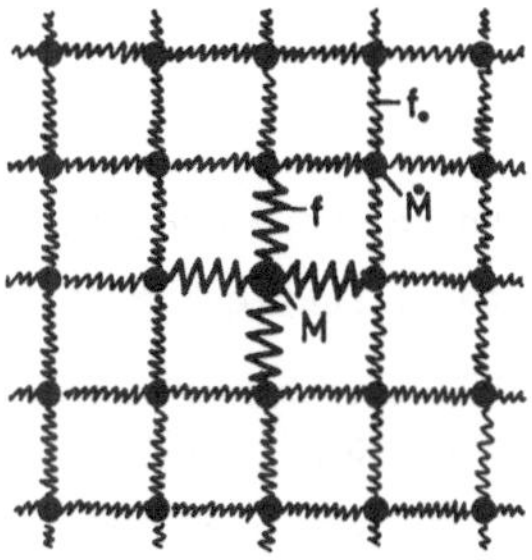

Fig. 9. Schematical representation of a substitutional defect of mass M and force constant f to its nearest neighbours. The mass of the host atoms is $\overset{\circ}{M}$, the host-host coupling is f_o

its nearest neighbours with springs f instead of the ideal spring $\overset{\circ}{f}$. In general we write

$$\phi = \overset{\circ}{\phi} + \varphi \quad , \quad M = \overset{\circ}{M} + \Delta M \tag{3.1}$$

where φ and ΔM refer to the change of the coupling constants and the masses near the defect. For the following it is most important that these changes φ, ΔM are locally restricted to a small three-dimensional region bounded in all directions and embedded in an otherwise ideal crystal. Thus the following discussion is not only valid for single point defects, but also for pairs and agglomerates of point defects, whereas "one-" or "two-dimensional" defects such as dislocations and stacking faults are not included.

The equation of motion for an eigenvibration of frequency ω

$$(\phi - M\omega^2)\ \underline{s}(\omega) = 0 \tag{3.2}$$

can be written as

$$(\overset{\circ}{\phi} - \overset{\circ}{M}\omega^2)\ \underline{s}(\omega) = -V(\omega)\ \underline{s}(\omega) \quad \text{with} \quad V(\omega) = \varphi - \Delta M\omega^2 \ . \tag{3.3}$$

Regarding the "perturbation" $V(\omega)\underline{s}(\omega)$ as an inhomogeneity, the general solution is the sum of a solution $\overset{\circ}{\underline{s}}(\omega)$ of the homogeneous equation, if such a solution exists for a given ω,

$$(\overset{\circ}{\phi} - \overset{\circ}{M}\omega^2)\ \overset{\circ}{\underline{s}}(\omega) = 0 \tag{3.4}$$

and a partial solution with the inhomogeneity $-V(\omega)\underline{s}(\omega)$:

$$\underline{s} = \overset{\circ}{\underline{s}} - \overset{\circ}{G}(\omega)\ V(\omega)\ \underline{s} \qquad \text{with}\ \overset{\circ}{G}(\omega) = \frac{1}{\overset{\circ}{\phi} - \overset{\circ}{M}(\omega + i\eta)^2} \ . \tag{3.5}$$

For frequencies within the allowed spectrum $0 \leq \omega \leq \overset{\circ}{\omega}_{max}$ of the ideal crystal such homogeneous solutions always exist representing plane waves $\underline{k}\sigma\rangle$ of frequency $\omega = \omega_\sigma(\underline{k})$ incident on the "perturbation potential" $V(\omega)$:

$$\overset{o}{s}{}^{\underline{m}}_{i}(\underline{k}\sigma) = \frac{1}{\sqrt{V_B}} \exp(i\underline{k}\underline{R}^{\underline{m}})\, e_i(\underline{k}\sigma) \quad \text{with} \quad \omega_\sigma(\underline{k}) = \omega \, .$$

Scattering by $V(\omega)$ produces an outgoing scattered wave $-\overset{o}{G}V\underline{s}$. The retarded Green's function $\overset{o}{G}(\omega)$ takes care of the radiation condition. Thus the total scattering state $\underline{s} = \underline{s}(\underline{k}\sigma)$ can be labelled by the $\underline{k}\sigma$ value of the incident phonon.

For frequencies ω larger than the maximal frequency $\overset{o}{\omega}_{max}$ of the ideal lattice there exist no eigensolutions of the homogeneous equation. Thus we have a homogeneous equation

$$\underline{s}(\omega) = -\overset{o}{G}(\omega)\, V(\omega)\, \underline{s}(\omega) \tag{3.6}$$

which has only solutions for those discrete frequencies ω_ℓ , $\ell = 1,2,\ldots$ for which the determinant vanishes

$$\text{Det} \left| 1 + \overset{o}{G}(\omega)\, V(\omega) \right| = 0 \, . \tag{3.7}$$

Because the Green's function $\overset{o}{G}(\omega)$ decreases exponentially with distance for $\omega > \overset{o}{\omega}_{max}$ the amplitudes $\underline{s}(\ell)$ are localized at the defect and the vibrational modes are therefore called "localized states".

Thus for an isolated point defect the allowed frequencies of the eigenvibrations consist of a continuous spectrum $0 \leq \omega \leq \overset{o}{\omega}_{max}$ coinciding with the frequencies of the ideal crystal and possibly of some discrete frequencies $\omega_\ell > \overset{o}{\omega}_{max}$ of the localized modes. According to (2.13) the eigenfunctions $\underline{\xi}(\alpha) = \sqrt{M}\,\underline{s}(\alpha)$ form a complete and orthonormalized set, so that we get for the amplitudes $\underline{s}(\underline{k}\sigma)$, $\underline{s}(\ell)$ of the eigenvibrations

$$\sum_{\underline{m}i} M^{\underline{m}} s^{*\underline{m}}_{i}(\underline{k}\sigma)\, s^{\underline{m}}_{i}(\underline{k}'\sigma') = \delta(\underline{k}-\underline{k}')\, \delta_{\sigma\sigma'} \quad , \quad \sum_{\underline{m}i} M^{\underline{m}} s^{*\underline{m}}_{i}(\underline{k}\sigma)\, s^{\underline{m}}_{i}(\ell) = 0 \tag{3.8}$$

and

$$\sum_{\sigma} \int \frac{d\underline{k}}{V_B}\, s^{\underline{m}}_{i}(\underline{k}\sigma)\, s^{*\underline{n}}_{j}(\underline{k}\sigma) + \sum_{\ell} s^{\underline{m}}_{i}(\ell)\, s^{\underline{n}}_{j}(\ell) = \delta^{\underline{m}\underline{n}}_{ij} / \sqrt{(M^{\underline{m}} M^{\underline{n}})} \, . \tag{3.9}$$

Note that unlike the plane waves $\overset{o}{\underline{s}}(\underline{k}\sigma)$, the scattering states alone do not form a complete set, except when there are no localized modes.

The problem has a direct analogy in quantum theory: The eigenstates of a free electron are plane waves $\left(\rightarrow \overset{o}{\underline{s}}(\underline{k}\sigma)\right)$ with energies $E = \hbar^2 k^2/2m > 0$ $\left(\rightarrow \omega = \omega_\sigma(\underline{k}) \leq \overset{o}{\omega}_{max}\right)$. In the field of a potential $V(\underline{r})$ the eigenstates are either scattering states with $E > 0$ $(\rightarrow \omega < \overset{o}{\omega}_{max})$ or bound states with discrete energies $E_\alpha < 0$ $(\rightarrow \omega_\ell > \overset{o}{\omega}_{max})$ which decrease exponentially for large distances similar to localized states.

The "perturbation" matrix $V(\omega)$ is strongly localized in space and has nonvanishing matrix elements only for the defect site itself and the adjacent sites affected by the perturbation. Thus (3.5) expresses the displacements $\underline{s}$ of all atoms in terms of the displacements of the small number of atoms in the "defect" subspace where $V \neq 0$. We can see this more clearly by matrix partition technique.

Let us introduce a projection operator P on the defect subspace of V and a second projector Q on the complementary unperturbed subspace of the surrounding ideal lattice. Then we have for the projectors

$$P^2 = P \quad , \quad Q^2 = Q \quad , \quad P + Q = 1 \tag{3.10}$$

and in particular for V

$$V = PVP = \begin{Bmatrix} V_{PP} & 0 \\ 0 & 0 \end{Bmatrix} \tag{3.11}$$

since all submatrices V_{PQ}, V_{QP}, V_{QQ} are connected with atoms of the unperturbed lattice and are therefore null matrices. Then (3.5) reads

$$\underline{s} = P\underline{s} + Q\underline{s} = \begin{Bmatrix} \underline{s}_P \\ \text{----} \\ \underline{s}_Q \end{Bmatrix} = \begin{Bmatrix} \overset{\circ}{\underline{s}}_P \\ \text{----} \\ \overset{\circ}{\underline{s}}_Q \end{Bmatrix} - \left\{\begin{array}{c|c} \overset{\circ}{G}_{PP}V_{PP} & 0 \\ \hline \overset{\circ}{G}_{QP}V_{PP} & 0 \end{array}\right\} \begin{Bmatrix} \underline{s}_P \\ \text{----} \\ \underline{s}_Q \end{Bmatrix} \tag{3.12}$$

and the vectors $\underline{s}_P$, $\underline{s}_Q$ in the subspaces P and Q are determined by

$$\underline{s}_P = \overset{\circ}{\underline{s}}_P - \overset{\circ}{G}_{PP}V_{PP}\underline{s}_P \qquad \underline{s}_Q = \overset{\circ}{\underline{s}}_Q - \overset{\circ}{G}_{QP}V_{PP}\underline{s}_P \ . \tag{3.13}$$

Thus in order to obtain $\underline{s}$ in the whole space, one has only to calculate $\underline{s}_P$, provided $\overset{\circ}{G}(\omega)$ is known. Since $\underline{s}_P$ is given by

$$\underline{s}_P = \frac{1}{1 + \overset{\circ}{G}_{PP}V_{PP}} \overset{\circ}{\underline{s}}_P \tag{3.14}$$

we have only to invert the finite matrix $1 + \overset{\circ}{G}_{PP}V_{PP}$ in the defect subspace P. Similar arguments hold for the localized mode. For instance, the condition (3.7) for localized modes is only a condition in the subspace P, since due to general rules for determinants (expansion into subdeterminants)

$$\mathrm{Det}\left|1 + \overset{\circ}{G}(\omega)\ V(\omega)\right| \equiv \mathrm{Det}\left|1 + \overset{\circ}{G}_{PP}(\omega)\ V_{PP}(\omega)\right| = 0 \ . \tag{3.15}$$

The defect Green's function can be expanded into the eigenfunctions $\underline{\xi}(\alpha)$ of the dynamical matrix (2.17) leading in our case to

$$G^{\mathfrak{mn}}_{ij}(\omega) = \sum_{\sigma} \int_{V_B} \frac{d\underline{k}}{V_B} \frac{s^{\mathfrak{m}}_i(\underline{k}\sigma)\ s^{*\mathfrak{n}}_j(\underline{k}\sigma)}{\omega^2(\underline{k}) - (\omega + i\eta)^2} + \sum_{\ell} \frac{s^{\mathfrak{m}}_i(\ell)\ s^{*\mathfrak{n}}_j(\ell)}{\omega^2_{\ell} - (\omega + i\eta)^2} \ . \tag{3.16}$$

As a function of ω^2, it has a branch cut along the real axis from 0 to $\overset{\circ}{\omega}{}^2_{max}$ and simple poles at ω_ℓ^2.

In order to calculate the Green's function in practical cases, this representation is not very useful. Instead we can derive directly an equation for G by means of the ideal Green's function $\overset{\circ}{G}$. Starting from

$$(\phi - M\omega^2)\, G(\omega) = 1 \quad \text{or} \quad (\overset{\circ}{\phi} - \overset{\circ}{M}\omega^2)\, G(\omega) = 1 - V(\omega)\, G(\omega) \tag{3.17}$$

we obtain

$$G(\omega) = \overset{\circ}{G}(\omega) - \overset{\circ}{G}(\omega)\, V(\omega)\, G(\omega) \qquad \text{with} \quad \overset{\circ}{G}(\omega) = \frac{1}{\overset{\circ}{\phi} - \overset{\circ}{M}(\omega + i\eta)^2} . \tag{3.18}$$

This equation can be solved directly with respect to $G(\omega)$

$$G(\omega) = \frac{1}{1 + \overset{\circ}{G}V}\,\overset{\circ}{G} = \overset{\circ}{G}\,\frac{1}{1 + V\overset{\circ}{G}} . \tag{3.19}$$

Inserting this into the right side of (3.18) yields

$$G = \overset{\circ}{G} - \overset{\circ}{G}\, t(\omega)\, \overset{\circ}{G} \qquad \text{with} \quad t = V\,\frac{1}{1 + \overset{\circ}{G}V} . \tag{3.20}$$

Thus the Green's function is directly given in terms of the ideal lattice Green's function, once the t matrix is known. Like $V(\omega)$, also $t(\omega)$ is restricted to the defect subspace P and has the form

$$t(\omega) = \begin{Bmatrix} t_{PP}(\omega) & 0 \\ 0 & 0 \end{Bmatrix} \qquad \text{with} \; t_{PP} = V_{PP}\,\frac{1}{1 + \overset{\circ}{G}_{PP}V_{PP}} . \tag{3.21}$$

Thus in order to calculate t, we have only to invert the finite matrix $1 + \overset{\circ}{G}_{PP}V_{PP}$ of dimension 3n if n atoms are affected by the perturbation. The defect Green's function is then according to (3.20) obtained by multiplying t from left and right with $\overset{\circ}{G}$.

For the t matrix, one can give various equivalent representations. First

$$t = V\,\frac{1}{1 + \overset{\circ}{G}V} = \frac{1}{1 + V\overset{\circ}{G}}\,V . \tag{3.22}$$

Using the identity $1/(1+x) = 1 - x/(1+x)$ we obtain from this

$$t = V - V\overset{\circ}{G}V\,\frac{1}{1 + \overset{\circ}{G}V} = V - V\overset{\circ}{G}\,\frac{1}{1 + V\overset{\circ}{G}}\,V \tag{3.23}$$

and by comparison with (3.22) and (3.19)

$$t = V - V\overset{\circ}{G}t = V - VGV . \tag{3.24}$$

Whereas the Green's function $G(\omega)$ has poles at the localized mode frequencies ω_ℓ, it is finite for frequencies within the spectrum of the ideal crystal. However $G(\omega)$ may here become especially large if for some particular frequencies

$$\text{Det}\left|1 + \overset{o}{G}(\omega)V(\omega)\right| \approx 0 \ . \tag{3.25}$$

(Note that the determinant cannot vanish exactly for $\omega < \overset{o}{\omega}_{max}$).

For instance, the determinant will be very small, if

$$\text{Det}\left|1 + \text{Re}\{G(\omega)\}V(\omega)\right| = 0 \tag{3.26}$$

for such frequencies ω, for which $\text{Im}\{G(\omega)\}$ is very small, e.g., near the band edges $\omega \approx 0$ and $\omega \approx \omega_{max}$. Then both $G(\omega)$ and $t(\omega)$ exhibit a resonance behaviour which manifests itself in a very strong scattering and in a peak in the local frequency spectrum.

Such resonances may have various physical reasons. The simplest example is a heavy isotope with mass $M \gg \overset{o}{M}$, discussed in detail in the following section. For large M a very narrow resonance peak in the spectrum appears which moves to $\omega \approx 0$ for $M \to \infty$. Then the isotope vibrates only with frequencies ≈ 0 and is completely decoupled from the host atoms. Other cases are weakly bound defects (see Chapter 5) or nearly unstable defects such as self interstitials (see Chapter 6). In both cases low frequency resonance modes appear. For a small negative change of the local coupling parameters the resonance frequency moves through zero. The configuration is then no longer stable and decays via a "decay mode". Similar resonances can occur at the upper end of the spectrum. For instance, if the force constants are only slightly too weak to form a localized mode with $\omega > \overset{o}{\omega}_{max}$, a resonance mode with a frequency slightly smaller than $\overset{o}{\omega}_{max}$ will occur which becomes a localized mode for a slight strengthening of the force constants.

3.2 The Isotopic Defect

The simplest defect one can imagine is a substitutional isotopic defect where one of the host atoms is replaced by an isotope of the host element. Since the isotope is chemically equivalent to the host atoms, the force constants are not changed, only the mass M of the isotope differs from $\overset{o}{M}$. Thus for an isotopic defect at side $\underline{d}$:

$$V_{ij}^{\underline{mn}} = -(M - \overset{o}{M})\omega^2 \, \delta^{\underline{md}} \, \delta^{\underline{nd}} \, \delta_{ij} \ . \tag{3.27}$$

From the sum rule (2.50) we obtain for the ratio of the averaged square frequencies of the isotope and of a host atom in the ideal lattice

$$<\omega^2>^{\text{isotope}} : <\omega^2>^{\text{host}} = \overset{\circ}{M}/M . \tag{3.28}$$

Thus we expect localized modes to occur for a very slight isotope with $M \ll \overset{\circ}{M}$, since in this case $<\omega^2>$ for the isotope will be much larger than $\overset{\circ}{\omega}{}^2_{max}$. On the other hand, for a very heavy isotope we have $<\omega^2>^{\text{isotope}} \ll \overset{\circ}{\omega}{}^2_{max}$ and most of the local spectrum has to be centered at very small frequencies giving rise to resonance modes.

Let us first discuss the occurrence of localized modes, i.e., solutions of (3.6),

$$s_i^{\underline{m}}(\omega) = \sum_j \overset{\circ}{G}{}_{ij}^{\underline{m}\underline{d}}(\omega)\ (M - \overset{\circ}{M})\omega^2\ s_j^{\underline{d}}(\omega) \tag{3.29}$$

which is a homogeneous equation for the amplitudes of the isotope.

All other amplitudes follow by multiplication of $s_j^{\underline{d}}$ by $\overset{\circ}{G}{}_{ij}^{(\underline{m}-\underline{d})}$. Thus the condition for a localized state is

$$\text{Det}\left|\delta_{ij} - \overset{\circ}{G}{}_{ij}^{(0)}(\omega)\ (M - \overset{\circ}{M})\omega^2\right| = 0 \tag{3.30}$$

or

$$1 = \overset{\circ}{G}{}_{xx}^{(0)}(\omega)\ (M - \overset{\circ}{M})\omega^2 \tag{3.31}$$

since for cubic crystals $\overset{\circ}{G}{}_{ij}^{(0)}(\omega) = \overset{\circ}{G}{}_{xx}^{(0)}(\omega)\delta_{ij}$. According to (2.19) the Green's function $\overset{\circ}{G}{}_{xx}^{(0)}(\omega)$ is for $\omega > \overset{\circ}{\omega}_{max}$ given by

$$\overset{\circ}{G}{}_{xx}^{(0)}(\omega) = \frac{1}{3\overset{\circ}{M}} \sum_\sigma \int_{V_B} \frac{d\underline{k}}{V_B} \frac{1}{\omega_\sigma^2(\underline{k}) - \omega^2} < 0 . \tag{3.32}$$

Consequently localized states exist only if $(M - \overset{\circ}{M}) < 0$ or $M < \overset{\circ}{M}$ as we expect from the sum rule (2.50). However not every isotope with $M < \overset{\circ}{M}$ will lead to a localized mode, but rather the mass has to be less than a critical value M_{cr}, which can be seen as follows.

First $-\overset{\circ}{G}{}_{xx}^{(0)}(\omega)$ is a monotonically decreasing function of ω^2 for $\omega > \omega_{max}$

$$\partial_{\omega^2}\left\{-\overset{\circ}{G}{}_{xx}^{(0)}(\omega)\right\} = \frac{-1}{3\overset{\circ}{M}} \sum_\sigma \int \frac{d\underline{k}}{V_B} \frac{1}{\left(\omega_\sigma^2(\underline{k}) - \omega^2\right)^2} < 0 \tag{3.33}$$

and approaches for large ω^2 the asymptotic value $1/\overset{\circ}{M}\omega^2$ from above, since $1/\omega^2 < 1/\left(\omega^2 - \omega_\sigma^2(\underline{k})\right)$. In Fig. 10 we have plotted $\overset{\circ}{G}{}_{xx}^{(0)}(\omega)$ qualitatively as a function of ω^2. As discussed in Sect. 2.6 it approaches $(\sim \sqrt{\omega^2 - \omega_c^2})$ a finite value at the band edge $\overset{\circ}{\omega}{}^2_{max}$. According to (3.31) the intersection $-\overset{\circ}{G}{}_{xx}^{(0)}(\omega)$ with $1/(\overset{\circ}{M} - M)\omega^2$ gives the frequency ω_ℓ of the localized mode. Evidently we only get an intersection if

$$-\overset{\circ}{G}{}_{xx}^{(0)}(\overset{\circ}{\omega}_{max}) > \frac{1}{(\overset{\circ}{M} - M)\omega^2_{max}} \quad \text{or if} \quad M < M_{cr} = \overset{\circ}{M}\left\{1 - \frac{1}{\overset{\circ}{M}\overset{\circ}{\omega}{}^2_{max}\left|\overset{\circ}{G}{}_{xx}^{(0)}(\omega_{max})\right|}\right\} . \tag{3.34}$$

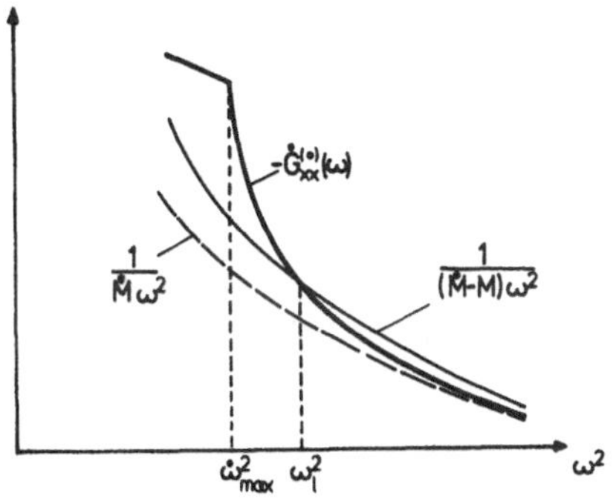

Fig. 10. Graphical construction of the localized mode frequency ω_ℓ^2 according to (3.31): ω_ℓ is determined by the intersection of $-\mathring{G}^{(0)}_{xx}(\omega)$ with $1/(\mathring{M}-M)\omega^2$

Thus we obtain exactly one triple degenerate localized mode if $M < M_{cr}$. For the example of a fcc lattice with nearest neighbour interaction with one force constant f_o the critical mass is $M_{cr} = 0.76\mathring{M}$. For this model Fig. 11 shows the resulting localized mode frequency as a function of the mass change $(\mathring{M}-M)/\mathring{M}$.

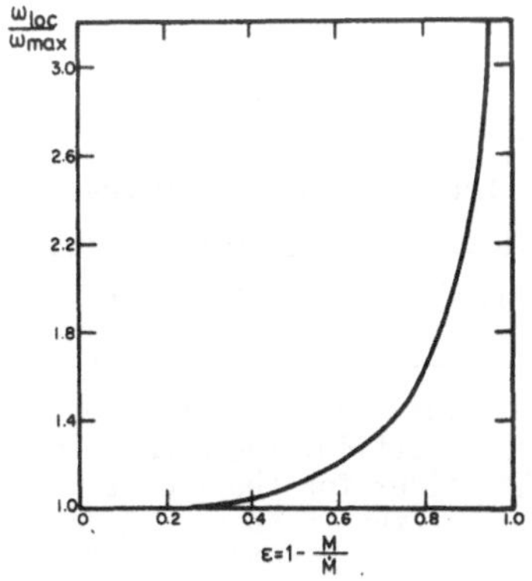

Fig. 11. The localized mode frequency $\omega_{loc}/\omega_{max}$ versus $\varepsilon = 1 - M/\mathring{M}$ for an isotopic defect in fcc (central force n.n. model)

For a very light impurity one can give an analytical solution. Since the localized mode frequency will be very high, we use the expansion (2.60) of $\mathring{G}(\omega)$ for high frequencies $\omega > \mathring{\omega}_{max}$:

$$\mathring{G}(\omega) = -\frac{1}{\mathring{M}\omega^2} - \frac{\mathring{\phi}}{(\mathring{M}\omega^2)^2} - \dots \tag{3.35}$$

and obtain from (3.31)

$$\omega_\ell^2 = \frac{\mathring{\phi}^{(0)}_{xx}}{M\mathring{M}/(\mathring{M}-M)} \cong \frac{\mathring{\phi}^{(0)}_{xx}}{M} = \langle\omega^2\rangle = \omega^2_{Einstein} \; . \tag{3.36}$$

Thus the localized mode frequency approaches the Einstein frequency of the isotope. From the Einstein model it is clear that the neighbouring atoms will practically not take part in the localized vibration. This is plausible from the asymptotic expansion of $\overset{\circ}{G}(\omega)$ for large distances giving a strong exponential decrease for large frequencies.

The defect Green's function can be calculated very easily for the isotopic defect. First we obtain from (3.18) with the inhomogeneity V of (3.27)

$$G^{\underline{m}\underline{n}}_{ij}(\omega) = \overset{\circ}{G}{}^{\underline{m}\underline{n}}_{ij}(\omega) + \sum_k \overset{\circ}{G}{}^{(\underline{m}-\underline{d})}_{ik} (M - \overset{\circ}{M})\omega^2 \, G^{\underline{d}\underline{n}}_{kj}(\omega) \, . \tag{3.37}$$

For the Green's function of the isotope $G^{\underline{d}\underline{d}}_{ij} = G^{\underline{d}\underline{d}}_{xx}\delta_{ij}$ we obtain immediately

$$G^{\underline{d}\underline{d}}_{xx}(\omega) = \overset{\circ}{G}{}^{(0)}_{xx}(\omega) + \overset{\circ}{G}{}^{(0)}_{xx}(\omega) \, (M - \overset{\circ}{M})\omega^2 \, G^{\underline{d}\underline{d}}_{xx}(\omega) \tag{3.38}$$

$$G^{\underline{d}\underline{d}}_{xx}(\omega) = \frac{1}{1/\overset{\circ}{G}{}^{(0)}_{xx}(\omega) - (M - \overset{\circ}{M})\omega^2} \, . \tag{3.39}$$

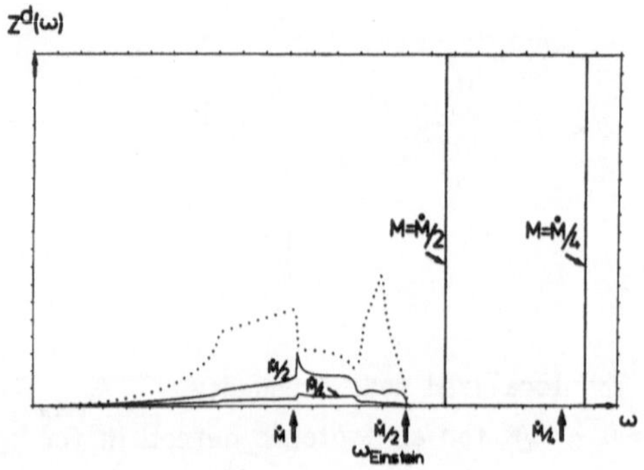

Fig. 12. Local frequency spectra $z^d(\omega)$ of light isotopes with masses $M = \overset{\circ}{M}/2$ and $M = \overset{\circ}{M}/4$ for the central force n.n. model in fcc. The dotted line indicates the spectrum of the ideal lattice. The arrows indicate the values for the Einstein frequencies of the isotopes

The local frequency spectrum of the isotope is given by $z^{\underline{d}}_x(\omega) = (2M\omega/\pi)\,\mathrm{Im}\{G^{\underline{d}\underline{d}}_{xx}(\omega)\}$ and is plotted in Fig. 12 for $M = \overset{\circ}{M}/2$ and $M = \overset{\circ}{M}/4$. For $\overset{\circ}{G}{}^{(0)}_{xx}(\omega)$ the nearest neighbour model for the fcc crystal with one longitudinal force constant is employed. For comparison the ideal spectrum is also shown for this model (thin line). For $M = \overset{\circ}{M}/2$ one gets a localized mode slightly above $\overset{\circ}{\omega}_{max}$. About 1/4 of the intensity remains in the range of the ideal spectrum $0 \leqslant \omega \leqslant \overset{\circ}{\omega}_{max}$. Here one sees also square root singularities appearing at the same critical frequencies as in the ideal lattice. For $M = \overset{\circ}{M}/4$ the localized mode frequency is appreciably higher and only little intensity remains in the continuous spectrum. The arrows give the positions of the Einstein frequencies of the isotope and of the ideal lattice. For $M = \overset{\circ}{M}/4$, $\omega_{\mathrm{Einstein}}$ agrees pretty well with ω_ℓ. Since $\omega^2_{\mathrm{Einstein}} = \langle\omega^2\rangle$, this means that nearly no intensity is left in the continuous spectrum and that the isotope vibrates only with the localized mode frequency.

Fig. 13 shows the local spectra for heavy isotopes with masses $M = 2\mathring{M}$, $4\mathring{M}$ and $8\mathring{M}$. The arrows again indicate the values of the Einstein frequencies. The spectra are Lorentzian shaped around a resonance frequency which is approximately given by the Einstein frequency. Due to the relatively small width, a very heavy isotope can

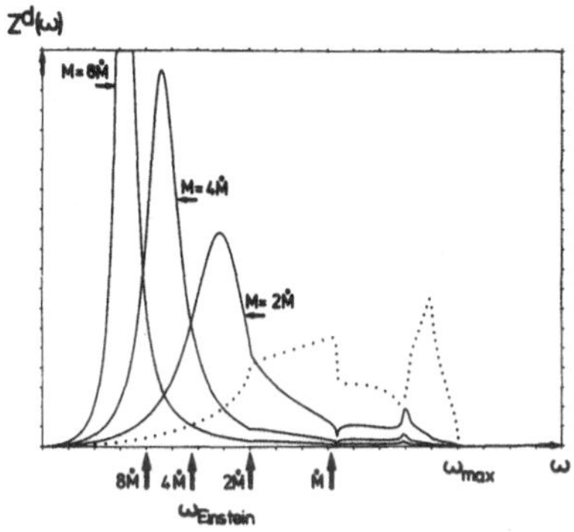

Fig. 13. Local spectra $z^d(\omega)$ for heavy isotopes with masses $M = 2\mathring{M}$, $4\mathring{M}$ and $8\mathring{M}$ compared with the ideal spectrum (dotted) for the central force n.n. model in fcc. The arrows give the values of the Einstein frequencies of the isotopes

only vibrate with frequencies nearly coinciding with its resonance frequency. In this sense it behaves similar to a very light isotope vibrating only with a localized mode.

In the next chapter we will give a more general discussion of localized and resonance modes which also includes the case of force-constant changes.

A very interesting effect occurs for masses M nearly coinciding with the critical mass $M_{cr} = 0.76\,\mathring{M}$ for occurrence of a localized mode. For M slightly larger than M_{cr}, e.g., for $M = 0.82\,\mathring{M}$, as shown in Fig. 14a, a resonance mode occurs just below the band edge ω_{max}. At the threshold ($M = 0.76\,\mathring{M}$) $z^d(\omega)$ diverges at ω_{max} (Fig. 14b). For $M < M_{cr}$ a localized mode appears. However, appreciable intensity can still be left in a resonance below ω_{max}, as is shown in Fig. 14c for $M = 0.70\,\mathring{M}$. The occurrence of such resonances near the threshold of a localized mode are discussed in more detail in the next chapter.

The dynamics of the neighbouring atoms of the isotope is not changed dramatically, since in the extreme cases of a high-frequency localized mode or a low-frequency resonant mode the neighbours do not participate in the motion of the isotope. This is suggested also by the moments (2.50,52) of the local frequency spectra of the neighbours. Both the moment $\langle\omega^2\rangle$ of (2.50) and the reciprocal moment $\langle 1/\omega^2\rangle$ (2.52) are not changed for neighbouring atoms, so that the gross features of the spectra should be retained.

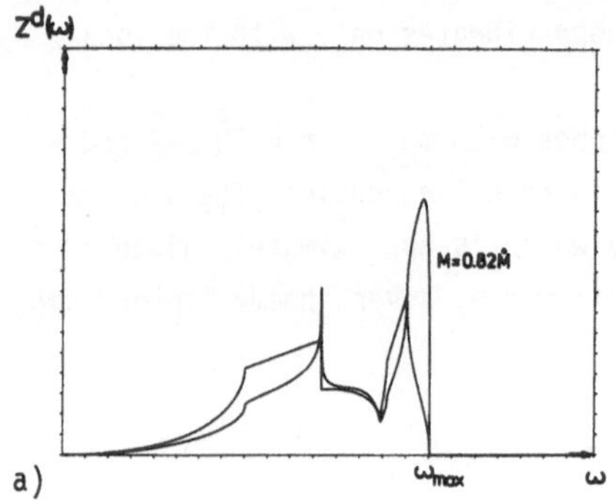

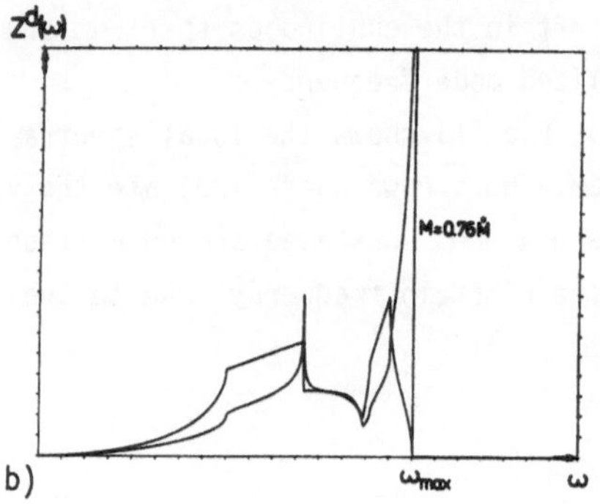

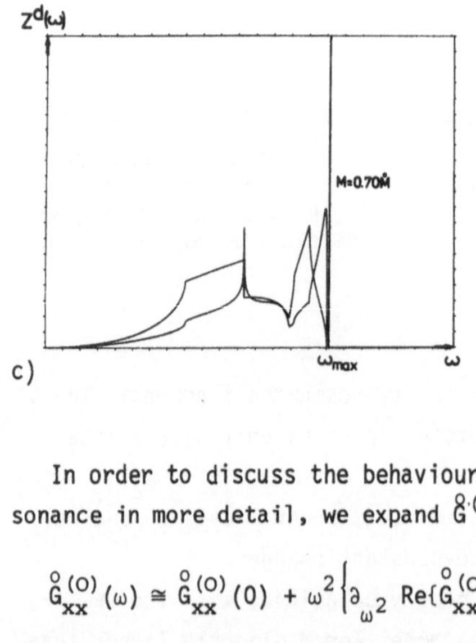

Fig. 14a-c. Local frequency spectra $z^d(\omega)$ for isotopes with masses near the critical mass M_{cr} for the occurrence of a localized mode
a) $M = 0.82\mathring{M} > M_{cr} = 0.76\mathring{M}$: resonant mode slightly below ω_{max}
b) $M = M_{cr} = 0.76\mathring{M}$: $z^d(\omega)$ diverges at ω_{max}
c) $M = 0.70\mathring{M} < M_{cr}$: localized mode above ω_{max} and resonant mode below ω_{max}

In order to discuss the behaviour of $G^{dd}(\omega)$ for the case of a low frequency resonance in more detail, we expand $\mathring{G}^{(0)}(\omega)$ for small frequencies (compare Sect. 2.4)

$$\mathring{G}^{(0)}_{xx}(\omega) \cong \mathring{G}^{(0)}_{xx}(0) + \omega^2 \left\{\partial_{\omega^2} \mathrm{Re}\{\mathring{G}^{(0)}_{xx}(\omega)\}\right\}_{\omega^2 \to +0} + i\,\frac{3\pi}{2\mathring{M}\omega_D^3}\,\omega + \dots \qquad (3.40)$$

and obtain for the isotope Green's function:

$$\mathring{G}^{dd}_{xx}(\omega) = \frac{1}{\mathring{f}_{eff} - M_{eff}\,\omega^2 - i\Gamma\omega} = \frac{1}{M_{eff}(\omega^2_{res} - \omega^2 - i\gamma\omega)}\,. \qquad (3.41)$$

This corresponds to the Green's function of a one-dimensional oscillator with mass M_{eff} and force constant f_{eff} which is the static restoring force of an atom of the ideal lattice, if all neighbouring atoms are allowed to relax. M_{eff} has the meaning of an effective mass related to the mass- and displacement-weighted number of atoms participating in the vibration (see Chap. 4)

$$\mathring{f}_{eff} = \frac{1}{\mathring{G}^{(0)}_{xx}(\omega = 0)}\,, \qquad M_{eff} = M - \mathring{M} + \left\{\partial_{\omega^2}\,\mathrm{Re}\left\{\frac{1}{\mathring{G}^{(0)}_{xx}(\omega)}\right\}\right\}_{\omega^2 \to +0} \,. \qquad (3.42)$$

The term $\Gamma\omega$ describes a velocity-dependent damping

$$\Gamma \equiv M_{eff}\gamma = f_{eff}^2 \frac{3\pi}{2\overset{o}{M}\omega_D^3} . \tag{3.43}$$

For small γ, the Green's function has a quasi-pole at the resonance frequency $\omega_{res}^2 = \overset{o}{f}_{eff}/M_{eff}$.

The physical meaning of $\overset{o}{f}_{eff}$, M_{eff} and Γ becomes more evident when we write down the equation of motion in the above approximation for the time-dependent Green's function $G_{\mathbf{xx}}^{\underline{dd}}(t)$ being the Fourier transform of $G_{\mathbf{xx}}^{\underline{dd}}(\omega)$. Since each factor $-i\omega$ corresponds to a time derivative ∂_t of $G_{\mathbf{xx}}^{\underline{dd}}(t)$, we obtain

$$M_{eff}\, \ddot{G}_{\mathbf{xx}}^{\underline{dd}}(t) + \Gamma\, \dot{G}_{\mathbf{xx}}^{\underline{dd}}(t) + \overset{o}{f}_{eff}\, G_{\mathbf{xx}}^{\underline{dd}}(t) = \delta(t) \tag{3.44}$$

with the solution

$$G_{\mathbf{xx}}^{\underline{dd}}(t) = \frac{1}{M_{eff}} \frac{\sin \tilde{\omega} t}{\tilde{\omega}} \exp(-\frac{\gamma t}{2})\, \theta(t) \quad , \quad \tilde{\omega} = \sqrt{\omega_{res}^2 - \gamma^2/4} \tag{3.45}$$

with $\theta(t) = 1$ for $t > 0$ and 0 for $t < 0$.

For small γ this solution represents a slightly damped oscillatory motion of frequency ω_{res}. In the same approximation the frequency spectrum of the isotope is

$$z_{\mathbf{x}}^{\underline{d}}(\omega) = \frac{M}{M_{eff}} \frac{1}{\pi} \frac{2\gamma\omega}{(\omega_{res}^2 - \omega^2)^2 + (\gamma\omega)^2} \tag{3.46}$$

which is an asymmetric Lorentzian centered at ω_{res} with a relative halfwidth

$$\frac{\Delta\omega}{\omega_{res}} = \frac{\gamma}{\omega_{res}} = \frac{3\pi}{2\overset{o}{M}\omega_D^3} \frac{\sqrt{\overset{o}{f}{}_{eff}^3}}{\sqrt{M_{eff}}} \quad \text{with} \quad \omega_{res} = \sqrt{\frac{\overset{o}{f}_{eff}}{M_{eff}}} . \tag{3.47}$$

For a very heavy isotope the effective mass is approximately given by the isotopic mass $M_{eff} \cong M$. Thus we obtain a low resonance frequency $\omega_{res} \sim 1/\sqrt{M}$ with an even smaller halfwidth $\Delta\omega \sim 1/M$. This behaviour can be seen directly from Fig. 13 by comparing the spectra for $M = 2\overset{o}{M}$, $4\overset{o}{M}$ and $8\overset{o}{M}$.

The quantities $\overset{o}{f}_{eff}$, ω_D, M_{eff} can be calculated from numerical results for the Green's function. For the fcc model with nearest neighbour interaction via a longitudinal force constant $\overset{o}{f}$ we have

$$\overset{o}{M}\, \overset{o}{\omega}{}_{max}^2 = 2\overset{o}{M}\, \omega_{Einstein}^2 = 8\overset{o}{f} \quad , \quad \overset{o}{f}_{eff} = 2.4\overset{o}{f} \quad ,$$
$$\overset{o}{M}\, \omega_D^2 = 8.5\overset{o}{f} \quad , \quad M_{eff} = M - 0.37\overset{o}{M} \quad . \tag{3.48}$$

In Chapter 4 we will see in detail, that this expansion for the isotopic Green's function near the resonance can be generalized and extended to an arbitrary point defect. It works not only for resonance vibrations, but also for localized vibrations and gives a convenient description of a characteristic defect vibration by simple physical quantities.

Whereas the occurrence of localized vibration was known since the work of LIFSHITZ in the years 1942-1943 /1.1/, the existence and importance of resonance vibrations was pointed out in 1962-1963 by BROUT and VISSCHER /3.1/, KAGAN and ISOLEVSKII /3.2/, and TAKENO /3.3/.

3.3 Variational Method for Localized States

The equation of motion for an arbitrary system

$$\phi \underline{s} - \omega M \underline{s} = 0 \tag{3.49}$$

can be derived by means of a variational principle. Consider the expression

$$R[\tilde{\underline{s}}] = \frac{(\tilde{\underline{s}}\,,\phi\tilde{\underline{s}})}{(\tilde{\underline{s}}\,,M\tilde{\underline{s}})} \tag{3.50}$$

as a functional of $\tilde{\underline{s}}$. Then the vectors $\underline{s}$ extremalizing $R[\tilde{\underline{s}}]$ are the exact eigenvectors of (3.49) and the extrema of $R[\tilde{\underline{s}}]$, i.e., $R[\underline{s}]$, are the exact eigenvalues ω^2. For the condition $\delta R[\tilde{\underline{s}}] = 0$ gives as Euler's equations of the vibrational principle

$$\phi\underline{s} - \frac{(\underline{s}\,,\phi\underline{s})}{(\underline{s}\,,M\underline{s})} M\underline{s} = 0 \quad \text{or} \quad \omega^2 = \frac{(\underline{s}\,,\phi\underline{s})}{(\underline{s}\,,M\underline{s})} \tag{3.51}$$

which proves the above statement. Since the eigenvalues ω^2 are the extrema of $R[\tilde{\underline{s}}]$, they are insensitive against small changes $\delta\underline{s}$ of the eigenvectors. Thus one can choose a suitable ansatz $\tilde{\underline{s}}$ with free parameters a,b,... which are determined by variation, i.e., differentiation of R(a,b,...) with respect to a,b,... . The resulting extremal value of R is the optimal approximation for ω^2 for the chosen ansatz.

Expression (3.50) is called Rayleigh's quotient in continuum theory and is quite analogous to Ritz's variational principle in quantum theory. Its major advantages are the minimal and maximal properties of $R[\tilde{\underline{s}}]$, because it approximates the lowest eigenvalue ω^2_{min} from above and the highest ω^2_{max} from below, since we have by expanding ϕ into eigenfunctions $\underline{s}(\alpha)$

$$\omega^2_{\text{min}} \leqslant \frac{(\tilde{\underline{s}}\,,\phi\tilde{\underline{s}})}{(\tilde{\underline{s}}\,,M\tilde{\underline{s}})} = \frac{\sum\limits_{\alpha} \omega_\alpha^2 \left|\left(\tilde{\underline{s}}\,,M\underline{s}(\alpha)\right)\right|^2}{\sum\limits_{\alpha'} \left|\left(\tilde{\underline{s}}\,,M\underline{s}(\alpha')\right)\right|^2} \leqslant \omega^2_{\text{max}} \ . \tag{3.52}$$

For an isolated point defect this method can only be used for localized modes. The continuous frequencies of the scattering states are determined by the incident plane wave. Thus the highest localized mode frequency can be approximated from below. This statement can also be made for lower localized mode frequencies provided the ansatz $\tilde{s}$ is orthogonal to all exact eigenvectors with higher frequency. Here group theory is of great help and allows us to derive a lower bound for the highest frequency in each irreducible subspace since the different irreducible subspaces are automatically orthogonal on each other.

The following simple method first used by LENGELER and LUDWIG /3.4/ leads to a suitable ansatz for the displacements $\tilde{\underline{s}}$. Because of the exponential decrease of the amplitudes for large distances, one allows only nonzero displacements within a certain subspace near the defect, containing, e.g., the defect and its nearest neighbours. This "molecule" of moveable atoms is coupled to the rigid outer lattice. The eigenvectors of this system can be inserted into the variational principle. The resulting frequencies ω^2 are identical with the eigenfrequencies of the molecule which can be calculated

$$\text{Det}\left|\phi_{ij}^{\underline{MN}} - M^{\underline{M}}\omega^2\delta_{ij}\right| = 0$$

where the atoms of the molecule are labelled $\underline{M}$ and $\underline{N}$. They are lower bounds for the exact frequencies. By enlarging the molecule, these approximations can be improved step by step.

Many applications of the variational principle are discussed in /1.2/ which we will not repeat here. We will come back to the problem of finding bounds for the localized mode frequencies in the next chapter, where we will derive both lower and upper bounds by applying a projection technique.

4. Description of Resonant and Localized Defect Vibrations

An alternative to the standard Green's function method of Chap. 2.1 has been proposed by LITZMANN and ROSZA /4.1/, KRUMHANSL and MATTHEW /4.2/, and others /4.3 - 7/. This method is especially useful since it yields a direct description of the local vibrational properties of the defect. Though this description has the simple structure of the Einstein approximation, it is exact and has the further advantage of also being applicable to interstitials which introduce additional degrees of freedom into the lattice.

By applying this method to resonant and localized defect vibrations, we obtain a very simple and physically evident description of the local vibrational behaviour of the defect. Low frequency resonances are treated in Sect. 4.2. The defect behaves as a simple damped Einstein oscillator, characterized by an effective force constant f^{eff} describing the static response, by an effective mass M^{eff}, a measure of the participation of the surrounding atoms in the vibration, and by a damping constant γ. In Sect. 4.3 we derive approximate results for the characteristic quantities f^{eff} and M^{eff}. An exact equation for the damping γ in terms of f^{eff} and M^{eff} is given in Sect. 4.4.

As a further application upper and lower bounds for the frequencies of localized modes can be derived (Sect. 4.5) including the previous lower bounds of DETTMANN and LUDWIG /4.8 , 1.2/, and upper bounds of DEAN /4.9/, and FUJITA /4.10/.

Finally we discuss in Sect. 4.6 resonance vibrations just below the maximum frequency of the perfect crystal. These resonances occur in situations where, e.g., the force constants are just below a critical value needed for the existence of a localized mode. The shape of these resonance lines is quite different from the one of low-frequency resonances.

4.1 Method of Krumhansl and Matthew

Let us introduce projection operators on the different subspaces of the lattice.

1) A *projector* P_C projecting on a *central subspace* C consisting, e.g., of the defect and its nearest neighbours. In the following it will contain in most cases the defect alone.

2) A *projector* P_R projecting on the *rest lattice* R of all other atoms, so that

$$P_C = P_C^2 \, , \quad P_R = P_R^2 \, , \quad P_C + P_R = 1 \, . \tag{4.1}$$

The subspace R can be split up further into a *neighbour region* N, consisting of all atoms which are either directly coupled to the central region (e.g., to the defect) or for which force constants are changed, and into an *unperturbed subspace* U of "ideal" atoms

$$P_R = P_N + P_U \, , \quad P_C + P_N + P_U = 1 \, . \tag{4.2}$$

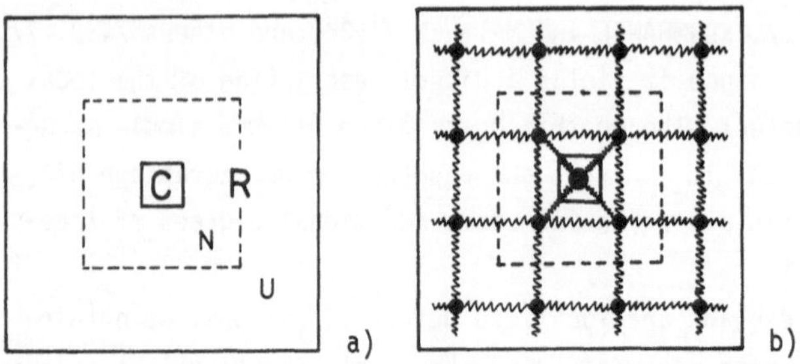

Fig. 15a,b. Schematic representation a) of the "central subspace" C and the "rest lattice subspace" R, which can be split up in a "unperturbed subspace" U and in a neighbouring subspace N containing all atoms directly coupled to C or with changed force constants. For the example in b) the subspace C consists of an interstitial and N contains its nearest neighbours

These subspaces are schematically sketched in Fig. 15a,b for the example of an interstitial in a planar lattice.

By using these projectors the coupling matrix can be partitioned using $P_C + P_R = 1$:

$$\phi = P_C \phi P_C + P_C \phi P_R + P_R \phi P_C + P_R \phi P_R \tag{4.3}$$

or in matrix form:

$$\phi = \begin{Bmatrix} \phi_{CC} & \phi_{CR} \\ \phi_{RC} & \phi_{RR} \end{Bmatrix} = \begin{Bmatrix} \phi_{CC} & \phi_{CN} & 0 \\ \phi_{NC} & \phi_{NN} & \overset{\circ}{\phi}_{NU} \\ 0 & \overset{\circ}{\phi}_{UN} & \overset{\circ}{\phi}_{UU} \end{Bmatrix} . \tag{4.4}$$

Here ϕ_{CC} describes the Einstein coupling of the central region (the defect) for a fixed rest lattice, ϕ_{RR} is the coupling of the atoms in the rest lattice and ϕ_{CR} the coupling of the defect with the neighbours in the rest lattice ($P_C \phi P_R \equiv P_C \phi P_N$). In the unperturbed subspace U the coupling constants are not changed, i.e., $\phi_{NU} = \overset{\circ}{\phi}_{NU}$, $\phi_{UU} = \overset{\circ}{\phi}_{UU}$.

In this whole chapter our central interest will be focused on the properties of the Green's function $G_{CC}(\omega)$ of the defect, i.e., the displacement of the defect due to a unit force on the defect. By partitioning the Green's function $G(\omega)$ in the same way as the coupling matrix, we will derive an equation for $G_{CC}(\omega)$ alone. First the equations of motion are

$$\begin{Bmatrix} \phi_{CC} - M_{CC}\omega^2 & \phi_{CR} \\ \phi_{RC} & \phi_{RR} - M_{RR}\omega^2 \end{Bmatrix} \begin{Bmatrix} G_{CC}(\omega) & G_{CR}(\omega) \\ G_{RC}(\omega) & G_{RR}(\omega) \end{Bmatrix} = \begin{Bmatrix} 1 & 0 \\ 0 & 1 \end{Bmatrix} \tag{4.5}$$

where 1 is a unit matrix in C space, R space, respectively. Then we obtain two equations coupling the Green's function $G_{CC}(\omega)$, with the Green's function $G_{RC}(\omega)$, i.e., the atomic displacements in the rest lattice due to a force on the defect.

$$(\phi_{CC} - M_{CC}\omega^2)\, G_{CC}(\omega) + \phi_{CR}\, G_{RC}(\omega) = 1 \tag{4.6a}$$

$$\phi_{RC}\, G_{CC}(\omega) + (\phi_{RR} - M_{RR}\omega^2)\, G_{RC}(\omega) = 0 \tag{4.6b}$$

By eliminating G_{RC}, we obtain an equation for the defect Green's function G_{CC} in terms of a frequency-dependent effective force constant $\phi_{CC}^{eff}(\omega)$

$$G_{CC}(\omega) = \frac{1}{\phi_{CC}^{eff}(\omega) - M_{CC}\omega^2} = \frac{1}{\phi_{CC} - \phi_{CR}\, \hat{G}_{RR}(\omega)\, \phi_{RC} - M_{CC}\omega^2} . \tag{4.7}$$

with

$$\hat{G}_{RR}(\omega) = \frac{1}{\phi_{RR} - M_{RR}(\omega + i\eta)^2} . \tag{4.8}$$

Equation (4.7), though rigorously valid, has the form of the Einstein approximation for the central region, which follows if $\phi_{CC}^{eff} \cong \phi_{CC}$. Thus the second contribution $\phi_{CR}\hat{G}_{RR}\phi_{RC}$ to ϕ_{CC}^{eff} describes the effect of the dynamic relaxations of the R atoms on the defect. This term can be understood quite easily by comparison with (4.6b): $-\phi_{RC}$ represents the forces on the rest lattice atoms due to a unit displacement of the defect and produces displacements $-\hat{G}_{RR}\phi_{RC}$. These displacements act back on the defect by the force $\phi_{CR}\hat{G}_{RR}\phi_{RC}$ which adds to the Einstein restoring force ϕ_{CC}. Note that ϕ_{CC}^{eff} is complex with its imaginary part describing the damping of the central region.

The remaining problem is the determination of $\hat{G}_{RR}$, i.e., the Einstein Green's function of the rest lattice for a fixed defect. For the case of an interstitial, the subspace R contains all the atoms of the ideal lattice. Then $\hat{G}_{RR}$ can be calculated by the standard Green's function method

$$\hat{G}_{RR} = \mathring{G} - \mathring{G}\, V_{RR}\, \hat{G}_{RR} = \mathring{G} - \mathring{G}\, t_{RR}\, \mathring{G} \tag{4.9}$$

with

$$t_{RR} = V_{RR} \frac{1}{1 + \mathring{G}V_{RR}} \equiv t_{NN} \,, \qquad V_{RR} = \phi_{RR} - \mathring{\phi} - (M_{RR} - \mathring{M})\omega^2 \equiv V_{NN} \,.$$

Here the perturbation $V_{RR} = V_{NN}$ is localized to the neighbouring subspace N. To calculate t_{NN} one has to invert a matrix of dimension 3 times the number of perturbed atoms in subspace N.

For a substitutional defect the space R contains less degrees of freedom than the perfect lattice. In this case we can also apply (4.9) if we replace $\mathring{G}$ by the Green's function $\mathring{G}_{RR}(\omega)$ being the inverse of $\mathring{\phi}_{RR} - \mathring{M}_{RR}\omega^2$ in the subspace R.

Since the defect is fixed, we can calculate $\hat{\mathring{G}}$ with the Green's function method, if we introduce an infinitely strong Einstein spring $V_{CC} \to \infty$ which pins the defect. Then $\hat{\mathring{G}}$ is given by

$$\hat{\mathring{G}} = \mathring{G} - \mathring{G}\, V_{CC}\, \hat{\mathring{G}} = \mathring{G} - \mathring{G}\, t_{CC}\, \mathring{G} \tag{4.10}$$

where

$$t_{CC} = V_{CC} \frac{1}{1 + \mathring{G}_{CC}\, V_{CC}} \cong \frac{1}{\mathring{G}_{CC}} \quad \text{for } V_{CC} \to \infty \,,$$

herefore we obtain

$$\hat{\mathring{G}}_{RR} = \mathring{G}_{RR} - \mathring{G}_{RC} \frac{1}{\mathring{G}_{CC}} \mathring{G}_{CR} \,. \tag{4.11}$$

Whereas so far we were only concerned with the Green's function $G_{CC}(\omega)$ of the defect, the basic equation (4.5) yields, besides (4.6a,b) for G_{CC} and G_{RC}, two additional equations for G_{CR} and G_{RR}, from which the Green's function G_{RR} for the rest lattice can be calculated.

$$G_{RR}(\omega) = \left\{\phi_{RR} - \phi_{RC}\,\frac{1}{\phi_{CC} - M_{CC}(\omega + i\eta)^2}\,\phi_{CR} - M_{RR}(\omega + i\eta)^2\right\}^{-1} . \tag{4.12}$$

This equation is especially useful for the case of an interstitial, since it shows that the additional degrees of freedom can be eliminated leading to an effective coupling between the lattice atoms. Thus G_{RR} can be calculated by the Green's function method

$$G_{RR}(\omega) = \frac{1}{\overset{o}{\phi} - \overset{o}{M}(\omega + i\eta)^2 - \tilde{V}_{NN}} = \overset{o}{G} - \overset{o}{G}\,\tilde{V}_{NN}\,G_{RR} \tag{4.13}$$

if we add to the perturbation V_{NN} the interstitial-induced interaction between the host atoms

$$\tilde{V}_{NN} = V_{NN} - \phi_{NC}\,\frac{1}{\phi_{CC} - M_{CC}(\omega + i\eta)^2}\,\phi_{CN} .$$

Compared with the standard Green's function method, the resulting equations (4.7) for $G_{CC}(\omega)$, (4.9) for $\hat{G}_{RR}(\omega)$ and (4.12) for $G_{RR}(\omega)$ look formally more complicated. However, the actual numerical effort is not larger since in both cases essentially the same matrix has to be inverted. The major advantage of the present method consists in the very suggestive description of the defect Green's function $G_{CC}(\omega)$ by an Einstein-like force constant $\phi_{CC}^{eff}(\omega)$ which proves to be most useful in the following sections.

4.2 Resonant Modes

To describe resonant vibrations of the defect we have to know the Green's function $G_{CC}(\omega) = (\phi_{CC}^{eff} - M_{CC}\omega^2)^{-1}$ for small frequencies. Since in the case of a resonance, G_{CC} has a quasi-pole at small frequencies we expand ϕ_{CC}^{eff},

$$\phi_{CC}^{eff} = \phi_{CC} - \phi_{CR}\,\hat{G}_{RR}(\omega)\,\phi_{RC} \tag{4.14}$$

and thus $\hat{G}_{RR}(\omega)$, in powers of ω, separately for the real and imaginary part. For simplicity we assume that the central space C contains only the defect which is coupled to its neighbouring atoms in subspace N contained in R. Then $G_{CC}(\omega)$ and ϕ_{CC}^{eff} are three-dimensional matrices which are, e.g., for cubic crystals proportional to δ_{ij}, so that $G_{CC}(\omega) = G_{xx}^{dd}(\omega)\,\delta_{ij}$ and $\phi_{CC}^{eff} = \phi_{xx}^{dd}\,\delta_{ij}$ can be treated as scalar quantities in the following.

In analogy to (2.62) the expansion of $\mathrm{Re}\{\hat{G}_{RR}(\omega)\}$ yields a constant and a quadratic term in ω:

$$\mathrm{Re}\{\hat{G}_{RR}(\omega)\} \cong \hat{G}_{RR}(0) + \partial_{\omega^2} \mathrm{Re}\{\hat{G}_{RR}(\omega)\}\Big|_{\omega^2\to+0} \cdot \omega^2 + \dots$$
$$\cong \frac{1}{\phi_{RR}} + \tilde{P}\left(\frac{1}{\phi_{RR} - \overset{\circ}{M}\omega^2}\right)^2\Bigg|_{\omega^2\to+0} \cdot \overset{\circ}{M}\omega^2 \tag{4.15}$$

with $\tilde{P}\left(\frac{1}{x^2}\right) = -\partial_x P\left(\frac{1}{x}\right) = \frac{x^2-\eta^2}{(x^2+\eta^2)^2}\Bigg|_{\eta\to 0}$.

Here we assume that all atoms in N space have the ideal mass $\overset{\circ}{M}$. The limit $\eta\to 0$ has to be evaluated before the limit $\omega^2\to +0$ is taken (note that according to (2.88) the limit $\omega^2\to -0$ does not exist.)

The first term of (4.15) leads to an effective static spring constant of the defect.

$$f^{eff} = \phi_{CC}^{eff}(0) = \phi_{CC} - \phi_{CR}\,\hat{G}_{RR}(0)\,\phi_{RC} = \phi_{CC} - \phi_{CR}\frac{1}{\phi_{RR}}\phi_{RC} \tag{4.16}$$

The quadratic term in ω of ϕ_{CC}^{eff} has the dimension of a mass and can be taken together with the defect mass M into an effective mass M^{eff}.

$$M^{eff} = M + \phi_{CR}\left\{\partial_{\omega^2} \mathrm{Re}\{\hat{G}_{RR}(\omega)\}\right\}\Big|_{\omega^2\to+0}\phi_{RC}$$
$$= M + \overset{\circ}{M}\phi_{CR}\left\{\tilde{P}\left(\frac{1}{\phi_{RR} - \overset{\circ}{M}\omega^2}\right)^2\right\}\Bigg|_{\omega^2\to+0}\phi_{RC}\ . \tag{4.17}$$

The expansion of the imaginary part of $\hat{G}_{NN}(\omega)$ gives - similar to the expansion (2.68) of the ideal Green's function $\overset{\circ}{G}(\omega)$ - a term linear in ω which will be calculated and discussed in detail in Sect. 4.4. The result is [see (4.65,68)]

$$\mathrm{Im}\{\phi_{CC}^{eff}(\omega)\} \cong -\Gamma\omega = -M^{eff}\ \gamma\omega \quad \text{with } \Gamma = \frac{3\pi}{2\overset{\circ}{M}\omega_D^3}(f^{eff})^2\ . \tag{4.18}$$

It is remarkable that Γ depends on the defect quantities only through f^{eff} since ω_D is the Debye frequency of the ideal lattice (2.70).

Thus for cubic symmetry the Green's function of the defect behaves for small ω as

$$G_{xx}^{dd}(\omega) = \frac{1}{f^{eff} - M^{eff}\omega^2 - i\Gamma\omega} \tag{4.19}$$

which represents the Green's function of a linear oscillator with velocity-dependent damping. The corresponding time-dependent equations of motion are

$$\left(M^{eff}\frac{d^2}{dt^2} + \Gamma\frac{d}{dt} + f^{eff}\right) G_{xx}^{dd}(t) = \delta(t)\ . \tag{4.20}$$

Thus compared with the Einstein approximation, the embedding of the defect into the lattice results in a renormalization of the Einstein force constant to an effective force constant, and of the defect mass to an effective mass and in addition to a velocity-dependent damping.

The resonance frequency of the defect is

$$\omega_{res}^2 = \frac{f^{eff}}{M^{eff}} . \tag{4.21}$$

Thus we can distinguish two kinds of resonances, "spring resonances" with small f^{eff} and "mass resonances" with large M^{eff}. Only the former give a large static response $G_{xx}^{dd}(\omega = 0) = 1/f^{eff}$, which leads to a change of the elastic behaviour of the crystal (see Sect. 7.6), $f^{eff} = 0$ yields the stability limit of the considered defect configuration which for $f^{eff} < 0$ is instable and decays exponentially in time into a different, stable configuration.

The vibrational spectrum of the defect is in the above approximation

$$z^{d}(\omega) \cong \frac{M}{M^{eff}} \frac{2}{\pi} \frac{\gamma\omega^2}{(\omega^2 - \omega_{res}^2)^2 + \gamma^2\omega^2} \tag{4.22}$$

representing an asymmetric Lorentzian centered at ω_{res} which is plotted in Fig. 16.

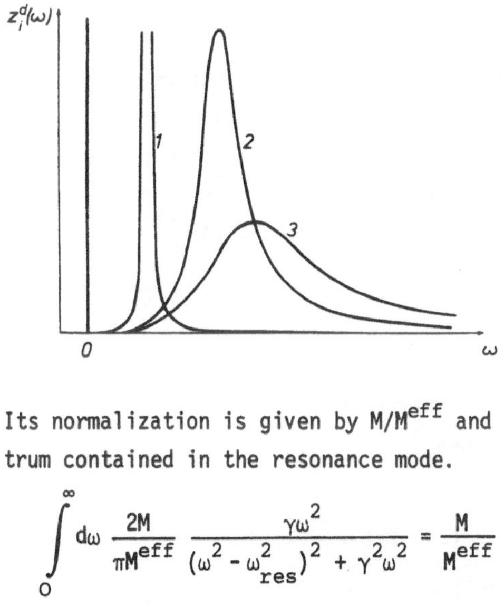

Fig. 16. Local frequency spectrum near a resonance mode. For $\omega_{res} \to 0$, the relative halfwidths $\Delta\omega/\omega_{res} \to 0$, so that the curve gets narrower and degenerates into a δ-function for $\omega_{res} = 0$

Its normalization is given by M/M^{eff} and represents the fraction of the total spectrum contained in the resonance mode.

$$\int_0^\infty d\omega \frac{2M}{\pi M^{eff}} \frac{\gamma\omega^2}{(\omega^2 - \omega_{res}^2)^2 + \gamma^2\omega^2} = \frac{M}{M^{eff}} . \tag{4.23}$$

The halfwidth of the resonance is with (4.64) of Sect. 4.4

$$\Delta\omega = \gamma = \frac{3\pi}{2\overset{o}{M}\omega_D^3} \frac{(f^{eff})^2}{M^{eff}} \quad \text{or} \quad \frac{\Delta\omega}{\omega_{res}} = \frac{3\pi}{2\overset{o}{M}\omega_D^3} \frac{\sqrt{(f^{eff})^3}}{\sqrt{M^{eff}}} . \tag{4.24}$$

Thus both for a mass resonance (M^{eff} large) as well as for a spring resonance with small f^{eff} the relative width goes to zero with $\omega_{res} \to 0$ resulting in sharper and sharper peaks in the spectrum which for $\omega_{res} = 0$ degenerates into a δ-function. Due to the 3/2 power, spring resonances are much weaker damped than mass resonances and give much sharper resonances. Examples are given in Sect. 5.1.

For frequencies $\omega \ll \omega_{res}$ we obtain for $\mathrm{Im}\{G^{\underline{dd}}_{xx}\}$ or $z^{\underline{d}}_{x}(\omega)$ with Γ from (4.18)

$$\mathrm{Im}\{G^{\underline{dd}}_{xx}(\omega)\} \cong \frac{\Gamma\omega}{(f^{eff})^2} = \frac{3\pi}{2\overset{\circ}{M}\omega_D^3}\,\omega = \mathrm{Im}\{\overset{\circ}{G}{}^{(0)}_{xx}(\omega)\} \tag{4.25}$$

$$z^{\underline{d}}_{x}(\omega) \cong \frac{M}{\overset{\circ}{M}}\frac{3\omega^2}{\omega_D^3} = \frac{M}{\overset{\circ}{M}}\overset{\circ}{z}(\omega) \ . \tag{4.26}$$

Thus the imaginary part of the defect Green's function is identical with the ideal one and the spectra agree up to a factor $M/\overset{\circ}{M}$. Equation (4.24) for $\Delta\omega$ and (4.25) are equivalent statements and it is sufficient to prove either (4.24) or (4.25) as will be done in Sect. 4.4.

All the above results have been derived under the assumption that the expansion (4.15) for $\hat{G}_{NN}(\omega)$ converges sufficiently fast, so that higher order terms can be neglected. As a consequence we obtain for a central region with one atom and for cubic symmetry at most one resonance, being threefold degenerate. If this is not the case, e.g., if more than one resonance exists, the Green's function $\hat{G}_{NN}(\omega)$ for a fixed defect is no longer slowly varying. Therefore the central region has to be enlarged. On the other hand for a large enough central region the Green's function $\hat{G}_{NN}(\omega)$ depends only on properties of the ideal lattice and is therefore slowly varying for small ω, so that the previous expansion can be carried out. As an example we treat the 100-dumbbell in the fcc lattice in Chap. 6. Here also symmetry suggests to take both atoms together as the central region. In such a case the expansion is very similar, the difference is only that the quantities M^{eff}, f^{eff} and Γ are tensors in C space

$$G_{CC}(\omega) = \frac{1}{f^{eff}_{CC} - M^{eff}_{CC}\,\omega^2 - i\Gamma_{CC}\omega} \tag{4.27}$$

which can be decomposed into irreducible representations.

If $\underline{M}$, $\underline{N}$, label the atoms of the C space, then [3]

$$G^{\underline{MN}}_{ij}(\omega) = \sum_{\Gamma}\ \sum_{\gamma,\gamma'=1}^{\sigma_\Gamma}\ \sum_{\mu=1}^{d_\Gamma} s^{\underline{M}}_{i}(\Gamma;\gamma;\mu)\ G^{\Gamma}_{\gamma\gamma'}(\omega)\ s^{\underline{N}}_{j}(\Gamma;\gamma';\mu) \ . \tag{4.28}$$

Here σ_Γ is the multiplicity of the irreducible representation Γ and d_Γ is its dimension determining the degeneracy of the eigenvalues. The basis vector $s^{\underline{M}}_{i}(\Gamma;\gamma;\mu)$ with $\gamma = 1,\ldots,\sigma_\Gamma$; $\mu = 1,\ldots,d_\Gamma$ spans the irreducible subspace corresponding to

---

[3] As is commonly used, we have labelled the irreducible representation by Γ. Since this Γ is an index only, a confusion with the halfwidth Γ in (4.27) and (4.29) is not possible.

the irreducible representation Γ. If all irreducible representations occur only once, $\sigma_\Gamma = 1$, as it is the case, e.g., for the dumbbell interstitial in Chap. 6, then the indexes γ, γ' can be left out and the problem reduces for each subspace to a scalar Green's function $G^\Gamma(\omega)$ which has the same general form as the Green's function $G^{\underline{dd}}_{\mathbf{xx}}(\omega)$ of (4.19)

$$G^\Gamma(\omega) = \frac{1}{f^{eff}_\Gamma - M^{eff}_\Gamma\,\omega^2 - i\Gamma_\Gamma\omega} \quad . \tag{4.29}$$

4.3 Effective Force Constant and Effective Mass

According to the last section the resonant mode is completely determined by the effective force constant f^{eff} and the effective mass M^{eff} since the damping Γ can be obtained from these quantities (4.24). According to (4.16,17) f^{eff} and M^{eff} can be calculated by means of the Green's function $\overset{o}{G}$ of the ideal lattice. But, since in general only numerical results for $\overset{o}{G}$ are available, we will give other equivalent expressions for these effective quantities, which enable us to derive approximations without explicity using the Green's function. For simplicity we restrict the discussion as before to the case of a subspace C containing one defect with cubic symmetry.

First we will give an alternative and equivalent derivation for the low frequency behaviour of $G^{\underline{dd}}_{\mathbf{xx}}(\omega)$ (4.19). The expansion of ϕ^{eff}_{CC} for small frequencies means an expansion of $1/G_{CC}(\omega)$ in powers of ω. Thus in the expansion

$$G^{\underline{dd}}_{\mathbf{xx}}(\omega) = \frac{1}{f^{eff} - M^{eff}\,\omega^2 - i\Gamma\omega} \tag{4.30}$$

the quantities f^{eff}, M^{eff} and Γ can directly be expressed by $G^{\underline{dd}}_{\mathbf{xx}}$ itself

$$f^{eff} = \frac{1}{G^{\underline{dd}}_{\mathbf{xx}}(0)} \quad , \qquad M^{eff} = -\left\{\partial_{\omega^2}\,\mathrm{Re}\,\frac{1}{G^{\underline{dd}}_{\mathbf{xx}}(\omega)}\right\}_{\omega^2\to+0} \tag{4.31}$$

$$\Gamma = -\left\{\partial_\omega\,\mathrm{Im}\,\frac{1}{G^{\underline{dd}}_{\mathbf{xx}}(\omega)}\right\}_{\omega\to 0} = (f^{eff})^2\left\{\partial_\omega\,\mathrm{Im}\{G^{\underline{dd}}_{\mathbf{xx}}(\omega)\}\right\}_{\omega\to 0} . \tag{4.32}$$

One has to be cautious in performing the limit $\omega^2 \to 0$ due to the square root behaviour of $G(\omega)$ at the critical point $\omega = 0$. Since

$$\partial_{\omega^2}\,G(\omega) \sim i/\sqrt{\omega^2} \tag{4.33}$$

the derivative $\partial_{\omega^2}\,\mathrm{Re}\{G^{\underline{dd}}_{\mathbf{xx}}(\omega)\}$ exists only for $\omega^2 \to +0$, but not for $\omega^2 \to -0$.

f^{eff} and M^{eff} can also be expressed in terms of a resonance mode $\underline{u}(\omega)$ given by

$$u_i^{\underline{m}}(\omega) = \frac{G_{ix}^{\underline{md}}(\omega)}{G_{xx}^{\underline{dd}}(\omega)} \qquad \text{with } u_i^{\underline{d}}(\omega) = \delta_{ix} \,. \tag{4.34}$$

The displacements $u_i^{\underline{m}}$ are for $\underline{m} \neq \underline{d}$ determined by the force-free equation of motion

$$\sum_{\underline{n}j} \phi_{ij}^{\underline{mn}} \, u_j^{\underline{n}}(\omega) - M^{\underline{m}} \, \omega^2 \, u_i^{\underline{m}}(\omega) = 0 \text{ for } \underline{m} \neq \underline{d} \,. \tag{4.35}$$

For instance, we obtain for f^{eff} by multiplying the static equation

$$\sum_{\underline{n}j} \phi_{ij}^{\underline{mn}} \, G_{jx}^{\underline{nd}}(0) = \delta^{\underline{md}} \, \delta_{ix} \tag{4.36}$$

from left with $G_{xi}^{\underline{dm}}(\omega)$, summing over $\underline{m}$ and i and then dividing both sides by $(f^{eff})^2$

$$f^{eff} = \sum_{\substack{\underline{m}\,\underline{n}\\ i\,j}} u_i^{\underline{m}}(0) \, \phi_{ij}^{\underline{mn}} \, u_j^{\underline{n}}(0) = \sum_{\underline{n}j} \phi_{xj}^{\underline{dn}} \, u_j^{\underline{n}}(0) \tag{4.37}$$

where the last part of the equation follows (4.35) for $\omega = 0$. This expression can be derived somewhat more straightforward by considering the elastic energy due to a static force $F_i^{\underline{n}} = \delta^{\underline{nd}} \, \delta_{ix} \, \mathring{F}$ on the defect. First we have

$$E = \frac{1}{2} \, (\underline{s}, \underline{F}) = \frac{1}{2} \, G_{xx}^{\underline{dd}}(0) \, \mathring{F}^2 = \frac{1}{2} \, f^{eff} (s_x^{\underline{d}})^2 \quad \text{with } s_x^{\underline{d}} = G_{xx}^{\underline{dd}}(0) \, \mathring{F} = \frac{\mathring{F}}{f^{eff}} \,. \tag{4.38}$$

This is just the energy of a linear oscillator with force constant f^{eff}. On the other hand the energy is also given by

$$E = \frac{1}{2} \, (\underline{s}, \underline{\phi} \, \underline{s}) = \frac{1}{2} \left(\underline{u}(0), \underline{\phi} \, \underline{u}(0) \right) (s_x^{\underline{d}})^2 \quad \text{since } s_i^{\underline{m}} = G_{ix}^{\underline{md}}(0) \, \mathring{F} = u_i^{\underline{m}}(0) \, s_x^{\underline{d}} \,. \tag{4.39}$$

By comparison of (4.38) and (4.39), (4.37) follows immediately.

A similar expression can be obtained for M^{eff} by using the identity (2.64) for the matrix $G(\omega)$

$$\partial_{\omega^2} \, G(\omega) = G(\omega) \, M \, G(\omega) \,. \tag{4.40}$$

The result is

$$M^{eff} = \left\{ \mathrm{Re} \sum_{\underline{m}i} u_i^{\underline{m}}(\omega) \, M^{\underline{m}} \, u_i^{\underline{m}}(\omega) \right\}_{\omega^2 \to +0} = M + \mathring{M} \left\{ \mathrm{Re} \sum_{\underline{m}i} \left(u_i^{\underline{m}}(\omega) \right)^2 \right\}_{\omega^2 \to +0} \,. \tag{4.41}$$

Thus the effective mass consists of the defect mass M itself and of contributions from the masses of the host atoms weighted with the square of their amplitude. The resonance frequency ω_{res} can be written in the form

$$\omega^2_{res} = \frac{\left(\underline{u}(0), \underline{\phi}\, \underline{u}(0)\right)}{\left\{\mathrm{Re}\left(\underline{u}(\omega), M\,\underline{u}(\omega)\right)\right\}_{\omega^2 \to +0}} \tag{4.42}$$

similar to Rayleigh's quotient (3.51) for localized modes.

From the expressions (4.40,41) we see that the divergence of $\partial_{\omega^2}G(\omega)$ for $\omega \to 0$ arises because the sum $\sum_{\underline{m}} \left(u_i^{\underline{m}}(0)\right)^2$ diverges: each term of the sum gives a finite contribution, but for large distances $G_{ix}^{\underline{md}}(0)$ varies as $1/|\underline{R}^{\underline{m}} - \underline{R}^{\underline{d}}|$ and the sum diverges.

Similar to the Rayleigh quotient, the expression (4.42) for ω^2_{res} is insensitive against small changes of $\underline{u}(\omega)$. Thus the resonance frequencies of a system characterized by force constants $\tilde{\phi} + \delta\phi$ and masses $\tilde{M} + \delta M$ can be calculated within first-order perturbation theory from the amplitudes $\tilde{\underline{u}}(\omega)$ for the reference system $\tilde{\phi}$, $\tilde{M}$

$$\omega^2_{res} \cong \frac{\tilde{f}^{eff} + \left(\tilde{\underline{u}}(0), \delta\underline{\phi}\, \tilde{\underline{u}}(0)\right)}{\tilde{M}^{eff} + \left(\tilde{\underline{u}}(0), \delta M\, \tilde{\underline{u}}(0)\right)} \tag{4.43}$$

For instance, for a spring resonance with $f^{eff} > 0$, but small, we can take as reference a system for which $f^{eff} = 0$ due to a small additional negative force constant. This system is then on the brink of being instable and $\tilde{u}_i^m(0)$ represents the amplitude distribution of the mode with which the defect configuration "decays" into stable configuration. Therefore $\tilde{\underline{u}}(0)$ has been called "decay mode" by PAGE /4.11/.

The use of Green's functions can be avoided when the effective force constants are calculated approximately, by allowing only a finite number of atoms around the defect to relax. In this way one obtains always upper bounds for f^{eff}, which can be seen as follows. From (4.16) we obtain

$$f^{eff} = \phi_{CC} - \phi_{CR} \frac{1}{\phi_{RR}} \phi_{RC} \leq \phi_{CC} = f^{eff}_{Einstein} \tag{4.44}$$

since the second term is, together with ϕ_{RR}, positive definite. Thus the Einstein approximation gives a too high value for f^{eff}. As the next step we can derive an exact expression for $1/\phi_{RR}$. In (4.44) we only need the matrix elements $\{1/\phi_{RR}\}_{NN}$ in the neighbour subspace N directly coupled to the C space. By dividing the R space into the N space and a U space, so that R = N + U, we obtain for $\{1/\phi_{RR}\}_{NN}$ similar to the expression (4.44) for $(1/\phi)_{CC} = 1/f^{eff}$

$$\left(\frac{1}{\phi_{RR}}\right)_{NN} = \frac{1}{\phi_{NN} - \phi_{NU} \frac{1}{\phi_{UU}} \phi_{UN}} . \tag{4.45}$$

Thus

$$f^{eff} = \phi_{CC} - \phi_{CN}\left(\frac{1}{\phi_{RR}}\right)_{NN} \phi_{NC} \leq \phi_{CC} - \phi_{CN} \frac{1}{\phi_{NN}} \phi_{NC} \leq \phi_{CC} , \tag{4.46}$$

i.e., the Einstein approximation for the space (C + N) gives a better bound than the usual Einstein approximation. In this way we obtain, by allowing more and more atoms to relax statically, better and better upper bounds for f^{eff} which finally coincide with the exact value.

Similar approximations for M^{eff} cannot be obtained due to the convergence difficulties discussed previously. However, we can give an alternative expression for (4.41) by noting, that for large distances and small frequencies the Green's function $G^{\underline{md}}_{\underline{ix}}(\omega)$ approaches the ideal lattice Green's function $\overset{\circ}{G}{}^{\underline{md}}_{\underline{ix}}(\omega)$.

$$G^{\underline{md}}_{\underline{ix}}(\omega) = \overset{\circ}{G}{}^{\underline{md}}_{\underline{ix}} - \sum_{\substack{\underline{np}\\ \underline{jk}}} \overset{\circ}{G}{}^{\underline{mn}}_{\underline{ij}} V^{\underline{np}}_{\underline{jk}} G^{\underline{pd}}_{\underline{kx}} \tag{4.47}$$

For large distances ($R^{\underline{m}} \gg a$) we replace $\overset{\circ}{G}{}^{\underline{mn}}$ by the elastic Green's function (2.75) which we may expand for $R^{\underline{n}} \ll R^{\underline{m}}$ due to the short range of $V^{\underline{np}}_{\underline{jk}}$

$$\overset{\circ}{G}_{ij}(\underline{R}^{\underline{m}} - \underline{R}^{\underline{n}}, \omega) = \overset{\circ}{G}_{ij}(\underline{R}^{\underline{m}} - \underline{R}^{\underline{d}}, \omega) + (\underline{R}^{\underline{n}} - \underline{R}^{\underline{d}})\cdot\partial_{\underline{R}^{\underline{m}}} \overset{\circ}{G}_{ij}(\underline{R}^{\underline{m}} - \underline{R}^{\underline{d}}, \omega) + \dots \quad . \tag{4.48}$$

The first term gives no contribution (4.47) since $\sum_{\underline{n}} V^{\underline{np}}_{\underline{jk}} = 0$ if we have a change of the force constants only. (For a mass change the second term in (4.47) is at least proportional to ω^2.) The contribution from the next term in (4.48) can be neglected since due to the derivative it is smaller than $\overset{\circ}{G}{}^{\underline{md}}_{\underline{ix}}$ by a factor $R^{\underline{n}}/R^{\underline{m}} \ll 1$ or $R^{\underline{n}}\omega/c^{\sigma} \ll 1$. Thus for large distances and small frequencies we obtain the result

$$G^{\underline{md}}_{\underline{ix}}(\omega) \cong \overset{\circ}{G}{}^{\underline{md}}_{\underline{ix}}(\omega) \ . \tag{4.49a}$$

Thus for large $\underline{m}$ and small ω the amplitudes $u^{\underline{m}}_{i}(\omega)$ approach

$$u^{\underline{m}}_{i}(\omega) \cong \frac{\overset{\circ}{G}{}^{\underline{md}}_{\underline{ix}}(\omega)}{G^{\underline{dd}}_{\underline{ix}}(\omega)} \cong f^{eff}\, \overset{\circ}{G}{}^{\underline{md}}_{\underline{ix}}(\omega) \cong \frac{f^{eff}}{\overset{\circ}{f}{}^{eff}}\, \overset{\circ}{u}{}^{\underline{m}}_{i}(\omega) \tag{4.49b}$$

where $\overset{\circ}{f}{}^{eff}$ and $\overset{\circ}{u}{}^{\underline{m}}_{i}$ refer to the ideal lattice. Therefore we have the following expression for M^{eff}:

$$M^{eff} = \left(\frac{f^{eff}}{\overset{\circ}{f}{}^{eff}}\right)^2 \overset{\circ}{M}{}^{eff} + \sum_{\underline{m}i} \left\{ M^{\underline{m}}\, |u^{\underline{m}}_{i}(0)|^2 - \overset{\circ}{M}\left(\frac{f^{eff}}{\overset{\circ}{f}{}^{eff}}\right)^2 |\overset{\circ}{u}{}^{\underline{m}}_{i}(\omega)|^2 \right\} \tag{4.50}$$

where $\overset{\circ}{M}{}^{eff}$ is the effective mass of the ideal lattice, given by (3.42) for $M = \overset{\circ}{M}$. In this form no difficulty arises with the limit $\omega^2 \to 0$ due to the subtraction of the diverging parts. For a strong spring resonance ($f^{eff} \ll \overset{\circ}{f}{}^{eff}$) we may neglect the terms with $(f^{eff}/\overset{\circ}{f}{}^{eff})^2$ and obtain in a rough approximation

$$M^{eff} \approx \sum_{\underline{m}i} M^{\underline{m}}\, |u^{\underline{m}}_{i}(0)|^2 = M + \overset{\circ}{M} \sum_{\substack{\underline{m}(\neq\underline{d})\\ i}} |u^{\underline{m}}_{i}(0)|^2 \tag{4.51a}$$

provided that we allow only a rather small number of atoms near the defect to relax, since the sum diverges for the infinite crystal. In this approximation M^{eff} can be calculated just as easily as f^{eff} of (4.39) from the static displacements $u^{\underline{m}}_{i}(0)$.

On the other hand (4.51) is rather useless for mass resonances. For an isotopic defect we obtain $f^{eff} = \overset{o}{f}{}^{eff}$ and $u_i^{\underline{m}}(0) = \overset{o}{u}{}_i^{\underline{m}}(0)$ so that the effective mass is given by

$$M^{eff} = M + \overset{o}{M}{}^{eff} - \overset{o}{M} \quad . \tag{4.51b}$$

Numerical solutions for a fcc lattice with nearest neighbour interaction yield $\overset{o}{M}{}^{eff} = 0.63\ \overset{o}{M}$ and thus $M^{eff} < M$, whereas (4.51a) implies $M^{eff} > M$. As a consequence M^{eff} cannot be calculated by an evaluation of the equation of motion for a finite number of atoms.

Only in special cases, e.g., for resonant modes with even symmetry $u_i^{\underline{m}} = -u_i^{-\underline{m}}$ where $u_i^{\underline{m}}(0)$ decreases as $1/(R^{\underline{m}})^2$, the sum in (4.51) converges and obviously we obtain lower bounds for M^{eff} for an arbitrary finite region N. Such modes are of minor interest for the above considered case of a single atom defect since the defect remains at rest. However they are important, e.g., in the case of split interstitials (see Chap. 6).

So far we have concentrated on the Green's function $G_{CC}(\omega)$ of the defect only. However, knowing G_{CC}, the other matrix elements G_{RC}, G_{CR} and G_{RR} follow immediately. For instance, we obtain from (4.6b)

$$G_{RC}(\omega) = \frac{-1}{\phi_{RR} - \overset{o}{M}(\omega + i\eta)^2}\, \phi_{RC}\, G_{CC} = u_{RC}(\omega)\, G_{CC}(\omega) \tag{4.52}$$

and

$$G_{CR}(\omega) = G'_{RC}(\omega) = G_{CC}(\omega)\, u_{CR}(\omega) \tag{4.53}$$

where G'_{RC} is the transpose of G_{RC}. In the same way we obtain for G_{RR}

$$\begin{aligned} G_{RR}(\omega) &= \frac{1}{\phi_{RR} - \overset{o}{M}(\omega + i\eta)^2} + \frac{-1}{\phi_{RR} - (\omega + i\eta)^2}\, \phi_{RC}\, G_{CR}(\omega) \\ &\cong \frac{1}{\phi_{RR} - \overset{o}{M}(\omega + i\eta)^2} + u_{RC}(\omega)\, G_{CC}(\omega)\, u_{CR}(\omega) \quad . \end{aligned} \tag{4.54}$$

For frequencies near the resonance the second term is in general larger than the first one, so that

$$G_{ij}^{\underline{mn}}(\omega) \cong \frac{u_i^{\underline{m}}(\omega)\, u_j^{\underline{n}}(\omega)}{f^{eff} - M^{eff}\,\omega^2 - i\Gamma\omega} \cong \frac{u_i^{\underline{m}}(0)\, u_j^{\underline{n}}(0)}{M^{eff}\,(\omega_{res}^2 - \omega^2 - i\gamma\omega)} \, . \tag{4.55}$$

Here the ω dependence of $u_i^{\underline{m}}$ can be neglected if $\underline{m}$, $\underline{n}$ are not too large, so that the static displacements $u_i^{\underline{m}}(0)$ can be taken. For large distances on the other hand, the first term in (4.54) is always dominating. Expression (4.55) with the static displacements $\underline{u}(0)$ can, e.g., be used to calculate the t matrix (3.24) under resonance conditions (see Chap. 7)

$$t = V - V\,G\,V \cong - \frac{V\,|\underline{u}(0)\rangle\langle\underline{u}(0)|\,V}{M^{eff}\,(\omega_{res}^2 - \omega^2 - i\gamma\omega)} \;. \tag{4.56}$$

4.4 Damping of Resonant Modes

In order to calculate the damping of the resonance vibration we will proof first the identity (4.25) of $\mathrm{Im}\{G(\omega)\}$ and $\mathrm{Im}\{\overset{\circ}{G}(\omega)\}$ for low frequencies $\omega \to 0$. Starting with the standard equation (3.19) for G

$$G = \frac{1}{1 + \overset{\circ}{G}V}\,\overset{\circ}{G} \tag{4.57}$$

we obtain with the hermitian adjunct $G^+ = G^*$

$$\begin{aligned}\mathrm{Im}\{G(\omega)\} &= \frac{1}{2i}\left(G(\omega) - G^+(\omega)\right) = \frac{1}{2i}\left\{\frac{1}{1+\overset{\circ}{G}V}\,\overset{\circ}{G} - \overset{\circ}{G}{}^+\,\frac{1}{(1+\overset{\circ}{G}V)^+}\right\}\\ &= \frac{1}{2i}\,\frac{1}{1+\overset{\circ}{G}V}\left\{\overset{\circ}{G} + \overset{\circ}{G}V^+\overset{\circ}{G}{}^+ - \overset{\circ}{G}{}^+ - \overset{\circ}{G}V\overset{\circ}{G}{}^+\right\}\frac{1}{(1+\overset{\circ}{G}V)^+}\;.\end{aligned} \tag{4.58}$$

Since $V^+ = V$ the result is

$$\mathrm{Im}\{G(\omega)\} = \frac{1}{1+\overset{\circ}{G}V}\,\mathrm{Im}\{\overset{\circ}{G}\}\,\frac{1}{(1+\overset{\circ}{G}V)^+} = (1 - GV)\,\mathrm{Im}\{\overset{\circ}{G}\}(1 - VG^*)\;. \tag{4.59}$$

For small frequencies the imaginary part of $\overset{\circ}{G}{}^{\underline{m}\underline{n}}_{ij}(\omega)$ is proportional to ω and independent of $\underline{m}$ and $\underline{n}$ (2.68). To obtain the leading term of $\mathrm{Im}\{G(\omega)\}$ for small ω we can therefore replace $G(\omega)$ by $G(0)$ and $V(\omega)$ by $V(0) = \varphi$

$$\mathrm{Im}\{G^{\underline{m}\underline{n}}_{ij}(\omega)\} \cong \sum_{\substack{\underline{p}\underline{q}\\ k}}\left(\delta^{\underline{m}\underline{p}}_{ik} - \sum_{\underline{r}\ell} G^{\underline{m}\underline{r}}_{i\ell}(0)\,\varphi^{\underline{r}\underline{p}}_{\ell k}\right)\frac{3\pi\omega}{2\overset{\circ}{M}\omega_D^3}\left(\delta^{\underline{q}\underline{n}}_{kj} - \sum_{\underline{s}\ell'}\varphi^{\underline{q}\underline{s}}_{k\ell'}\,G^{\underline{s}\underline{n}}_{\ell' j}(0)\right). \tag{4.60}$$

Now due to the translation symmetry of the coupling parameters

$$\sum_{\underline{m}} \varphi^{\underline{m}\underline{n}}_{ij} = 0 = \sum_{\underline{n}} \varphi^{\underline{m}\underline{n}}_{ij} \tag{4.61}$$

we obtain finally

$$\mathrm{Im}\{G^{\underline{m}\underline{n}}_{ij}(\omega)\} \cong \mathrm{Im}\{\overset{\circ}{G}{}^{(\underline{m}-\underline{n})}_{ij}(\omega)\} \cong \frac{3\pi}{2\overset{\circ}{M}\omega_D^3}\,\omega\,\delta_{ij}\;, \tag{4.62}$$

i.e., for small ω, $\mathrm{Im}\{G\}$ does not depend on the structure and properties of the defect. The comparison with (4.25) for $\omega \to 0$

$$\mathrm{Im}\{G^{\underline{dd}}_{xx}(\omega)\} \cong \frac{\Gamma\omega}{(f^{eff})^2} \tag{4.63}$$

yields the following result for Γ and the relative halfwidth $\Delta\omega/\omega_{res}$ of the resonance

$$\Gamma \equiv M^{eff}\gamma = \frac{3\pi}{2\overset{o}{M}\omega_D^3}(f^{eff})^2 \;, \quad \frac{\Delta\omega}{\omega_{res}} = \frac{\gamma}{\omega_{res}} = \frac{3\pi}{2\overset{o}{M}\omega_D^3}\frac{\sqrt{(f^{eff})^3}}{\sqrt{M^{eff}}} \;. \tag{4.64}$$

Instead of comparing the low-frequency behaviour of $\mathrm{Im}\{G^{\underline{dd}}_{xx}(\omega)\}$, we can also calculate the damping constant Γ directly by going back to the definition (4.18)

$$\Gamma\omega \cong -\,\mathrm{Im}\{\phi^{eff}_{CC}(\omega)\} = \phi_{CR}\;\mathrm{Im}\{\hat{G}_{RR}(\omega)\}\;\phi_{RC} \quad \text{for } \omega \to 0 \;. \tag{4.65}$$

This method has the advantage of being applicable also for interstitial-type defects, for which the previous derivation fails due to the additional degrees of freedom. For interstitials we obtain for $\hat{G}(\omega)$ from (4.59)

$$\mathrm{Im}\{\hat{G}^{\underline{mn}}_{ij}(\omega)\} \cong \sum_{\substack{\underline{pq}\\ k}} \left\{\delta^{\underline{mp}}_{ik} - \sum_{\underline{r}\ell} \hat{G}^{\underline{mr}}_{i\ell}(0)\;(\varphi_{RR})^{\underline{rp}}_{\ell k}\right\} \frac{3\pi\omega}{2\overset{o}{M}\omega_D^3} \left\{\delta^{\underline{qn}}_{kj} - \sum_{\underline{s}\ell'} (\varphi_{RR})^{\underline{qs}}_{k\ell'}\;\hat{G}^{\underline{sn}}_{\ell' j}(0)\right\} . \tag{4.66}$$

The indices $\underline{m}\ldots\underline{r}\;\underline{s}$ refer to all atoms of the ideal lattice spanning the R space and φ_{RR} is the change of the coupling matrices of these atoms for a fixed defect d. Thus due to the translation symmetry of the coupling parameters we have

$$\sum_{\underline{n}(\neq\underline{d})} \varphi^{\underline{mn}}_{ij} + \phi^{\underline{md}}_{ij} = 0 \;, \quad \phi^{\underline{dd}}_{ij} + \sum_{\underline{n}(\neq\underline{d})} \phi^{\underline{dn}}_{ij} = 0 \quad . \tag{4.67}$$

Therefore by summing over $\underline{p}$ and $\underline{q}$ in (4.66) we obtain for Γ

$$\Gamma = \frac{3\pi}{2\overset{o}{M}\omega_D^3}(f^{eff})^2 \tag{4.68}$$

with f^{eff} given by (4.16). Thus the above formula for the damping is also valid for interstitial-type defects.

To see the intuitive meaning of this result we consider the work done on the crystal by given forces $F^{\underline{m}}_i(t) = f^{\underline{m}}_i \cos\omega_o t$ applied to the defect and the neighbouring atoms

$$W(t) = \sum_{\underline{m}i} F^{\underline{m}}_i(t)\;\dot{s}^{\underline{m}}_i(t) \tag{4.69}$$

Analogous to (2.58), $s^{\underline{m}}_i$ is given by

$$\begin{aligned} s^{\underline{m}}_i(t) &= \sum_{\underline{n}j}\left(G^{\underline{mn}}_{ij}(\omega_o)\;f^{\underline{n}}_j\,\frac{1}{2}\exp(i\omega_o t) + G^{\underline{mn}}_{ij}(-\omega_o)\;f^{\underline{n}}_j\,\frac{1}{2}\exp(-i\omega_o t)\right) \\ &= \sum_{\underline{n}j}\left(\mathrm{Re}\{G^{\underline{mn}}_{ij}(\omega_o)\}\cos\omega_o t + \mathrm{Im}\{G^{\underline{mn}}_{ij}(\omega_o)\}\sin\omega_o t\right) f^{\underline{n}}_j \;. \end{aligned} \tag{4.70}$$

Averaging over one period of the motion we obtain for the average work per sec

$$\overline{W(t)} = \frac{\omega_o}{2} \sum_{\substack{\underline{m}\underline{n} \\ ij}} f_i^{\underline{m}} \, \mathrm{Im}\{G_{ij}^{\underline{m}\underline{n}}(\omega)\} \, f_j^{\underline{n}} \, . \tag{4.71}$$

By combining with equation (4.62), $\overline{W(t)}$ is for small frequencies given by

$$\overline{W(t)} \cong \frac{3\pi\omega_o^2}{2M\omega_D^3} (\underline{f})^2 \quad \text{with } f_i = \sum_{\underline{m}} f_i^{\underline{m}} \, . \tag{4.72}$$

Thus the average energy transferred to the system is independent of the details of the defect structure, but depends only on the magnitude of the total force $\underline{f}$ and on the properties of the surrounding ideal lattice, i.e., its density of states for small ω's.

A more general proof of this statement, valid for an arbitrarily disordered but finite region embedded into an otherwise ideal crystal, is given in /4.12/.

For a force distribution with vanishing total force $\underline{f}$, as it is, for instance, the case for a distribution with even ("gerade") symmetry

$$f_i^{\underline{m}} = -f_i^{-\underline{m}} \quad \text{and} \quad f_i = \sum_i f_i^{\underline{m}} = 0 \tag{4.73}$$

expression (4.72) vanishes and the average work is proportional to ω_o^4. This can be seen by taking the ω^3 term of $\mathrm{Im}\{\overset{o}{G}(\omega)\}$ (2.68) with the tensor $T_{ij,k\ell}$ into account. The result can be written as

$$\overline{W(t)} = \frac{1}{2} \omega_o^4 \sum_{ij,k\ell} P_{ik} \, T_{ij,k\ell} \, P_{j\ell} \tag{4.74}$$

where P_{ij} is the dipole moment of the forces $\underline{K}$

$$P_{ij} = \sum_{\underline{m}} R_i^{\underline{m}} K_j^{\underline{m}} \, , \quad \underline{K} = \left(1 - \varphi \, G(0)\right) \underline{f} \tag{4.75}$$

which relate to the forces $\underline{f}$ as the Kanzaki forces relate to the "original" forces $\underline{f}$ in lattice statics /2.4/. Thus in this case the work $\overline{W(t)}$ does depend on the defect structure via the changed coupling parameters φ.

Equation (4.74) for the work per sec is analogous to the radiated energy of an Hertzian dipole with moment $\underline{P}$ in electrodynamics, which is also proportional to ω_o^4

$$W(t) = \frac{\omega_o^4}{3c^3} (\underline{P})^2 \, . \tag{4.76}$$

There is no analogy to (4.72), since in electrodynamics time varying charges (corresponding to the forces $f_i^{\underline{n}}(t)$) do not exist.

Due to the identity of $\mathrm{Im}\{G(\omega)\}$ and $\mathrm{Im}\{\overset{o}{G}(\omega)\}$ for $\omega \to 0$ the local spectra are equal up to a factor

$$z_x^{\underline{m}}(\omega) \cong \frac{M^{\underline{m}}}{\overset{\circ}{M}} \overset{\circ}{z}(\omega) \cong \frac{M^{\underline{m}}}{\overset{\circ}{M}} \frac{3\omega^2}{\omega_D^3} \,. \tag{4.77}$$

This result may lead to some confusion, when connected with the average frequency spectrum of the crystal

$$z(\omega) = \frac{1}{3N} \sum_{\underline{m}i} z_i^{\underline{m}}(\omega) \,. \tag{4.78}$$

It would mean that the average spectrum depends for $\omega \to 0$ only on the masses of the defect, but not on force constant changes so that for equal masses the spectrum is not changed at all for small frequencies. As we will see in Chap. 7, this is not true at all; force constant changes are very important and lead to changes of the elastic constants of the crystal, thus influencing the average low frequency spectrum. The solution of the puzzle is that the limit $\omega \to 0$ in (4.78) cannot be exchanged with the sum over all atoms $\underline{m}$. The expression (2.68) for $\mathrm{Im}\{\overset{\circ}{G}(\omega)\}$ is only valid for $\omega \ll c_\sigma/R^{(\underline{m}-\underline{n})}$. Therefore (4.62) for $\mathrm{Im}\{G(\omega)\}$ is then valid if both $\omega \ll c_\sigma/R^{\underline{m}}$ and $\omega \ll c_\sigma/R^{\underline{n}}$, assuming the defect to be located near the origin. Thus for large $R^{\underline{m}}$ or $R^{\underline{n}}$ ω has to be extremely small. Then it is plausible, that by summing (4.78) for a finite ω over all $\underline{m}$ we obtain a finite contribution from the very distant atoms with $c_\sigma/R^{\underline{m}} \lesssim \omega$, which is connected with the changed elastic constants. Details are discussed in Sect. 7.6.

4.5 Localized Modes

The concept of an effective mass can also be used for localized vibrations, i.e., eigenvibrations with $\omega > \overset{\circ}{\omega}_{max}$. For simplicity we assume that $\mathrm{Im}\{\hat{G}_{RR}(\omega)\} = 0$ for $\omega > \overset{\circ}{\omega}_{max}$ which means that there are no localized vibrations, if the defect itself is fixed, see (4.8,9).

Then the localized mode frequencies ω_ℓ are determined by

$$M_{CC}\, \omega_\ell^2 = \phi_{CC}^{eff}(\omega_\ell) = \phi_{CC} - \phi_{CR}\, \hat{G}_{RR}(\omega_\ell)\, \phi_{RC} \,. \tag{4.79}$$

The behaviour of the Green's function near ω_ℓ is obtained by a Taylor expansion of ϕ_{CC}^{eff}. If the retardation is taken into account by an infinitesimal damping $i\eta\omega$ we obtain

$$G_{xx}^{\underline{d}\underline{d}}(\omega) \cong \frac{1}{M_\ell^{eff}(\omega_\ell^2 - \omega^2 - i\eta\omega)} \quad \text{and} \quad z_x^{\underline{d}}(\omega) = \frac{M}{M_\ell^{eff}}\, \delta(\omega - \omega_\ell) \,. \tag{4.80}$$

Thus M/M_ℓ^{eff} gives the fractional contribution of the localized mode to the vibrations of the defect in x direction. The effective mass of the mode is in analogy to the case of a resonant mode determined by

$$M_\ell^{eff} = M + \phi_{CR}\, \partial_{\omega^2} \hat{G}_{RR}(\omega)\Big|_{\omega=\omega_\ell} \phi_{RC} = \partial_{\omega^2} \frac{1}{G_{CC}(\omega)}\Big|_{\omega=\omega_\ell} . \tag{4.81}$$

By making use of (4.40) for $\partial_{\omega^2}G(\omega)$ we can again introduce mode amplitudes $\underline{u}(\omega_\ell)$ by

$$u_i^{\underline{m}}(\omega) = G_{ix}^{\underline{md}}(\omega) \Big/ G_{xx}^{\underline{dd}}(\omega) \tag{4.82}$$

so that

$$M_\ell^{eff} = M + \mathring{M} \sum_{\substack{\underline{m}(\neq\underline{d}) \\ i}} \left| u_i^{\underline{m}}(\omega_\ell) \right|^2 > M . \tag{4.83}$$

Here we are not faced with a convergence problem, since the Green's function decreases exponentially for large distances.

Near the resonance the total Green's function can be written as

$$G_{ij}^{\underline{mn}}(\omega) \cong \frac{1}{\omega_\ell^2 - \omega^2 - i\eta\omega} \frac{1}{M_\ell^{eff}} u_i^{\underline{m}}(\omega_\ell)\, u_j^{\underline{n}}(\omega_\ell) = \frac{1}{\omega_\ell^2 - \omega^2 - i\eta\omega} s_i^{\underline{m}}(\omega_\ell)\, s_j^{\underline{n}}(\omega_\ell) . \tag{4.84}$$

By comparing with (3.16) we see that $u_i^{\underline{m}} / \sqrt{M^{eff}}$ is just the normalized amplitude $s_i^{\underline{m}}$ of atom $\underline{m}$ in the localized mode ℓ.

Equation (4.79) for the localized mode frequency ω_ℓ is very convenient for calculating *upper and lower bounds* for ω_ℓ without using the numerical lattice Green's functions. The condition for a localized mode is

$$M\omega^2 = (\phi^{eff})_{xx}^{\underline{dd}}(\omega) = \phi_{CC} + \phi_{CR} \frac{1}{\mathring{M}\omega^2 - \phi_{RR}} \phi_{RC} \tag{4.85}$$

which is graphically sketched in Fig. 17. As before we assume that the crystal has no localized mode if the defect is at rest. Then the maximal eigenfrequency of ϕ_{RR} is identical with $\mathring{\omega}_{max}$ of the ideal crystal. The matrix $\mathring{M}\omega^2 - \phi_{RR}$ is for $\omega > \omega_{max}$ positive definite, so that $\phi_{CC}^{eff}(\omega) > \phi_{CC}$ in this range. Further we see that $\phi_{CC}^{eff}(\omega)$ is monotonically decreasing, since $\partial_{\omega^2}\phi_{CC}^{eff}(\omega) < 0$, and approaches the limiting

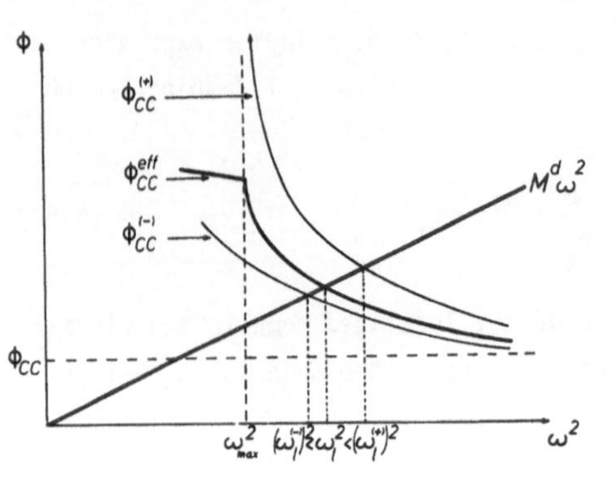

Fig. 17. Construction of bounds for the localized mode frequency ω_ℓ. The intersection of each approximation $\phi_{CC}^{(+)}(\omega) > \phi_{CC}^{eff}(\omega)$ with $M^d\omega^2$ gives an upper bound $\omega_\ell^{(+)}$, whereas each approximation $\phi_{CC}^{(-)}(\omega) < \phi_{CC}^{eff}(\omega)$ leads to a lower bound $\omega_\ell^{(-)}$

value ϕ_{CC} as $1/\omega^2$. At $\omega \gtrsim \omega_{max}$, $\phi_{CC}^{eff}(\omega)$ shows a square root behaviour.

From Fig. 17 it is obvious that whenever we have an approximation $\phi_{CC}^{(+)}(\omega)$ or $\phi_{CC}^{(-)}(\omega)$ with

$$\phi_{CC}^{eff}(\omega) < \phi_{CC}^{(+)}(\omega) \quad \text{or} \quad \phi_{CC}^{eff}(\omega) > \phi_{CC}^{(-)}(\omega) \quad \text{for all } \omega > \omega_{max} \tag{4.86}$$

then the intersection of this function with $M\omega^2$ gives always an upper bound $\omega_\ell^{(+)}$ or lower bound $\omega_\ell^{(-)}$ for the exact frequency ω_ℓ

$$\omega_\ell^{(-)} < \omega_\ell < \omega_\ell^{(+)} \; . \tag{4.87}$$

Examples for this are:

Lower bounds: A Taylor expansion of $(\overset{o}{M}\omega^2 - \phi_{RR})^{-1}$ in powers of $1/\omega^2$ yields

$$\begin{aligned}\phi_{CC}^{eff}(\omega) &= \phi_{CC} + \frac{1}{\overset{o}{M}\omega^2}\phi_{CR}\,\phi_{RC} + \frac{1}{(\overset{o}{M}\omega^2)^2}\phi_{CR}\,\phi_{RR}\,\phi_{RC} + \dots \\ &\geqslant \phi_{CC} + \frac{1}{\overset{o}{M}\omega^2}\phi_{CR}\,\phi_{RC} \geqslant \phi_{CC}\end{aligned} \tag{4.88}$$

since all terms in the expansion are positive. In this way we obtain an infinite series of lower bounds, approaching ω_ℓ^2 from below. The most simple one gives the Einstein frequency

$$\frac{\phi_{xx}^{\underline{dd}}}{M} = \omega_E^2 < \omega_\ell^2 \tag{4.89}$$

whereas the second one is given by

$$\omega_E^2\left(\frac{1}{2} + \frac{1}{2}\sqrt{1+\alpha}\right) < \omega_\ell^2 \tag{4.90}$$

with

$$\alpha = 4\frac{M}{\overset{o}{M}}\sum_{\underline{n}_j(\neq \underline{d})} (\phi_{xj}^{\underline{dn}})^2 \quad (\phi_{xx}^{\underline{dd}})^{-2} = 4\,\frac{\langle\omega^4\rangle - \omega_E^4}{\omega_E^4}\,, \quad \omega_E^2 = \langle\omega^2\rangle$$

where $\langle\omega^4\rangle$, $\langle\omega^2\rangle$ are the moments of the local frequency spectrum of the defect.

Somewhat better lower bounds can be obtained as follows: Since in (4.85) we need the inverse of $\overset{o}{M}\omega^2 - \phi_{RR}$ only in the subspace N, we can subdivide R into R = N + U, so that analogously to (4.45)

$$\left(\frac{1}{\overset{o}{M}\omega^2 - \phi_{RR}}\right)_{NN} = \frac{1}{\overset{o}{M}\omega^2 - \phi_{NN} - \phi_{NU}\dfrac{1}{\overset{o}{M}\omega^2 - \phi_{UU}}\phi_{UN}} \cong \frac{1}{\overset{o}{M}\omega^2 - \phi_{NN}} \tag{4.91}$$

which is in first approximation the Einstein approximation for the N space. However, since the correction matrix in the denominator is for $\omega > \overset{o}{\omega}_{max}$ positive definite, we have

$$\phi_{CC}^{eff}(\omega) > \phi_{CC} + \phi_{CN} \frac{1}{\overset{\circ}{M}\omega^2 - \phi_{NN}} \phi_{NC} \ . \tag{4.92}$$

This means that the Einstein approximation for the combined spaces C + N gives a lower bound for ω_ℓ, a result we obtained already in Sect. 3.3 from the variational principle.

Upper bounds: Due to our previous assumption, the maximal eigenvalue of ϕ_{RR} is $\overset{\circ}{M}\overset{\circ}{\omega}{}^2_{max}$. Thus

$$\phi_{CC}^{eff}(\omega) \leq \phi_{CC} + \frac{1}{\overset{\circ}{M}(\omega^2 - \overset{\circ}{\omega}{}^2_{max})} \phi_{CR}\, \phi_{RC} \tag{4.93}$$

which yields the upper bound

$$\omega_\ell^2 < \frac{\overset{\circ}{\omega}{}^2_{max} + \omega_E^2}{2} + \left\{ \left(\frac{\overset{\circ}{\omega}{}^2_{max} - \omega_E^2}{2} \right)^2 + \langle \omega^4 \rangle - \omega_E^4 \right\}^{\frac{1}{2}} \ . \tag{4.94}$$

For $\omega_E > \overset{\circ}{\omega}_{max}$ this can be further estimated due to $\sqrt{1+x} \leq 1 + \frac{x}{2}$ for $x > 0$

$$\omega_\ell^2 < \omega_E^2 + \frac{\langle \omega^4 \rangle - \omega_E^4}{\omega_E^2 - \overset{\circ}{\omega}{}^2_{max}} \ . \tag{4.95}$$

This result can also be obtained directly from (4.93) by underestimating $\omega^2 - \overset{\circ}{\omega}{}^2_{max}$ by the lower bound $\omega_E^2 - \overset{\circ}{\omega}{}^2_{max}$.

The method can be further improved by writing

$$\frac{1}{\overset{\circ}{M}\omega^2 - \phi_{RR}} = \frac{1}{\overset{\circ}{M}(\omega^2 - \overset{\circ}{\omega}{}^2_{max}) + (\overset{\circ}{M}\overset{\circ}{\omega}{}^2_{max} - \phi_{RR})} = \frac{1}{\overset{\circ}{M}(\omega^2 - \overset{\circ}{\omega}{}^2_{max})} \{1 - X + X^2 - X^3 + - \dots\} \tag{4.96}$$

with

$X = (\overset{\circ}{M}\overset{\circ}{\omega}{}^2_{max} - \phi_{RR}) / \overset{\circ}{M}(\omega^2 - \overset{\circ}{\omega}{}^2_{max})$ positive definite.

Now we have the following inequality valid for $X > 0$ and n integer

$$1 - X + X^2 - \dots - X^{2n-1} = \frac{1 - X^{2n}}{1 + X} < \frac{1}{1 + X} < \frac{1 + X^{2n+1}}{1 + X} = 1 - X + X^2 - \dots + X^{2n} \ . \tag{4.97}$$

Thus by taking the 1st, 3rd, 5th, etc., terms in the expansion (4.96), we always get upper bounds, whereas 2nd, and 4th terms give lower bounds.

Such upper bounds for the localized mode frequencies have been derived by a somewhat different method by DEAN /4.9/, and also by FUJITA /4.10/, in particular for simple cubic lattices. We believe that the present derivation is more straight forward. Unfortunately all upper bounds yield less useful approximations, if ω_ℓ approaches $\overset{\circ}{\omega}_{max}$, since all $\phi_{CC}^{(+)}$ diverge at $\overset{\circ}{\omega}_{max}$.

4.6 Resonance and Localized Modes near the Band Edge ω_{max}

As discussed in Sect. 3.1, resonance vibrations can occur for those frequencies of the continuous spectrum $0 \leq \omega \leq \overset{\circ}{\omega}_{max}$, for which $\mathrm{Im}\{\overset{\circ}{G}(\omega)\}$ is very small. Besides for $\omega \cong 0$ this is also the case at the upper band edge[4] where $\mathrm{Im}\{\overset{\circ}{G}(\omega)\}$ vanishes as $\sqrt{\omega_{max}^2 - \omega^2}$ (see Sect. 2.6). Physically these resonances are connected with the formation of a localized mode, since a slight strengthening of the force constants or a slight decrease of the defect mass leads to the appearance of a localized mode just above ω_{max}. This behaviour is similar to the low-frequency resonance modes discussed in the previous sections, which for a slight increase of some negative force constants go over into localized "decay modes" with $\omega^2 < 0$. However there are some peculiar differences which we discuss in the following. The existence of this type of resonance vibrations, though obvious from the above point of view, has to our knowledge been discussed only recently /4.13/.

We start with the Green's function $G_{xx}^{\underline{dd}}(\omega)$ of the defect

$$G_{xx}^{\underline{dd}}(\omega) = \frac{1}{\phi_{CC}^{eff}(\omega) - M\omega^2} \quad . \tag{4.98}$$

In the case of a resonance near $\omega_m = \omega_{max}$, $G_{xx}^{\underline{dd}}(\omega_m)$ is very large. Thus we expand $1/G_{xx}^{\underline{dd}}$ or ϕ_{CC}^{eff} into powers of $\omega^2 - \omega_m^2$, leading in analogy to (4.19) to

$$G_{xx}^{\underline{dd}}(\omega) = \frac{1}{f^{eff} - M^{eff}(\omega^2 - \omega_m^2) - iM^{eff}\gamma\sqrt{\omega_m^2 - \omega^2}} \tag{4.99}$$

where f^{eff} is the effective restoring force at ω_m

$$f^{eff} \equiv f^{eff}(\omega_m) = \left(G_{xx}^{\underline{dd}}(\omega_m)\right)^{-1} = \phi_{CC}^{eff}(\omega_m) - M\omega_m^2 \quad . \tag{4.100}$$

From the representation (3.16) for $G(\omega)$, it follows that $f^{eff}(\omega_m) < 0$, as long as no localized mode exists, whereas, with the appearance of a localized mode at ω_m, $f^{eff}(\omega_m)$ goes through zero and gets positive. The effective mass is given by

$$\begin{aligned} M^{eff} &= M^{eff}(\omega_m) = -\partial_{\omega^2} \mathrm{Re}\left\{\frac{1}{G_{xx}^{\underline{dd}}(\omega)}\right\}\Bigg|_{\omega^2 \to \omega_m^2 - 0} \\ &= M + \sum_{\substack{\underline{mn}, ij \\ (\neq \underline{d})}} \phi_{xi}^{\underline{dm}} \, \partial_{\omega^2} \mathrm{Re}\{\hat{G}_{ij}^{\underline{mn}}(\omega)\}\Big|_{\omega^2 \to \omega_m^2 - 0} \phi_{ix}^{\underline{nd}} \quad . \end{aligned} \tag{4.101}$$

The damping constant γ is determined by the behaviour of $\mathrm{Im}\{G_{xx}^{\underline{dd}}(\omega)\}$ near ω_m

---

4 In nonprimitive lattices additional resonances can occur near bandgaps if a localized mode moves into the gap.

$$\mathrm{Im}\left\{\frac{-1}{G_{xx}^{dd}(\omega)}\right\} \cong \frac{\mathrm{Im}\{G_{xx}^{dd}(\omega)\}}{\left(G_{xx}^{dd}(\omega_m)\right)^2} \cong M^{eff}\ \gamma\sqrt{\omega_m^2-\omega^2}\ \theta(\omega_m^2-\omega^2)\ . \tag{4.102}$$

Note that $i\sqrt{\omega_m^2-\omega^2}$ is real for $\omega > \omega_m$. Thus we obtain for the local frequency spectrum as a function of ω^2 for $\omega^2 < \omega_m^2$

$$z_x^d(\omega^2) = \frac{M}{\pi}\ \mathrm{Im}\{G_{xx}^{dd}(\omega)\} = \frac{M}{M^{eff}}\frac{1}{\pi}\frac{\gamma\sqrt{\omega_m^2-\omega^2}}{\left((f^{eff}/M^{eff}) - (\omega^2-\omega_m^2)\right)^2 + \gamma^2(\omega_m^2-\omega^2)} \tag{4.103}$$

and for frequencies above the band edge ($\omega^2 > \omega_m^2$)

$$z_x^d(\omega) = \frac{M}{M^{eff}}\delta\left\{(f^{eff}/M^{eff}) - (\omega^2-\omega_m^2) - \gamma\sqrt{\omega^2-\omega_m^2}\right\}\ \text{for}\ \omega^2 > \omega_m^2\ . \tag{4.104}$$

For $f^{eff} < 0$, the δ function gives no contribution so that the spectrum $z_x^d(\omega)$ is an asymmetric Lorentzian centered at about $\omega_m^2 - \omega_{res}^2 \cong |f^{eff}|/M^{eff}$. Compared to the resonances at $\omega \approx 0$, the important difference is, as will be shown below, that the damping constant γ remains finite as $f^{eff} \to 0$ whereas $\gamma \sim (f^{eff})^2/M^{eff}$ for low-frequency resonances. As a result the resonance curve has always a finite width $\Delta\omega = \gamma$ contrary to the very sharp resonances for $\omega \approx 0$, which degenerate into a δ function as ω_{res}^2 approaches 0 (see Fig. 16).

On the contrary, for $f^{eff} \to 0$, the spectrum z_x^d diverges at the band edge as $(\omega_m^2-\omega^2)^{-\frac{1}{2}}$. This behaviour for a negative value of f^{eff} and for $f^{eff} = 0$ is shown in Fig. 18. The normalization of the resonance curve remains the same, namely

$$\int_0^{\omega_m^2} d\omega^2\ z_x^d(\omega^2) = \frac{M}{M^{eff}} \tag{4.105}$$

as can be shown by residuum calculus.

The different behaviour of the damping for resonances near $\omega = 0$ and ω_{max} can be traced back to the behaviour of $\mathrm{Im}\{\overset{o}{G}(\omega)\}$ at these points, since according to (4.59)

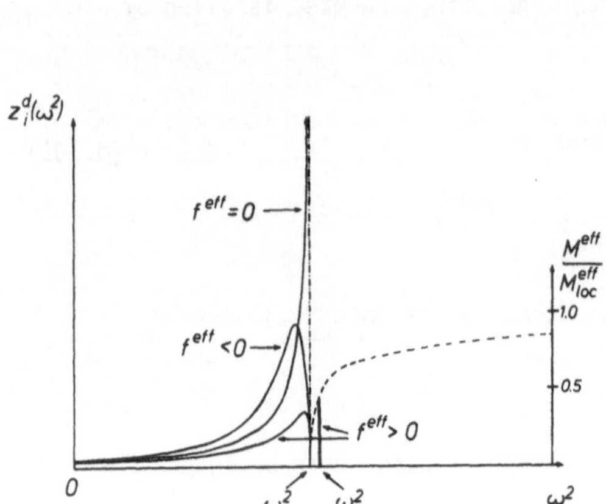

Fig. 18. Resonances and localized modes near the band edge ω_{max}. For $f^{eff} < 0$ we have a broad resonance below ω_{max}, which diverges as f^{eff} approaches zero. For $f^{eff} > 0$ a localized mode occurs above ω_{max}, however some intensity is still left in the resonance mode

$$\mathrm{Im}\{G(\omega)\} = \Big(1 - G(\omega)\, V(\omega)\Big)\, \mathrm{Im}\{\overset{\circ}{G}(\omega)\}\Big(1 - G(\omega)\, V(\omega)\Big)^{+}$$
$$\cong \Big(1 - \mathrm{Re}\{G(\omega_m)\}\, V(\omega_m)\Big)\, \mathrm{Im}\{\overset{\circ}{G}(\omega)\}\Big(1 - V(\omega_m)\, \mathrm{Re}\{G(\omega_m)\}\Big) \,. \tag{4.106}$$

Near ω_m, $\mathrm{Im}\{\overset{\circ}{G}\}$ shows a square root behaviour

$$\mathrm{Im}\{\overset{\circ}{G}{}_{ij}^{(\underline{m}-\underline{n})}(\omega) \cong \sqrt{\omega_m^2 - \omega^2}\, \theta(\omega_m^2 - \omega^2)\, g_{ij}^{(\underline{m}-\underline{n})} \tag{4.107}$$

where, contrary to the behaviour at $\omega \cong 0$, $g_{ij}^{(\underline{m}-\underline{n})}$ depends explicitly on $(\underline{m}-\underline{n})$. This follows from the expansion (2.84,85) near the critical point ω_c, which for the maximum ω_m gives

$$g_{ij}^{(\underline{m}-\underline{n})} = \frac{2\pi^2\, \mathrm{sign}(\omega)}{\overset{\circ}{M} V_B \sqrt{\lambda_1\lambda_2\lambda_3}} \sum_{\nu} e_i^{\sigma}(\underline{k}_\nu)\, e_j^{\sigma}(\underline{k}_\nu)\, \cos(\underline{k}_\nu \underline{R}^{(\underline{m}-\underline{n})}) \tag{4.108}$$

where we have to sum over the "star" of all critical points $\underline{k}_\nu$ corresponding to the maximal frequency ω_m. Due to the $\underline{m}$, $\underline{n}$ dependence of $\mathrm{Im}\{\overset{\circ}{G}\}$ arising from the finite value of $|\underline{k}_\nu|$, the V dependence in (4.106) no longer drops out, so that $\mathrm{Im}\{G(\omega)\} \neq \mathrm{Im}\{\overset{\circ}{G}(\omega)\}$. Thus the extremely weak damping of the low frequency resonances is due to the peculiar behaviour of the plane waves at $\underline{k} = 0$, which represent a common translation of all atoms. If $f^{eff}(\omega_m)$ approaches 0, $G_{xx}^{\underline{dd}}(\omega_m)$ and $\mathrm{Re}\{G(\omega_m)\}$ become infinitely large. However, the ratio (4.102) determining the damping remains finite, so that by combination of (4.102), (4.106) and (4.107) the damping constant γ is under resonance conditions given by

$$\gamma \cong \frac{1}{M^{eff}} \sum_{\underline{mn}} u_i^{\underline{m}}(\omega_m)\, V_{ii'}^{\underline{mm}'}(\omega_m)\, g_{i'j'}^{\underline{m}'\underline{n}'}\, V_{j'j}^{\underline{n}'\underline{n}}(\omega_m)\, u_j^{\underline{n}}(\omega_m) \tag{4.109}$$

where $u_i^{\underline{m}}(\omega_m)$ are the amplitudes of the resonance modes

$$u_i^{\underline{m}}(\omega_m) = \frac{G_{ix}^{\underline{md}}(\omega_m)}{G_{xx}^{\underline{dd}}(\omega_m)} \,. \tag{4.110}$$

For $f^{eff} > 0$, the δ function in (4.104) describes a localized mode with frequency $\omega_\ell > \omega_m$ given by

$$\omega_\ell^2 = \omega_m^2 + \left\{\sqrt{\left(\frac{\gamma}{2}\right)^2 + \frac{f^{eff}}{M^{eff}}} - \frac{\gamma}{2}\right\}^2 \cong \begin{cases} \omega_m^2 + \left(\dfrac{f^{eff}}{M^{eff}\gamma}\right)^2 & \text{for } \dfrac{f^{eff}}{M^{eff}} \ll \dfrac{\gamma^2}{4} \\ \omega_m^2 + \dfrac{f^{eff}}{M^{eff}} & \text{for } \dfrac{f^{eff}}{M^{eff}} \gg \dfrac{\gamma^2}{4} \end{cases} \,. \tag{4.111}$$

By expanding the argument of the δ function (4.104) around ω_ℓ we obtain

$$z_x^{\underline{d}}(\omega) = \frac{M}{M^{eff}}\, \delta\left((\omega^2 - \omega_\ell^2)\left(1 + \frac{\gamma/2}{\omega_\ell^2 - \omega_m^2}\right)\right) = \frac{M}{M_{\ell oc}^{eff}}\, \delta(\omega^2 - \omega_\ell^2) \tag{4.112}$$

where $M_{\ell oc}^{eff}$ is the effective mass of the localized vibration

$$M^{eff}_{\ell oc} = M^{eff}\left\{1 - \frac{\gamma/2}{\sqrt{(\gamma/2)^2 + (f^{eff}/M^{eff})}}\right\}^{-1} = M^{eff}\left\{1 + \frac{\gamma/2}{\sqrt{\omega_\ell^2 - \omega_m^2}}\right\}. \qquad (4.113)$$

Thus the effective mass of the localized vibration becomes infinite for $\omega_{\ell oc} \to \omega_{max}$ signalizing that more and more atoms take part in the vibration which gets more and more delocalized. On the other hand for $\omega_\ell^2 - \omega_m^2 \gg \gamma^2/4$ we obtain the "normal" contribution $M^{eff}_{\ell oc} = M^{eff}$. Thus in the local frequency spectrum the localized mode starts at $\omega_\ell \cong \omega_m$ with an infinitesimal intensity $M^{eff}/M^{eff}_{\ell oc} \approx 0$.

The remaining intensity is still in the continuous spectrum slightly below ω_m, since we obtain from (4.103) by residuum calculus for $f^{eff} > 0$

$$\int_0^{\omega_m^2} d\omega^2 \; z_x^d(\omega^2) = \frac{\gamma/2}{\sqrt{(\gamma/2)^2 + (f^{eff}/M^{eff})}} \equiv \frac{M}{M^{eff}} - \frac{M}{M^{eff}_{\ell oc}} \; . \qquad (4.114)$$

Thus the total intensity M/M^{eff} is split into a localized mode contribution $M/M^{eff}_{\ell oc}$ above ω_m and a resonance contribution below ω_m, which finally dissappears for $\omega_\ell^2 - \omega_m^2 \gg \gamma^2/4$. This behaviour of the spectrum for $f^{eff} > 0$ is shown in Fig. 18. The dashed line shows the factor $M^{eff}/M^{eff}_{\ell oc}$ as a function of the localized mode frequency ω_ℓ. According to (4.112,114) this gives the portion of the intensity contained in the localized mode.

Finally a short note about some papers reviewed in this chapter. The derivation of (4.7) for $G_{CC}(\omega)$ is due to KRUMHANSL and MATTHEW /4.2,3/. The technique is similar to the one used previously by LITZMANN and ROSZA /4.1/ and has also been used by KUNC /4.4/, MAHANTY /4.5/, and SACHDEW and MAHANTY /4.6/. The elimination of the additional degrees of freedom for $G_{RR}(\omega)$ is due to WAGNER /4.14/. The concept of an effective constant and an effective mass has been discussed previously by KLEIN /1.7/, AGRAWAL and RAM /4.15,16/, and STONEHAM /4.15/. However, the simplicity and the full advantage of this concept had not been used. Only recently PAGE /4.11/ has given a description of low-frequency resonances in terms of such effective quantities. His method is based on the Green's function technique and restricted to resonances caused by weakened force constants only. Our method is valid for all type of resonances. Both methods do not rely on numerical results for the ideal Green's function. Our presentation follows closely our work presented in /4.13/.

5. Dynamics of Substitutional Defects

After the general description of localized and resonant vibrations as presented in the preceding chapter we will now discuss in detail some simple models for a substitutional point defect. In the simplest case only the coupling between the defect and its neighbours is changed. For this model MANNHEIM /5.1-3/ obtained a very elegant result for the local frequency spectrum of the defect. We will give an elementary derivation of this model which is not based on group theory. In Sect. 5.2 the results of recent Mössbauer measurements for Fe, Sn and Au impurities in metals are reviewed and interpreted in terms of Mannheim's model. Finally in Sect. 5.3 we discuss the vibrational behaviour of vacancies as a special substitutional defect.

5.1 Nearest Neighbour Model for Substitutional Defects

Fig. 19a-c shows a substitutional defect and its nearest neighbours for the three cubic lattices. We assume, that compared to the ideal coupling, only the longitudinal spring f between the defect and its nearest neighbours is changed by $\Delta f = f - \overset{\circ}{f}$. While at present we make no assumption for the interaction in the ideal lattice, later on we will also assume nearest neighbour interaction in the ideal lattice (force constant $\overset{\circ}{f}$).

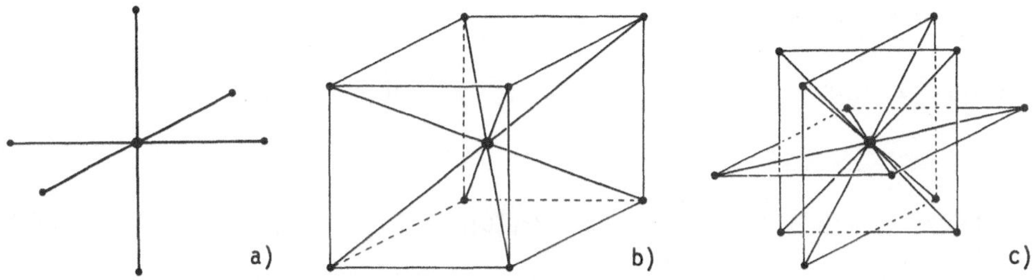

Fig. 19a-c. Schematic representation of a substitutional defect and its nearest neighbours for the three cubic lattices: a) sc, b) bcc, c) fcc

According to Sect. 3.1 the Green's function $G(\omega)$ for the defect lattice is determined by

$$G = \overset{\circ}{G} - \overset{\circ}{G} V G = \overset{\circ}{G} - \overset{\circ}{G} t \overset{\circ}{G} \tag{5.1}$$

with

$$V(\omega) = \varphi - \Delta M\omega^2 \quad \text{and} \quad t(\omega) = V \frac{1}{1 + \overset{\circ}{G} V} \quad . \tag{5.2}$$

In a first step we will assume, that the defect has the same mass as the host atoms ($M = \mathring{M}$). The generalization to both force constant and mass changes is then very easy. In direct analogy to the isotopic defect (Sect. 3.2) the Green's function $G^{dd}_{xx}(\omega;M)$ of a defect with mass M is related to the Green's function $G^{dd}_{xx}(\omega;\mathring{M})$ of a defect with mass $\mathring{M}$, but with identical coupling to its neighbour, by [see (3.38)]

$$G^{dd}_{xx}(\omega;M) = G^{dd}_{xx}(\omega;\mathring{M}) + G^{dd}_{xx}(\omega;\mathring{M})\ (M - \mathring{M})\omega^2\ G^{dd}_{xx}(\omega;M)\ , \tag{5.3}$$

thus

$$G^{dd}_{xx}(\omega;M) = \left\{\frac{1}{G^{dd}_{xx}(\omega;\mathring{M})} - (M - \mathring{M})\omega^2\right\}^{-1}\ . \tag{5.4}$$

For the defect in fcc, the matrix $V(\omega)$, coupling the defect with its 12 nearest neighbours, consists of $3\times 13 = 39$ lines and rows. To calculate $t(\omega)$ one has therefore to invert matrices of size 39×39.

In the following we will restrict ourselves to the Green's function $G^{dd}_{xx}(\omega)$ of the defect. By introducing 39-dimensional vectors $|\nu\rangle$, $G^{dd}_{xx}(\omega)$ can be written as

$$G^{dd}_{xx}(\omega) = \langle d|G(\omega)|d\rangle \qquad \text{where } |d\rangle = \begin{Bmatrix} 1\\ 0\\ \vdots\\ 0 \end{Bmatrix} \tag{5.5}$$

describes a unit displacement of the defect in x direction, while all other atoms are fixed (see Fig. 20a).

By introducing a complete set of 39 orthonormalized vectors $|\nu\rangle$, $\nu = 1,\ldots,39$ (5.1) can be written as

$$\langle d|G|d\rangle \ = \ \langle d|\mathring{G}|d\rangle \ - \sum_{\nu,\nu'=1}^{39} \langle d|\mathring{G}|\nu\rangle\ \langle\nu|t|\nu'\rangle\ \langle\nu'|\mathring{G}|d\rangle \tag{5.6}$$

and

$$\langle\nu|t|\nu'\rangle \ = \ \langle\nu|\varphi|\nu'\rangle \ - \sum_{\mu,\mu'=1}^{39} \langle\nu|\varphi|\mu\rangle\ \langle\mu|\mathring{G}|\mu'\rangle\ \langle\mu'|t|\nu'\rangle\ . \tag{5.7}$$

These equations can be greatly simplified by symmetry considerations. Only those vectors $|\nu\rangle$ and $|\nu'\rangle$ have to be taken into account in (5.6,7) which transform under cubic symmetry operations S in the same way as the vector $|d\rangle$ does. For instance if S is a symmetry operation which leaves $|d\rangle$ unchanged as e.g., a rotation around the x axis by 90° does (see Fig. 20), then

$$\langle d|\mathring{G}|\nu\rangle = \langle d|S\mathring{G}|\nu\rangle = \langle d|\mathring{G}S|\nu\rangle\ ,$$

since $S\,|d\rangle = |d\rangle$ and $S\mathring{G} = \mathring{G}S$ due to cubic symmetry.

Thus we conclude that either

$$S|\nu\rangle = |\nu\rangle \quad \text{or} \quad \langle d|\mathring{G}|\nu\rangle = 0\ .$$

In the latter case the vector $|\nu\rangle$ gives no contribution to the sum. A second example is an inversion I at the defect site. Since I $|d\rangle = -|d\rangle$ only those vectors $|\nu\rangle$ with uneven symmetry (I $|\nu\rangle = -|\nu\rangle$) have to be taken into account. Thus for fcc only 4 different and orthogonal vectors have to be considered, which are shown in Fig. 20a-d.

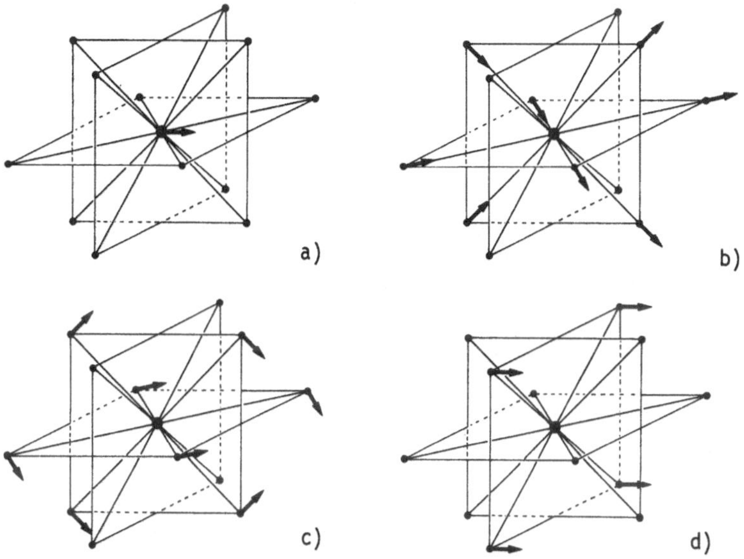

Fig. 20. a) The state $|d\rangle$ describing a unit displacement of the defect in x direction (fcc)
b) - d) The three 39-dimensional states $|1\rangle$ (Fig. b), $|2\rangle$ (Fig. c) and $|3\rangle$ (Fig. d) which are, due to symmetry, the only states which can couple to the defect displacement $|d\rangle$

In order to calculate the t matrix (5.7), an expansion of φ into eigenvectors would be very useful

$$\varphi = \sum_{\nu''} \varphi_{\nu''} \; |\nu''\rangle \langle\nu''| \quad .$$

For the above models, the vectors $|2\rangle$ and $|3\rangle$ are already eigenvectors with eigenvalue $\varphi_\nu = 0$, since the displacements of the neighbouring atoms are orthogonal to the changed spiral springs. A third eigenvector with eigenvalues $\varphi_\nu = 0$ is a common translation of the defect and the 12 nearest neighbours. Thus of the four vectors of Fig. 20 only one eigenvector with eigenvalue $\varphi_\nu \neq 0$ can be formed, being a linear combination of the vectors $|d\rangle$ and $|1\rangle$ and orthogonal to the translation. Figure 21a shows the displacements for this eigenvector $|0\rangle$, which, when properly normalized, is given by

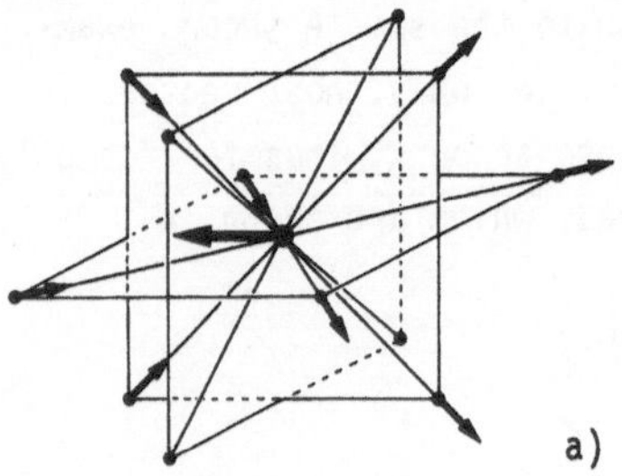

a)

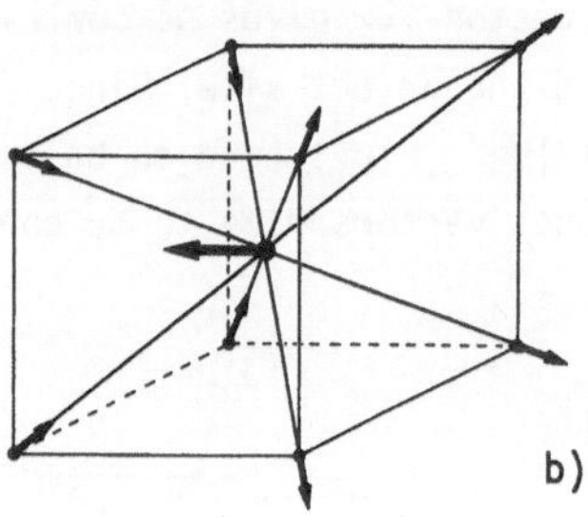

b)

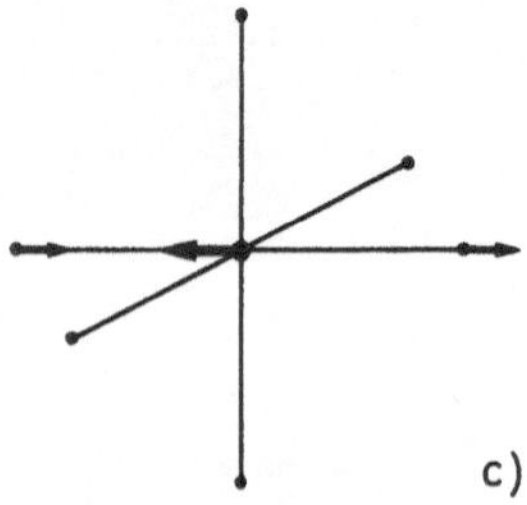

c)

Fig. 21a-c. The eigenstate $|0\rangle$ for the three cubic lattices, which is for a central force n.n. model the only mode which can couple to the defect state $|d\rangle$ of Fig. 20a a) fcc b) bcc c) sc

$$|0\rangle = \frac{1}{\sqrt{80}} \begin{Bmatrix} -8 \\ 0 \\ 0 \\ 1 \\ 1 \\ 0 \\ \vdots \end{Bmatrix} . \tag{5.9}$$

Thus we obtain from (5.7)

$$\langle d|G|d\rangle = \langle d|\mathring{G}|d\rangle - \langle d|\mathring{G}|0\rangle \, \langle 0|t|0\rangle \, \langle 0|\mathring{G}|d\rangle \tag{5.10}$$

$$\langle 0|t|0\rangle = \langle 0|\varphi|0\rangle - \langle 0|\varphi|0\rangle \, \langle 0|G|0\rangle \, \langle 0|t|0\rangle \tag{5.11}$$

or

$$G_{xx}^{dd}(\omega) = \mathring{G}_{xx}^{(o)}(\omega) - (\langle d|\mathring{G}|0\rangle)^2 \, \langle 0|\varphi|0\rangle \, \frac{1}{1 + \langle 0|\mathring{G}|0\rangle \, \langle 0|\varphi|0\rangle} . \tag{5.12}$$

For bcc and sc crystals, one obtains just the same result. The displacements of corresponding eigenvectors $|0\rangle$ are shown in Fig. 21b for bcc and Fig. 21c for sc. Note that the displacements of the nearest neighbours are always parallel to the changed spiral springs. The vector $|0\rangle$ is orthogonal to the translation and has to be properly normalized. The eigenvalues are for the different lattices

$$\varphi_o = \langle 0|\varphi|0\rangle = \begin{cases} 5\,(f - \mathring{f}) & \text{fcc} \\ \frac{11}{3}(f - \mathring{f}) \quad \text{for} & \text{bcc} \\ 3\,(f - \mathring{f}) & \text{sc} \end{cases} . \tag{5.13}$$

Similarly the matrix elements $\langle d|\mathring{G}|0\rangle$ and $\langle 0|\mathring{G}|0\rangle$ can be calculated, with the results

$$\langle d|\overset{o}{G}|0\rangle = \begin{cases} \frac{8}{\sqrt{80}}\left(-\overset{o}{G}{}^{(0)}_{xx} + \overset{o}{G}{}^{(110)}_{xx} - \overset{o}{G}{}^{(110)}_{xy}\right) & \text{for fcc} \\ \frac{8}{\sqrt{88}}\left(-\overset{o}{G}{}^{(0)}_{xx} + \overset{o}{G}{}^{(111)}_{xx} + 2\overset{o}{G}{}^{(111)}_{xy}\right) & \text{for bcc} \\ \frac{2}{\sqrt{6}}\left(-\overset{o}{G}{}^{(0)}_{xx} + \overset{o}{G}{}^{(100)}_{xx}\right) & \text{for sc} \end{cases} \tag{5.14}$$

$$\langle 0|\overset{o}{G}|0\rangle = \begin{cases} \overset{o}{G}{}^{(0)}_{xx} - \frac{1}{5}\Big(8\overset{o}{G}{}^{(110)}_{xx} - \overset{o}{G}{}^{(110)}_{zz} + 9\overset{o}{G}{}^{(110)}_{xy} - \overset{o}{G}{}^{(211)}_{xx} - 2\overset{o}{G}{}^{(211)}_{xy} \\ \qquad -\overset{o}{G}{}^{(211)}_{yz} - \overset{o}{G}{}^{(220)}_{xx} - \overset{o}{G}{}^{(220)}_{xy}\Big) & \text{for fcc} \\ \overset{o}{G}{}^{(0)}_{xx} - \frac{1}{11}\Big(16\overset{o}{G}{}^{(111)}_{xx} + 32\overset{o}{G}{}^{(111)}_{xy} + \overset{o}{G}{}^{(200)}_{xx} - 2\overset{o}{G}{}^{(200)}_{zz} - 2\overset{o}{G}{}^{(220)}_{xx} \\ \qquad -2\overset{o}{G}{}^{(220)}_{xy} + \overset{o}{G}{}^{(220)}_{zz} - 3\overset{o}{G}{}^{(222)}_{xx} - 6\overset{o}{G}{}^{(222)}_{xy}\Big) & \text{for bcc} \\ \overset{o}{G}{}^{(0)}_{xx} - \frac{4}{3}\,\overset{o}{G}{}^{(100)}_{xx} + \frac{1}{3}\,\overset{o}{G}{}^{(200)}_{xx} & \text{for sc} \end{cases} \tag{5.15}$$

The result (5.12) can be simplified much further, if in addition to the change φ of the coupling also the ideal lattice coupling $\overset{o}{\phi}$ consists only of a nearest neighbour interaction with force constant $\overset{o}{f}$. Then $G^{\underline{dd}}_{xx}$ can be expressed in terms of the single Green's function $\overset{o}{G}{}^{(0)}_{xx}$ alone, as has been noticed by MANNHEIM. By multiplying the basic equation for $\overset{o}{G}$ from left with $\langle d|$

$$\langle d|\,\overset{o}{\phi}\overset{o}{G} - \overset{o}{M}\omega^2 \langle d|\overset{o}{G} = \langle d| \tag{5.16}$$

we can introduce between $\overset{o}{\phi}$ and $\overset{o}{G}$ the sum $\sum_\nu |\nu\rangle\langle\nu|$, since the vectors $|\nu\rangle$ form a complete set in the space spanned by the defect and its nearest neighbours. By choosing the $|\nu\rangle$'s as eigenvectors to $\overset{o}{\phi}$ as before, only the vector $|0\rangle$ of Fig. 21 remains having the only nonzero eigenvalue $\langle 0|\overset{o}{\phi}|0\rangle$. Thus by multiplying (5.16) from the right with $|d\rangle$, and $|0\rangle$, respectively, we can express $\langle 0|\overset{o}{G}|d\rangle$ and $\langle 0|\overset{o}{G}|0\rangle$ in terms of $\langle d|\overset{o}{G}|d\rangle = \overset{o}{G}{}^{(0)}_{xx}$ alone

$$\langle 0|\overset{o}{G}|d\rangle = (1 + \overset{o}{M}\omega^2\overset{o}{G}{}^{(0)}_{xx})\,\frac{1}{\langle d|\overset{o}{\phi}|0\rangle} \tag{5.17}$$

$$\langle 0|\overset{o}{G}|0\rangle = (\langle d|0\rangle + \overset{o}{M}\omega^2\,\langle d|\overset{o}{G}|0\rangle)\,\frac{1}{\langle d|\overset{o}{\phi}|0\rangle}\;. \tag{5.18}$$

Similarly, the matrix elements $\langle d|\overset{o}{\phi}|0\rangle$ and $\langle 0|\overset{o}{\phi}|0\rangle$ can be expressed in terms of $\langle d|\overset{o}{\phi}|d\rangle = \overset{o}{f}_E$, the Einstein frequency of the ideal lattice

$$\langle d|\overset{o}{\phi}|0\rangle = \langle d|0\rangle\,\langle 0|\overset{o}{\phi}|0\rangle = \frac{1}{\langle d|0\rangle}\,\langle d|\overset{o}{\phi}|d\rangle \tag{5.19}$$

$$\langle d|\overset{o}{\phi}|d\rangle = \langle d|0\rangle^2\,\langle 0|\overset{o}{\phi}|0\rangle \tag{5.20}$$

where

$$\langle d|\overset{\circ}{\phi}|d\rangle = \overset{\circ}{\phi}{}^{(0)}_{xx} = \overset{\circ}{f}_{E} = \begin{cases} 4\overset{\circ}{f} & \text{for fcc} \\ \frac{8}{3}\overset{\circ}{f} & \text{for bcc} \\ 2\overset{\circ}{f} & \text{for sc} \end{cases} \quad . \tag{5.21}$$

The equivalent results hold for the coupling $\phi = \overset{\circ}{\phi} + \varphi$ in the defect lattice; one has only to replace in (5.21) $\overset{\circ}{f}_E$ by $\overset{\circ}{f}{}^d_E$ and $\overset{\circ}{f}$ by f so that in (5.12)

$$\langle 0|\varphi|0\rangle = \frac{1}{\langle d|0\rangle^2}\,(f^d_E - \overset{\circ}{f}_E) \ . \tag{5.22}$$

Thus in (5.12) all quantities can be expressed in terms of $\overset{\circ}{G}{}^{(0)}_{xx}(\omega)$, $\overset{\circ}{f}_E$ and f^d_E. After some lengthy, but straightforward calculations one obtains

$$\frac{1}{\frac{1}{G^{dd}_{xx}} + \overset{\circ}{M}\omega^2} = \frac{1}{\frac{1}{\overset{\circ}{G}{}^{(0)}_{xx}} + \overset{\circ}{M}\omega^2} + \frac{1}{f^d_E} - \frac{1}{\overset{\circ}{f}_E} \ . \tag{5.23}$$

At this stage it is convenient to allow also a mass change of the defect atom. By taking into account (5.4) we obtain then for the Green's function of the defect with mass M

$$\frac{1}{\frac{1}{G^{dd}_{xx}} + M\omega^2} - \frac{1}{f^d_E} = \frac{1}{\frac{1}{\overset{\circ}{G}{}^{(0)}_{xx}} + \overset{\circ}{M}\omega^2} - \frac{1}{\overset{\circ}{f}_E} \ . \tag{5.24}$$

Note the remarkable symmetry of this equation: the left hand side depends only on defect quantities, the right side on ideal lattice quantities. (The simplicity of this result suggests, that perhaps there should be an easier derivation as the one sketched here. However we did not find one.) Finally we obtain for G^{dd}_{xx}

$$G^{dd}_{xx}(\omega) = \left\{ \frac{1}{\frac{\overset{\circ}{G}{}^{(0)}_{xx}(\omega)}{1 + \overset{\circ}{M}\omega^2\,\overset{\circ}{G}{}^{(0)}_{xx}(\omega)} + \frac{1}{f^d_E} - \frac{1}{\overset{\circ}{f}_E}} - M\omega^2 \right\}^{-1} \ . \tag{5.25}$$

In deriving this result from (5.12), we explicitly assumed nearest neighbour interaction also in the ideal lattice. As is well known only the face-centered lattice is stable for this interaction. Nevertheless (5.25) is widely used for the interpretation of Mössbauer measurements (see sect. 5.3) and usually the exact host Green's function $\overset{\circ}{G}{}^{(0)}_{xx}(\omega)$ is put into (5.25) instead of the one for nearest neighbour interaction. This procedure would be much better justified for (3.12).

Let us take advantage of the simplicity of this result and calculate for this model the characteristic values f^{eff}, M^{eff} and γ for resonance vibrations, for which the general formulas were derived in Chap. 4. With the low-frequency expansion (4.19) for the defect

$$G^{dd}_{xx}(\omega) = \frac{1}{f^{eff} - M^{eff}\omega^2 - iM^{eff}\gamma\omega} \tag{5.26}$$

and the analogous expression for $\overset{o}{G}{}^{(0)}_{xx}(\omega)$ (with $\overset{o}{f}{}^{eff}$, $\overset{o}{M}{}^{eff}$ and $\overset{o}{\gamma}$), we obtain from (5.24) by comparing the different powers of ω

$$\frac{1}{f^{eff}} = \frac{1}{f^d_E} - \frac{1}{\overset{o}{f}_E} + \frac{1}{\overset{o}{f}{}^{eff}} \rightarrow \frac{1}{4f} - \frac{1}{4\overset{o}{f}} + \frac{1}{2.38\overset{o}{f}} \quad \text{for fcc} \tag{5.27}$$

$$\gamma = \overset{o}{\gamma}\frac{\overset{o}{M}{}^{eff}}{M^{eff}}\frac{(f^{eff})^2}{(\overset{o}{f}{}^{eff})^2} = \frac{3\pi}{2\overset{o}{M}\omega_D^3}\frac{(f^{eff})^2}{M^{eff}} \tag{5.28}$$

$$M^{eff} = M + \left(\frac{f^{eff}}{\overset{o}{f}{}^{eff}}\right)^2 \left\{\overset{o}{M}{}^{eff} - \overset{o}{M} + (\overset{o}{M}{}^{eff})^2\,\overset{o}{\gamma}{}^2\left(\frac{1}{f^{eff}} - \frac{1}{\overset{o}{f}{}^{eff}}\right)\right\}. \tag{5.29}$$

From (5.27) we see, that f^{eff} will get small, i.e., the defect becomes nearly unstable, if the coupling f with the nearest neighbours is small (the defect is weakly bound): $f^{eff} \cong f^d_E = 4f$ for fcc. Since in this limit $M^{eff} \cong M$, the defect vibrates as an Einstein oscillator $\omega^2_{res} \cong f^d_E/M$ with the neighbouring atoms at rest. However, in all other cases, the defect behaves quite different than the Einstein-picture says. For instance, if the defect is strongly bound ($f \rightarrow \infty$), f^{eff} becomes independent of f (for fcc: $f^{eff} \cong 5.9\overset{o}{f}$). Thus if M is sufficiently large, we still can have a resonance mode.

By, somewhat arbitrarily, requiring $\omega^2_{res} = f^{eff}/M^{eff} \leqslant \omega^2_{max}/3$ as the condition for a resonance mode, we have plotted in Fig. 22 the line $\omega_{res} = \omega_{max}/3$ in the M-f plane (for fcc). For M and f values on the left side of this line, we always have a resonance mode.

Quite analogously to the foregoing discussion we can also discuss the occurrence of resonant and localized modes near ω_{max}. For instance, by introducing [see (4.100)]

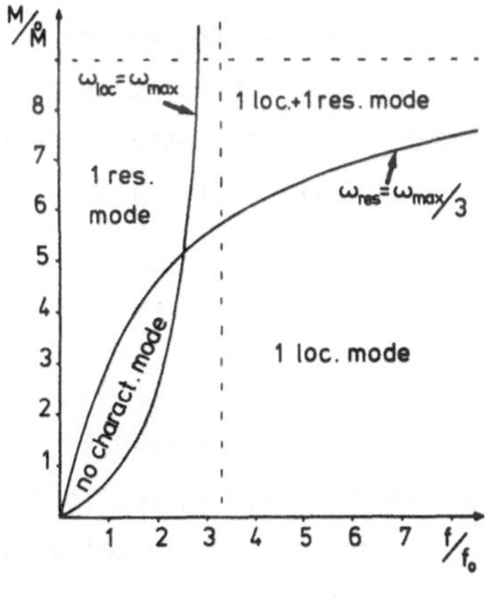

Fig. 22. Conditions for the occurrence of localized modes and resonant modes (central-force n.n. model in fcc). On the left of the curve $\omega_{res} = \omega_{max}/3$ a resonant mode occurs, on the right of the curve $\omega_{\ell oc} = \omega_{max}$ a localized mode exists

$G^{dd}_{xx}(\omega_m) = 1/f^{eff}(\omega_m)$ into (5.24), we obtain for $f^{eff}(\omega_m)$

$$\frac{1}{f^{eff}(\omega_m) + M\omega_m^2} - \frac{1}{f_E^d} = \frac{1}{\overset{\circ}{f}{}^{eff}(\omega_m) + \overset{\circ}{M}\omega_m} - \frac{1}{\overset{\circ}{f}_E} . \tag{5.30}$$

Since $f^{eff}(\omega_m) = 0$, when a localized mode occurs exactly at ω_{max}, we see that for fcc ($\overset{\circ}{M}\omega_m^2 = 8\overset{\circ}{f} = 2f_E^d$, $\overset{\circ}{f}{}^{eff}(\omega_m) = -2.27\overset{\circ}{f}$) we will always have a localized mode, if

$$\frac{M}{\overset{\circ}{M}} < \frac{M_{crit}}{\overset{\circ}{M}} = \left\{ 2\frac{f}{\overset{\circ}{f}} - 0.604 \right\}^{-1} . \tag{5.31}$$

By plotting $M_{crit}(f)$ in Fig. 22 for fcc, a localized mode occurs for values on the right side of this line. Thus we see, that the total M-f plane is divided into four regions: a region of moderate M-f values where we have no characteristic defect mode, a region with one resonant mode, a region with one localized mode, and, somewhat surprisingly, for large M and f values a region where we have both a resonant mode and a localized mode.

This peculiar behaviour can be understood qualitatively, if, in a generalized Einstein approximation, we allow the defect together with its nearest neighbours to move as shown in Fig. 20a,b. This leads to a quadratic equation in ω^2 for the allowed frequencies. For $f \gg \overset{\circ}{f}$ and $M \gg \overset{\circ}{M}$ we obtain one high frequency, corresponding to a localized mode and a low frequency for a resonant mode

$$\omega_{\ell oc}^2 \cong \frac{4\overset{\circ}{f} + f}{\overset{\circ}{M}} + \frac{4f}{M} \qquad \omega_{res}^2 \cong \frac{16\overset{\circ}{f}}{M + 4\overset{\circ}{M} + 4M\overset{\circ}{f}/f} . \tag{5.32}$$

In the localized mode the neighbouring atoms move practically alone, so that the large defect mass does not affect the frequency very much. On the other hand, in the resonance mode all atoms move in phase so that the strong springs f are practically not stretched: the restoring force is determined by the restoring forces of the rest lattice on the neighbouring atoms.

In the foregoing discussion we have strictly avoided the use of group theory. However, this is no longer possible, if we are not only interested in the motion of the defect alone, but also want to study the motion, e.g., of the nearest neighbours or if we want to study a more general nearest neighbour model with noncentral forces. In this case one has to start with (5.1) expressing $G(\omega)$ in the entire lattice in terms of the Green's functions $\overset{\circ}{G}(\omega)$ and the t matrix

$$G = \overset{\circ}{G} - \overset{\circ}{G}t\overset{\circ}{G} \quad \text{with} \quad t = V\frac{1}{1 + \overset{\circ}{G}V} \quad \text{and} \quad V = \varphi - \Delta M\omega^2 .$$

The essential problem is the calculation of the t matrix, which is restricted to the defect subspace, i.e., the defect and its nearest neighbours. Thus in fcc this requires the inversion of 39×39 matrices. For this we decompose all 39×39 matrices V, $\overset{\circ}{G}$ and t into their irreducible parts. For instance, V is written

$$V_{ij}^{mn} = \sum_{\Gamma} \sum_{\gamma,\gamma'=1}^{\sigma_\Gamma} \sum_{\mu=1}^{d_\Gamma} s_i^m(\Gamma;\gamma;\mu)\, V_{\gamma\gamma'}^{\Gamma}\, s_j^n(\Gamma;\gamma';\mu) \tag{5.33}$$

where σ_Γ denotes the number of times the total representation contains the irreducible representation Γ with dimension d_Γ. The basis vectors $s_i^m(\Gamma;\gamma;\mu)$ form an orthonormal and complete set. d_Γ gives the multiplicity of the distinct eigenvalues of V and σ_Γ the rank of the matrices in the remaining equations.

$$\begin{aligned} G_{\gamma\gamma'}^{\Gamma} &= \overset{o}{G}{}_{\gamma\gamma'}^{\Gamma} - \sum_{\nu,\nu'=1}^{\sigma_\Gamma} \overset{o}{G}{}_{\gamma\nu}^{\Gamma}\, t_{\nu\nu'}^{\Gamma}\, \overset{o}{G}{}_{\nu'\gamma'}^{\Gamma} \\ t_{\gamma\gamma'}^{\Gamma} &= V_{\gamma\gamma'}^{\Gamma} - \sum_{\nu,\nu'=1}^{\sigma_\Gamma} V_{\gamma\nu}^{\Gamma}\, \overset{o}{G}{}_{\nu\nu'}^{\Gamma}\, t_{\nu'\gamma'}^{\Gamma} \end{aligned} \tag{5.34}$$

For all important symmetries, the basis vectors have been worked out and tabulated by DETTMANN and LUDWIG /4.8/ (see also LUDWIG /1.2/). For the general nearest neighbour model in fcc the following decomposition of the total representation is obtained

$$\Gamma_{fcc} = A_{1g} + A_{2g} + 2E_g + 2F_{1g} + 2F_{2g} + A_{2u} + E_u + 4F_{1u} + 2F_{2u} \,. \tag{5.35a}$$

Here A denotes one-dimensional, E two-dimensional and F three-dimensional representations. The subscripts g and u refer to even and odd symmetry under inversion. The corresponding matrix elements $\overset{o}{G}{}_{\gamma\gamma'}^{\Gamma}$ are also, e.g., given by LAKATOS and KRUMHANSL /5.4/, and AGRAWAL /5.5/. Thus in order to calculate G, one has to invert only low-dimensional matrices in the different subspaces: a 4×4 matrix for F_{1u}, 2×2 matrices for the subspaces E_g, F_{1g}, F_{2g} and F_{2u} and only one-dimensional matrices for A_{1g}, A_{2g}, A_{2u} and E_u. The motion of the defect itself is described by the F_{1u} representation alone, the four independent basis vectors of which are plotted in Fig. 20a-d. Contrary to Mannheim's model, the basis vectors 2 and 3 can no longer be neglected in a general model due to transversal force constants.

For a substitutional defect in a bcc or sc lattice, the result is quite analogous /4.8, 1.2/. Here we give only the decomposition of the total representations

$$\Gamma_{bcc} = A_{1g} + E_g + F_{1g} + 2F_{2g} + A_{2u} + E_u + 3F_{1u} + F_{2u} \tag{5.35b}$$

$$\Gamma_{sc} = A_{1g} + E_g + F_{1g} + F_{2g} + 3F_{1u} + F_{2u} \,. \tag{5.35c}$$

5.2 Mössbauer Studies of Fe, Sn, and Au impurities

Very useful information about the dynamical behaviour of impurities has been obtained in the last ten years by Mössbauer technique. In this section we will discuss the theoretical background and then review the experimental results for Fe, Sn and Au impurities.

By the emission of a γ quantum, a momentum $\hbar\underline{k}$ is transferred to the Mössbauer atom $\underline{R}^{\underline{n}}$, so that

$$e^{i\underline{k}\underline{R}^{\underline{n}}}|i\rangle$$

is the wave function of the crystal atoms after the emission, if $|i\rangle$ describes the initial state of the crystal. The probability for a transition to a final crystal state $|f\rangle$ is then

$$|\langle f|e^{i\underline{k}\underline{R}^{\underline{n}}}|i\rangle|^2 .$$

Thus the probability $w_e(\underline{k};\omega)$ for the emission of a γ quantum with an energy $E_0 - \hbar\omega$ (E_0 = energy difference between excited and ground state of the nucleus) and momentum $\hbar\underline{k}$ ($k = E_0/\hbar c$) is given by

$$w_e(\omega,\underline{k}) = \sum_{i,f} P_i(T) \; |\langle f|e^{i\underline{k}\underline{R}^{\underline{n}}}|i\rangle|^2 \; \delta(E_i - E_f + \hbar\omega) \tag{5.36a}$$

where we have summed over all final states f and over all initial states i according to their thermal occupation $p_i(T)$ ($\sum_i p_i(T) = 1$). As has been shown by van HOVE /5.6/, this expression can be rewritten as a time integral over a correlation factor (see, e.g., /5.7/ or /1.3 - 5/).

$$w_e(\omega;\underline{k}) = \frac{1}{2\pi\hbar} \int_{-\infty}^{+\infty} dt \; e^{i\omega t} \; e^{-\gamma_M|t|} \; \langle e^{-i\underline{k}\underline{R}^{\underline{n}}(t)} \; e^{i\underline{k}\underline{R}^{\underline{n}}(0)}\rangle_T \tag{5.36b}$$

where $\underline{R}^{\underline{n}}(t)$ is the time-dependent position operator and where $\langle \;\rangle_T$ means the thermal average. In (5.36a) we assumed infinite lifetime of the excited state. To take into account the finite lifetime $\tau = 1/\gamma_M$, the δ function has to be replaced by a Lorentzian with halfwidth $\hbar\gamma_M$ so that in (5.36b) an additional time factor $e^{-\gamma_M|t|}$ appears /5.8/.

For large times $t \to \infty$, the impurity loses its memory and the coordinate $\underline{R}^{\underline{n}}(t)$ becomes uncorrelated with $\underline{R}^{\underline{n}}(0)$

$$\langle e^{-i\underline{k}\underline{R}^{\underline{n}}(t)} \; e^{i\underline{k}\underline{R}^{\underline{n}}(0)}\rangle_T \underset{t\to\infty}{\to} \langle e^{-i\underline{k}\underline{R}^{\underline{n}}(t)}\rangle_T \langle e^{i\underline{k}\underline{R}^{\underline{n}}(0)}\rangle_T = \left|\langle e^{i\underline{k}\underline{R}^{\underline{n}}(0)}\rangle_T\right|^2 = f(T) \tag{5.37}$$

This leads to an unbroadened Mössbauer line with natural linewidth γ_M, the "zero phonon line", the intensity of which is reduced by a Debye - Waller factor $f(T)$

$$w_e(\omega;\underline{k}) = f(T)\frac{2}{\pi}\frac{\gamma}{\omega^2+\gamma_M^2} . \tag{5.38}$$

In addition we get, due to phonon excitations, a very broad background with total intensity $1-f(T)$, distributed over a frequency interval comparable with an average phonon frequency. Therefore its intensity is too small to be detected experimentally. For instance, the decay time of the Mössbauer state of Fe^{57} is $T_{1/2} = 10^{-7}$s. Therefore the width of the background intensity is of the order of $10^5\gamma_M$ and its intensity a factor 10^{-5} smaller than the peak intensity of the Mössbauer line.

The transition of the correlation function (5.37) to its uncorrelated limit for large t can in general not be described correctly within the harmonic theory, as has been pointed out by KRIVOGLAZ /5.9/. Within harmonic theory correlations can persist infinitely, if localized modes exist, so that the harmonic theory leads to a wrong result for the intensity of the Mössbauer line, if (5.36b) is used instead of (5.38). This will be demonstrated in the following. Within the harmonic theory the following identity is valid (setting $\underline{R}^n = \overset{\circ}{\underline{R}}{}^n + \underline{s}^n$) (see, e.g., /5.7/ or /1.3-5/)

$$\begin{aligned} < e^{-i\underline{k}\underline{R}^n(t)}\, e^{i\underline{k}\underline{R}^n(0)} >_T &= < e^{-i\underline{k}\underline{s}^n(t)}\, e^{i\underline{k}\underline{s}^n(0)} >_T \\ &= e^{-<(\underline{k}\underline{s}^n)^2>}\, e^{<(\underline{k}\underline{s}^n(t))(\underline{k}\underline{s}^n(0))>} . \end{aligned} \tag{5.39}$$

The averages in the exponent have been calculated in Sect. 2.2 (2.39). In the case of a localized mode, $\omega\,\mathrm{Im}\{G_{ij}^{nn}(\omega)\}$ contains a δ function $\delta(\omega-\omega_\ell)$, so that we obtain a pure oscillating contribution

$$<\underline{k}\underline{s}^n(t)\ \underline{k}\underline{s}^n(0)> \sim \frac{\hbar}{2M^n\omega_\ell}\left\{\left(2n(\omega_\ell)+1\right)\cos\omega_\ell t - i\sin\omega_\ell t\right\} . \tag{5.40}$$

Thus the correlation function (5.39) does not approach a stationary value for $t\to\infty$, in contradiction to (5.37). In reality, however, the localized mode is damped due to anharmonic effects and the correlations (5.40) contain an additional damping factor $\sim e^{-\Gamma_a t}$ /5.9/, so that (5.37) is valid. Since typical lifetimes of localized modes are of the order of 10^{-10}s, the stationary value (5.37) is well approached within the lifetime of the Mössbauer state. For a more thorough discussion of anharmonic effects see /5.9/.

The Debye-Waller factor f(T) can be calculated by the methods of cumulants (see, e.g., /5.10/), i.e., by expanding ln f(T) in powers of $\underline{k}$. For example, we obtain in general for a fluctuating quantity x

$$<e^{ix}>_T = \exp\left\{\sum_{n=0}^{\infty}\frac{i^n}{n!}<x^n>_c\right\} . \tag{5.41}$$

Here $<x^n>_c$ are the so-called connected or cumulant averages, the lowest orders of which are

$$\langle x\rangle_c = \langle x\rangle \;;\quad \langle x^2\rangle_c = \langle x^2\rangle - \langle x\rangle^2 \;;\quad \langle x^3\rangle_c = \langle x^3\rangle - 3\langle x^2\rangle\langle x\rangle + 2\langle x\rangle^3$$
$$\langle x^4\rangle_c = \langle x^4\rangle - 4\langle x^3\rangle\langle x\rangle - 3\langle x^2\rangle^2 + 12\langle x^2\rangle\langle x\rangle^2 - 6\langle x\rangle^4 \,. \tag{5.42}$$

Thus by introducing the equilibrium positions $\overset{\circ}{\underline{R}}{}^{\underline{n}}$ in the averaged lattice we have

$$\underline{R}^{\underline{n}} = \overset{\circ}{\underline{R}}{}^{\underline{n}} + \underline{s}^{\underline{n}} \quad \text{with} \quad \langle \underline{R}^n\rangle = \overset{\circ}{\underline{R}}{}^n \,, \quad \langle \underline{s}^n\rangle = 0 \,.$$

Then we obtain

$$\langle e^{i\underline{k}\underline{R}^{\underline{n}}}\rangle = e^{i\underline{k}\overset{\circ}{\underline{R}}{}^n} e^{-M_{\underline{k}}} \quad \text{and} \quad f_T = e^{-2M_{\underline{k}}} \tag{5.43}$$

with $M_{\underline{k}}$ given by

$$M_{\underline{k}} = \frac{1}{2}\langle(\underline{k}\cdot\underline{s}^{\underline{n}})^2\rangle - \frac{1}{24}\left\{\langle(\underline{k}\cdot\underline{s}^{\underline{n}})^4\rangle - 3\langle(\underline{k}\cdot\underline{s}^{\underline{n}})^2\rangle^2\right\} + \dots \tag{5.44}$$

since for cubic crystals all uneven averages of $\underline{s}^{\underline{n}}$ vanish.

In the harmonic approximation the displacements $\underline{s}^{\underline{n}}$ are Gaussian distributed and all cumulants higher than the second order vanish /1.3 - 5/

$$M_{\underline{k}} = \frac{1}{2}\langle(\underline{k}\cdot\underline{s}^{\underline{n}})^2\rangle = \frac{1}{6}k^2\langle(\underline{s}^{\underline{n}})^2\rangle \quad \text{for cubic crystals} \tag{5.45}$$

where according to (2.54) the average $\langle(\underline{s}^{\underline{n}})^2\rangle$ is given by

$$\langle(\underline{s}^{\underline{n}})^2\rangle = 3\int d\omega\, z^{\underline{n}}(\omega)\,\frac{\varepsilon(\omega,T)}{M^{\underline{n}}\omega^2} \,. \tag{5.46}$$

Especially for high temperatures one has

$$\langle(\underline{s}^{\underline{n}})^2\rangle \cong 3\frac{kT}{M^{\underline{n}}}\langle\frac{1}{\omega^2}\rangle + \frac{\hbar^2}{12M^{\underline{n}}kT} + \dots \cong 3kTG^{\underline{n}\underline{n}}_{xx}(\omega=0) = 3\,\frac{kT}{f^{eff}} \tag{5.47}$$

where f^{eff} is the effective static force constant of the defect (see Sect. 4.3). The last equation can also be derived quite elementary: According to Sect. 4.3 the potential energy due to a given static displacement $\underline{s}^{\underline{d}}$ of the impurity is $E = (1/2)f^{eff}(\underline{s}^{\underline{d}})^2$. By averaging over $\underline{s}^{\underline{d}}$, we obtain $\langle E\rangle = (3/2)kT$, since the defect has three degrees of freedom (x,y,z), so that (5.47) follows immediately.

For the nearest neighbour model, f^{eff} is given by (5.27). From this we can give a very simple expression for the ratio of the high-T displacements of the impurity and of the host

$$\frac{\langle\underline{s}^2\rangle_{im}}{\langle\underline{s}^2\rangle_{host}} = \frac{\overset{\circ}{f}{}^{eff}}{f^{eff}} \cong 1 + \frac{\overset{\circ}{f}{}^{eff}}{f^d_E} - \frac{\overset{\circ}{f}{}^{eff}}{\overset{\circ}{f}_E} \Rightarrow 1 + 0.60\left(\frac{\overset{\circ}{f}}{f} - 1\right) \quad \text{for fcc.} \tag{5.48}$$

Thus this ratio is directly determined by the force constant ratio $f/\overset{\circ}{f}$. For $\overset{\circ}{f}{}^{eff}$ and f^{eff} we have used the nearest neighbour results ($\overset{\circ}{f}{}^{eff} = 2.38\overset{\circ}{f}$, $\overset{\circ}{f}_E = 4\overset{\circ}{f}$ for fcc).

Due to anharmonicity, there are two different corrections to the harmonic expression (5.45-47). The first and most important one is an anharmonic contribution to the quadratic average $<(\underline{k}\cdot\underline{s}^{n})^2>$ in (5.45) being proportional to $(kT)^2$ for high temperatures. The second correction comes from the fourth-order terms in (5.44). It is proportional to k^4, but also depends on the direction of $\underline{k}$, contrary to (5.45). For high temperatures, it is proportional to $(kT)^3$. Estimates of MARADUDIN and FLINN /5.11/ show that this term should be very small. Therefore the quadratic expression seems to be satisfactory for practical purposes.

Another item of information about the dynamics of the impurities, which can be extracted from Mössbauer measurements, is the second-order Doppler shift. Due to the emission of the γ quantum of energy E_o, the mass of the emitting impurity is decreased by $\delta M = -E_o/c^2$. Due to energy conservation this has to lead to a small shift ΔE of the γ energy, represented by the difference between the initial and final energy of the crystal. If $H^{(i)}$ is the initial Hamiltonian operator of the crystal and $H^{(f)}$ the final one, being different from $H^{(i)}$ only due to the mass change $\Delta M^{n} = -E_o/c^2$ of the impurity, we obtain

$$\Delta E = <H^{(i)} - H^{(f)}> = -\frac{\overset{o}{E}}{2(M^{n})^2c^2}<(\underline{p}^{n})^2>_T \ . \tag{5.49}$$

The thermal average $<(\underline{p}^{n})^2>$ has been calculated in (2.55). In the classical limit it is given by $3kT/M^{n}$. Thus nontrivial information about the impurities can only be obtained at low temperatures, where essentially the moment $<\omega>$ is measured.

This second-order Doppler shift is superimposed on the so-called isomer or chemical shift, which depends on the s-electron density at the position of the emitting nucleus. If this term is independent of temperature in a first approximation, the whole temperature dependence of the line shift is due to the second-order Doppler shift. However, due to thermal expansion, etc., also the isomer shift could become temperature dependent. For Fe^{57} this temperature dependence is small compared to the second-order Doppler shift. Contrary, for the very sharp line of Ta^{181} the dominating temperature dependence comes from the isomer shift /5.12,15/.

Most work has been done on Fe^{57} in various metal hosts. The characteristic values for this transition are: γ energy E = 14.4keV, wavelength λ = 0.86Å, decay time $T_{1/2} = 10^{-7}$s. An example of the measured Debye-Waller factors is shown in Fig. 23 for Fe^{57} in Cu as measured by SACHDEW and TEWARY /5.13/. Only recently the first measurements were reported for Au^{197} in Cu and Ag by PRINCE et al /5.14/. Due to the shorter wavelength of the γ quanta in Au^{197} (E = 77keV, λ = 0.16A, $T_{1/2} = 1.9\cdot10^{-9}$s) the f factor at T = 0 is rather small (see Fig. 24). The third Mössbauer atom which has been studied is Sn^{119} (E = 24keV, λ = 0.52A).

The experimental work for Fe^{57} has been summarized and analyzed by O'CONNOR et al /5.16/, and HOWARD and NUSSBAUM /5.17/. A recent critical analysis of GROW et al. /5.18/ summarizes and analyzes all known measurements for Fe^{57}, Sn^{119} and Au^{197} in

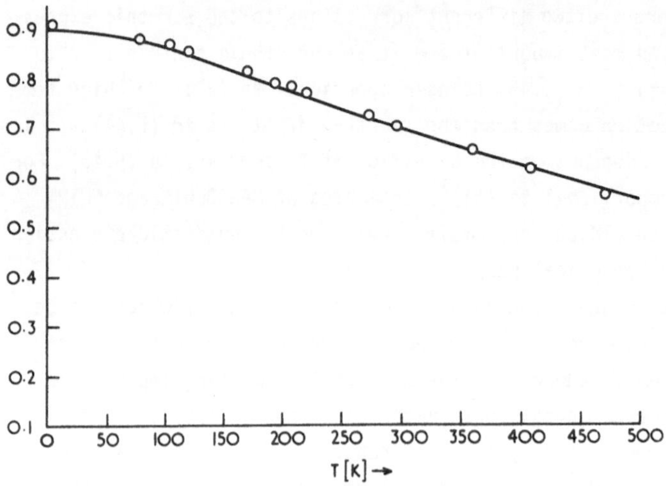

Fig. 23. Debye - Waller factor f(T) of Fe^{57} in Cu as measured by SACHDEW and TEWARY /5.17/

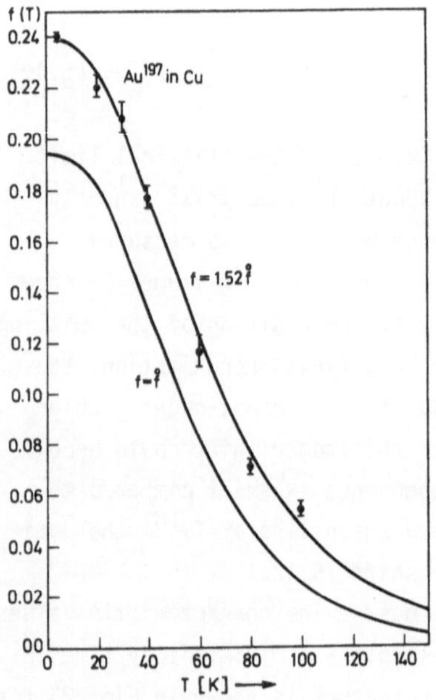

Fig. 24. Debye - Waller factor f(T) for Au^{197} in Cu according to PRINCE et al. /5.14/. The full lines refer to the Debye - Waller factor as calculated for an isotopic defect ($f = f_o$) and for a strengthened Au-Cu coupling ($f = 1.52f_o$)

terms of the Mannheim model. In the analysis, the experimental Debye-Waller factor f(T) and the second-order Doppler shift $\Delta E(T)$ are fitted with the theoretical expression (5.46) for f(T) and (5.49) for $\Delta E(T)$, where for the local impurity spectrum $z^d(\omega)$ the nearest neighbour approximation (5.25) of MANNHEIM is used $\left(z^d(\omega) = (2M\omega/\pi)\cdot \mathrm{Im}\{G^{dd}_{xx}(\omega)\}\right)$. For the ideal spectrum the exact values as calculated from neutron-scattering experiments are taken (or sometimes simply a Debye spectrum is used). The nearest neighbour force constant f is replaced by $8f = \overset{o}{M}\omega^2_{max}$ or $4\overset{o}{f} =$

$f^{o}_{Einstein}$, as it is valid for the nearest neighbour model in fcc.

The following table by GROW et al. /5.18/ summarizes the experimental results for Fe^{57}, Sn^{119} and Au^{197} in various metals. For more details about the evaluation of the data, the selection of the best experimental results, their uncertainties, etc., we refer to the original paper.

Table 1

Impurity	Host	M/M_o	f/f_o	Ref.	$\Delta V/V_c$
Fe^{57}	Al	2.12	0.62	5.17	*)
	Au	0.29	0.67	5.19	-0.20
	Cr	1.10	0.70	5.20	-0.02
	Cu	0.89	1.32	5.21	+0.04
	Mo	0.59	0.44	5.22	*)
	Nb	0.61	0.61	5.18	-0.38
	Ni	0.97	1.25	5.23	+0.10
	Pd	0.54	0.58	5.21	-0.12
	Pt	0.29	0.63	5.21	-0.10
	Ta	0.31	0.54	5.22	*)
	W	0.31	0.41	5.22	*)
Sn^{119}	Ag	0.10	1	5.24,25	
	Au	0.60	1	5.25	
	Pd	1.12	2	5.26	
	Pt	0.61	1	5.25	
Au^{197}	Cu	3.10	1.52	5.14	+0.48
	Ag	1.79	1.39	5.14	-0.02

*) no ΔV values reported, partly due to the extremely low solubility of the impurities in these hosts

For illustration, the local frequency spectrum of Fe^{57} in Cu and Fe^{57} in Al, as calculated by HOWARD and NUSSBAUM /5.17/, is shown in Fig. 25a,b. For Cu a localized mode exists slightly above ω_{max}, containing about 50% of the intensity. This high frequency is due to the combined action of the slightly reduced mass and the increased force constants (f/f_o = 1.39). In Al a large resonance appears at about $\omega_{max}/3$, which is due to the large Fe mass and the weakened force constants ($f = 0.61f_o$). Figure 25c shows the local spectrum of Au^{197} in Cu. Here the effect of the heavy mass (M/M_o = 3.10) overcomes the increased force constants (f/f_o = 1.52)

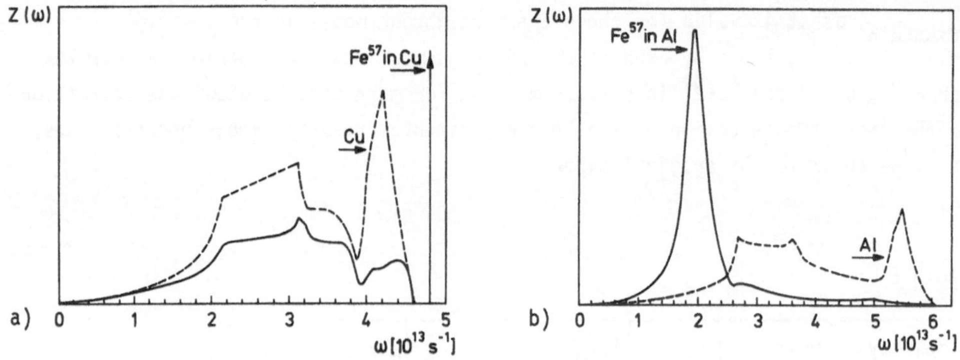

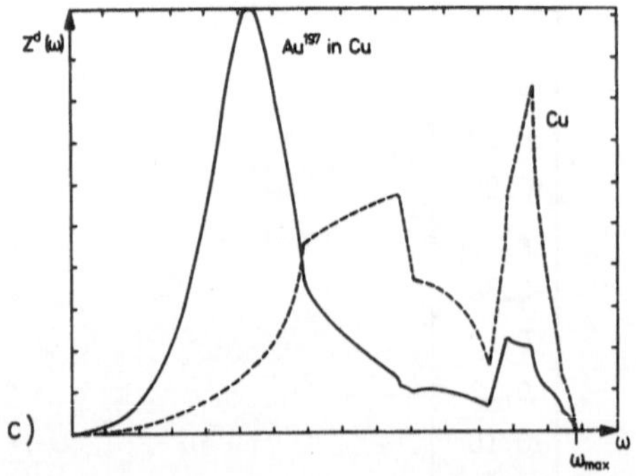

Fig. 25a-c. Local frequency spectra of impurities in fcc metals compared with the ideal host spectra (dashed)

a) Fe^{57} in Cu (f/f_o = 1.39 /5.17/, M/M_o = 0.89)

b) Fe^{57} in Al (f/f_o = 0.61 /5.17/, M/M_o = 2.12)

c) Au^{197} in Cu (f/f_o = 1.52 /5.14/, M/M_o = 3.10)

leading to a broad resonant mode.

The spectra for the two bcc hosts Cr and Nb are shown in Fig. 26a and b. In Nb a resonance mode appears slightly below ω_{max}, which for a slight increase of the force constant would lead to a localized mode.

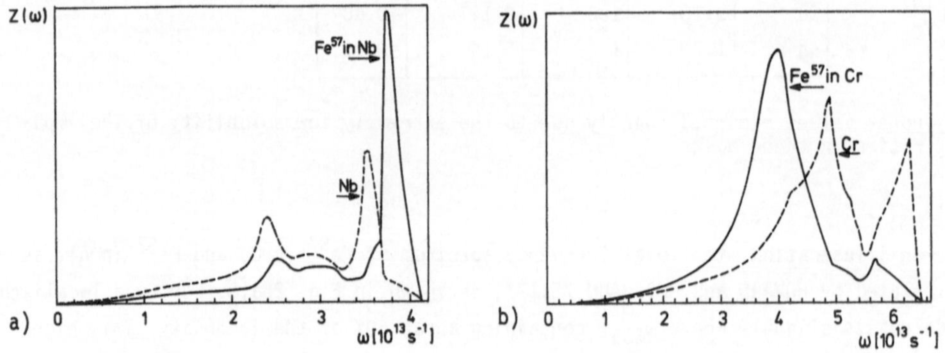

Fig. 26a,b. Local frequency spectra for impurities in bcc models

a) Fe^{57} in Nb (f/f_o = 0.61 /5.17/, M/M_o = 0.61)

b) Fe^{57} in Cr (f/f_o = 0.70 /5.17/, M/M_o = 1.10)

5.3 Dynamical Behaviour of Vacancies

In analogy to the subsitutional defect, a vacancy (see Fig. 27) can also be described in the nearest neighbour model by setting f = 0 and by disregarding the three translational modes with frequency $\omega = 0$ of the central atom (Fig. 20a). For the dynamics the most characteristic effects should be a slight softening of the "local frequencies" of the nearest neighbours due to the missing springs at the vacancy site. For instance, in fcc (Fig. 27) the Einstein frequency of a nearest neighbour

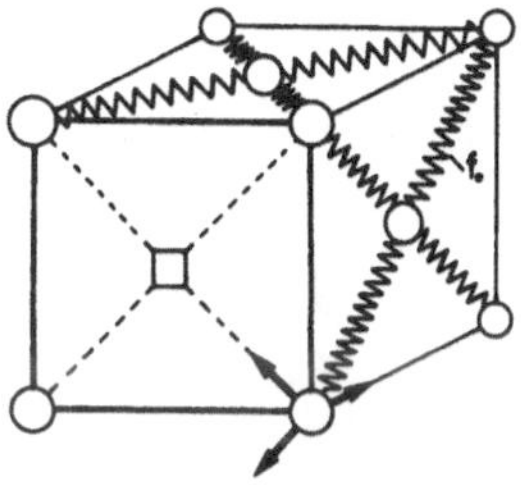

Fig. 27. Vacancy in fcc lattice. The dashed lines indicate for the central-force n.n. model the missing force constant f_o. The arrows at the nearest neighbour site indicate the three symmetry adapted directions of motions for the nearest neighbour

for a motion in the direction of the vacancy is reduced from its value $\langle\omega^2\rangle = 4f_o/M_o$ in the ideal crystal to the value $\langle\omega^2\rangle = 3f_o/M_o$. Contrary for motions perpendicular to the missing springs, the Einstein frequencies are unchanged, $\langle\omega^2\rangle = 4f_o/M_o$.

This trend of the Einstein approximation is also found in more detailed calculations /5.27/. These calculations are based on computer simulation. For the interaction short ranged Morse (Fig. 32) and Born - Mayer potentials have been used in order to calculate the static configuration of the vacancy. From the configuration and the given potential all relevant coupling parameters can be determined which are then used to calculate the vibrational behaviour via the Green's function method. The resulting local frequency spectra of a nearest neighbour in fcc are plotted in Fig. 28a-f both for the Morse potential (a,c,e) and the Born - Mayer potential (b,d,f). Figure 28a,b shows the spectrum for vibrations into the direction of the vacancy site. For both potentials the major effect is the formation of a quasi-resonant peak at somewhat lower frequencies than in the ideal spectrum which is slightly more pronounced for the Born - Mayer potential. The two perpendicular directions $(1\bar{1}0)$ and (001) show no such drastic effect, at least for the Morse potential. For the Born - Mayer potential the spectrum is shifted towards higher frequencies. This can be explained by the different role of static relaxations for both potentials. Whereas the Morse potential practically does not give rise to relaxations, these are quite important for the Born - Mayer potential. Due to the relaxations the distances between the nearest neighbours are smaller and the coupling parameter stronger than in the ideal lattice leading on the average to higher frequencies for motions into the direction of a nearest neighbour.

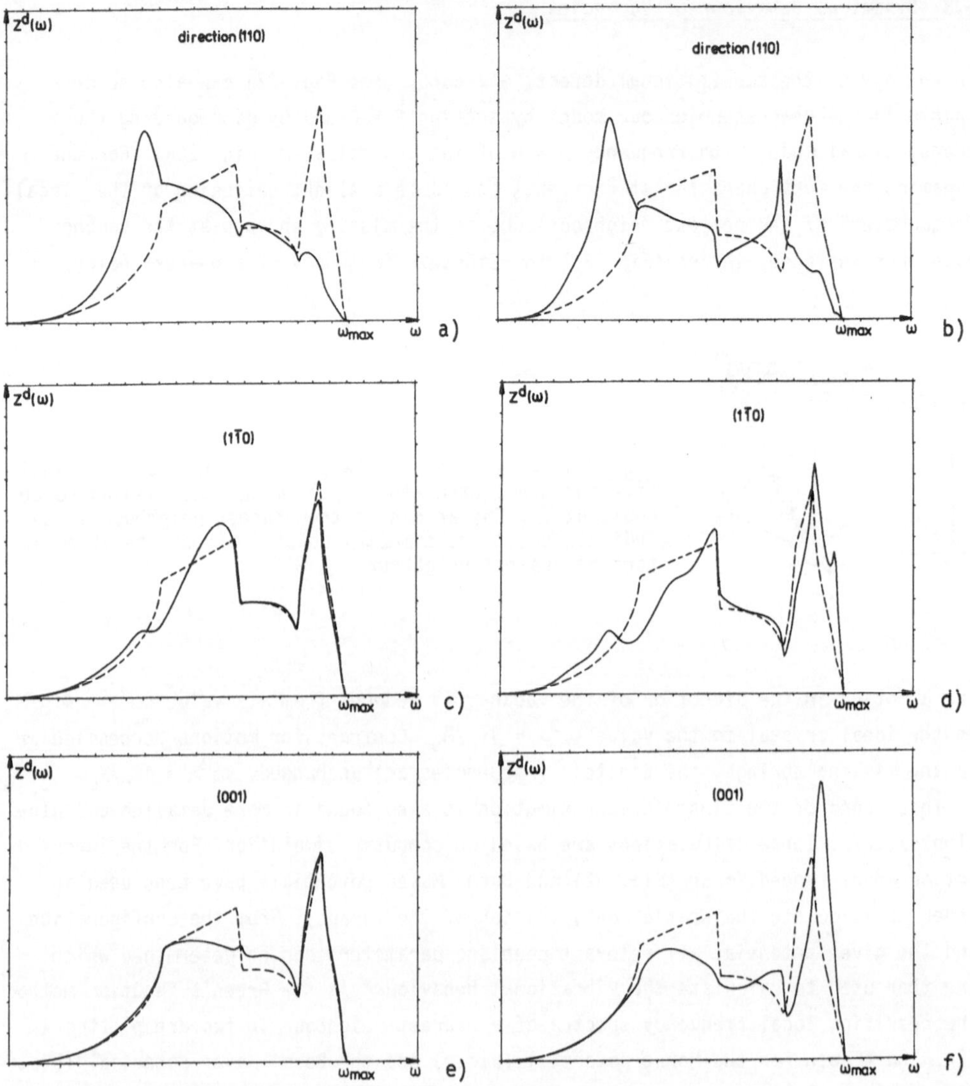

Fig. 28a-f. Local frequency spectra of a nearest neighbour (110) of a vacancy in fcc for vibrations in different directions

a) direction (110), i.e., towards the vacancy (Morse potential)
b) direction (110), i.e., towards the vacancy (Born - Mayer potential)
c) direction (1$\bar{1}$0) (Morse potential)
d) direction (1$\bar{1}$0) (Born - Mayer potential)
e) direction (001) (Morse potential)
f) direction (001) (Born - Mayer potential)

6. Vibrational Properties of Interstitials

In this chapter we will discuss the vibrational properties of interstitials. First, we briefly discuss the dynamics of light interstitials, especially of hydrogen atoms, which due to their small masses are characterized by localized vibrations. In the following sections we discuss in detail the dynamics of self-interstitials. Due to the strong and repulsive interaction between the interstitial and its nearest neighbour, the interstitial vibrates both with localized modes as well as with low-frequency resonant modes. The latter are typical for self-interstitials and can be understood as representing a tendency of the interstitial towards instability. Two special configurations are discussed in detail: the octahedral interstitial which is especially simple because of its cubic symmetry and the 100-dumbbell which according to recent experiments represents the interstitial configuration in Al and Cu (for a recent review see /6.1/).

6.1 Vibrations of H in Metals

Localized Vibrations of H:

Gases like H, O, N and also C have a high solubility in many metals. Due to their small size, they go into solution as interstitials. One generally assumes that they occupy either the octahedral or tetrahedral position, both for fcc and bcc crystals. Because of their small mass, their dynamics should be characterized by localized vibrations. In a first approximation, the localized frequency can be calculated by the Einstein model, i.e., the interstitial moves in the rigid lattice. For instance, the tetrahedral configuration in bcc crystals, shown in Fig. 29a, favours one direction, in the case shown the x axis. Thus we expect one double degenerate frequency for vibrations in y and z direction and a different frequency for vibrations in x direction. If we assume a longitudinal force constant f between the interstitial

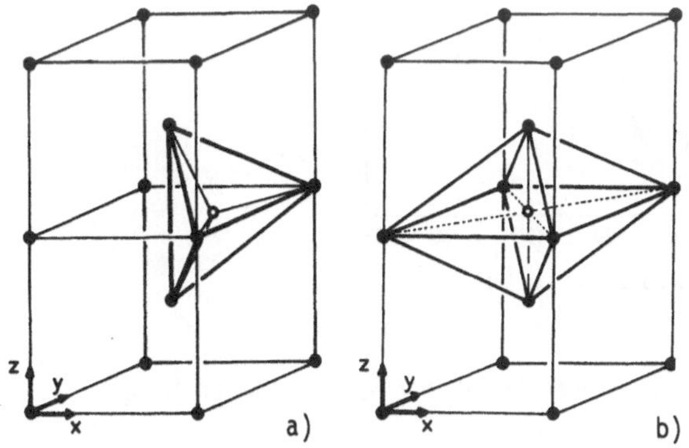

Fig. 29.
a) Tetrahedral configuration in bcc lattice (showing the interstitial (∘) and the neighbouring atoms forming an irregular tetrahedron)
b) Octahedral configuration in bcc lattice (showing the interstitial (∘) and the neighbouring atoms forming an irregular octahedron)

and its four nearest neighbours, we obtain in the Einstein approximation

$$\omega_x^2 = \frac{4}{5}\frac{f}{M} \quad , \quad \omega_y^2 = \omega_z^2 = \frac{8}{5}\frac{f}{M} \quad . \tag{6.1}$$

Thus the double degenerate frequency ω_y should be a factor $\sqrt{2}$ higher than ω_x.

Analogously also the octahedral position in bcc crystals (Fig. 29b) leads to two different frequencies. In a nearest neighbour model (Fig. 29b) we assume spiral springs f to the two nearest neighbours in z direction and springs f' to the four neighbours in the x-y plane. Thus the frequencies of the localized modes are $\omega_z^2 = 2f/M$ and $\omega_x^2 = \omega_y^2 = 2f'/M$. Since the z neighbours are appreciably nearer to the defect, one expects $f > f'$. Thus contrary to the tetrahedral configuration, the *lower* frequency is double degenerate.

In fcc crystals the octahedral position has cubic symmetry and therefore a triple degenerate mode with $\omega^2 = 2f/M$ if f denotes the springs to the nearest neighbours. For the tetrahedral position one has to note that contrary to bcc the neighbours form a regular tetrahedron. The only contribution to the local spectrum of the interstitial comes from a triple degenerate F representation. Thus effectively, one has cubic symmetry and therefore a triple degenerate mode, with a frequency of $\omega^2 = \frac{4}{3}\frac{f}{M}$.

Experimentally, only hydrogen in metals has been studied in detail by incoherent neutron scattering. For instance, VERDAN et al. /6.2/ have measured the scattering of $NbH_{0.05}$ and $VH_{0.04}$ by the time of flight method. The resulting frequency distribution in the vicinity of the local peaks is shown in Fig. 30a,b. Clearly two double

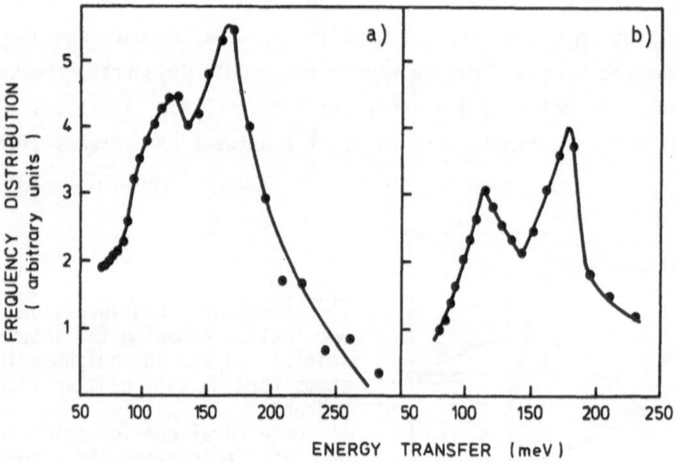

Fig. 30a,b. Local frequency spectra of H in metals according to time of flight measurements of VERDAN et al. /6.2/
a) $VH_{0.04}$ b) $NbH_{0.05}$

peaks can be seen, with more intensity in the higher peak, which is in agreement with the expectations for the tetrahedral position. For $NbH_{0.08}$ BIRCHALL and ROSS obtain essentially the same result /6.3/. The following table summarizes the results of the neutron measurements and gives the resonant frequencies for the various systems studied. For the bcc metals V, Nb, and Ta the ratio of the two localized

Table 2

System	$V-H_{0.04}^{+)}$	$Nb-H_{0.05}^{+)}$	$Ta-H_{0.15}^{+)}$	$Pd-H_{0.02}^{*)}$	$Th-H_{0.05}^{*)}$
$\hbar\omega_{\ell oc}$ [meV]	167 ± 6 123 ± 6	177 ± 6 114 ± 6	170 ± 5 120 ± 5	66	114 ± 5
Ref.	/6.2/	/6.2/	/6.4/	/6.5/	/6.6/

$^{+)}$bcc $\quad$ $^{*)}$fcc

frequencies agrees with $\sqrt{2}$ within the experimental errors, thus indicating the tetrahedral position. Recent channeling measurements /6.7/ show conclusively that in the related systems $NbD_{0.02}$ and $TaD_{0.07}$ deuterium occupies the tetrahedral position. For NbD, the same result has been obtained by diffuse neutron scattering /6.8/. By studying the line broadening due to the diffusion of hydrogen with quasi-elastic scattering, one can also obtain information about the hydrogen positions. For Pd (fcc) the results can be interpreted uniquely by the hydrogen atoms occupying octahedral positions /6.4/. However, for bcc metals, the results are not quite conclusive. A recent review by KEHR /6.9/ summarizes the diffusion aspect of hydrogen in metals. For a general review about hydrogen in metals see /6.10,11/.

In-Band Motion of Hydrogen:

In the Einstein approximation /6.1/ the defect Green's function $G^{dd}_{xx}(\omega)$ and the local frequency spectrum is given by

$$G^{dd}_{xx}(\omega) = \frac{1}{\phi^{dd}_{xx} - M\omega^2} \quad \text{and} \quad z^d_x(\omega^2) = \delta(\omega^2 - \frac{\phi^{dd}_{xx}}{M}) \ . \tag{6.2}$$

Thus the total intensity is contained in the localized mode and there is no intensity in the normal frequency band. However this is not quite true, as can be seen from (4.7) by approximating the Green's function $\hat{G}_{RR}(\omega)$ by its high-frequency limit

$$\hat{G}_{RR}(\omega) \cong -\frac{1}{\overset{\circ}{M}\omega^2} \ .$$

In this approximation one obtains

$$G^{dd}_{xx}(\omega) = \frac{1}{\phi^{dd}_{xx} + \frac{1}{\overset{o}{M}\omega^2}\sum\limits_{n=d}(\phi^{dn}_{xi})^2 - M\omega^2} \; . \tag{6.3}$$

Thus the localized mode frequency is given by

$$\omega^2_{\ell oc} = \frac{\phi^{dd}_{xx}}{M}\left\{1 + \frac{1}{\overset{o}{M}\omega^2}\sum_{n=d}(\phi^{dn}_{xi})^2/\phi^{dd}_{xx}\right\} \cong \frac{\phi^{dd}_{xx}}{M}(1 + \alpha\frac{M}{\overset{o}{M}})$$

$$\text{with} \quad \alpha = \sum_{\substack{n\neq d\\ i}} (\phi^{dn}_{xi})^2/(\phi^{dd}_{xx})^2 \; . \tag{6.4}$$

By expanding $G^{dd}_{xx}(\omega)$ near $\omega = \omega_{\ell oc}$, one obtains for the spectrum

$$z^d_x(\omega) = \frac{1}{1 + \alpha M/\overset{o}{M}}\,\delta(\omega - \omega_{\ell oc}) \cong (1 - \alpha M/\overset{o}{M})\;\delta(\omega - \omega_{\ell oc}) \quad . \tag{6.5}$$

Thus $\alpha M/\overset{o}{M}$ gives the intensity of the local spectrum in the range of the host frequency band. Due to the smallness of M, this is a very small fraction of the total intensity. Nevertheless, it can be very important for the dynamics of hydrogen, especially for its thermal fluctuations, as we will see below.

The total spectrum $z^d_x(\omega)$ can be written as

$$z^d_x(\omega) = \alpha\,\frac{M}{\overset{o}{M}}\,z^{band}_x(\omega) + (1 - \alpha\frac{M}{\overset{o}{M}})\;\delta(\omega - \omega_{\ell oc}) \tag{6.6}$$

where $z^{band}_x(\omega)$ is a normalized spectrum in the range of the host's band. The thermal displacements squared is according to (2.53) given by

$$\begin{aligned}\langle(s^d_x)\rangle &= \int d\omega\,\frac{\varepsilon(\omega,T)}{M\omega^2}\,z^d_x(\omega)\\ &= \alpha\int d\omega\,\frac{\varepsilon(\omega,T)}{\overset{o}{M}\omega^2}\,z^{band}_x(\omega) + (1 - \alpha\frac{M}{\overset{o}{M}})\,\frac{\varepsilon(\omega_{\ell oc},T)}{M\omega^2_{\ell oc}}\\ &\approx \alpha\,\langle s^2_x\rangle_{host} + \langle(s^d_x)^2\rangle_{\ell oc} \; .\end{aligned} \tag{6.7}$$

Here, in a rough approximation, $z^{band}_x(\omega)$ has been replaced by the ideal host spectrum $\overset{o}{z}(\omega)$, so that the first term yields besides the factor α the displacements squared of the host, whereas the second term is the contribution of the localized mode, for which $1 - \alpha M/M \approx 1$. It is just the division by $1/M\omega^2$ in (6.7), which makes the in-band motion of similar importance for $\langle s^2\rangle$ as the localized mode.

The factor α can easily be calculated in a nearest neighbour model with a longitudinal force constant f between the interstitial and its nearest neighbours. One obtains $\alpha = f/f_{Einstein}$, so that for the

$$\begin{array}{lll} \text{tetrahedral position in bcc:} & \alpha_x = \frac{5}{4} \text{ and } \alpha_y = \alpha_z = \frac{5}{8} & \\ \text{octahedral position in bcc :} & \alpha = \frac{1}{2} & \\ \text{tetrahedral position in fcc:} & \alpha = \frac{3}{4} & \text{for all directions} \\ \text{octahedral position in fcc :} & \alpha = \frac{1}{2} & \end{array} \tag{6.8}$$

As a simple example we will consider the octahedral position in fcc. For this the defect Green's function $G^{dd}_{xx}(\omega)$ will be derived in Sect. 6.3. For a pure longitudinal force constant f ($f_\parallel = f$, $f_\perp = 0$) we obtain from (6.12)

$$G^{dd}_{xx}(\omega) = \left\{ \frac{2f}{1 + \left[\overset{\circ}{G}{}^{(0)}_{xx}(\omega) + \overset{\circ}{G}{}^{(200)}_{xx}(\omega) \right] f} - M\omega^2 \right\}^{-1} . \tag{6.9}$$

Within the band, we can practically neglect the term $M\omega^2$, since $M \ll \overset{\circ}{M}$. Then

$$G^{dd}_{xx}(\omega) = \frac{1}{2f} + \frac{1}{2}\left(\overset{\circ}{G}{}^{(0)}_{xx}(\omega) + \overset{\circ}{G}{}^{(200)}_{xx}(\omega) \right) \tag{6.10}$$

$$z^{band}_x(\omega) = \overset{\circ}{z}{}^{(0)}_x(\omega) + \overset{\circ}{z}{}^{(200)}_x(\omega) , \quad \overset{\circ}{z}{}^{(m)}_x(\omega) = \frac{2\overset{\circ}{M}\omega}{\pi} \mathrm{Im}\{\overset{\circ}{G}{}^{(m)}_{xx}(\omega)\} .$$

Whereas the first term $\overset{\circ}{z}{}^{(0)}(\omega)$ is the ideal host spectrum, the second term $\overset{\circ}{z}{}^{(200)}(\omega)$ represents a correction to this. For very high temperatures $\langle (s^d_x)^2 \rangle$ is proportional to kT and determined by the static Green's function $G^{dd}_{xx}(0)$, for which we obtain from above

$$\langle (s^d_x)^2 \rangle = kT\, G^{dd}_{xx}(0) = kT \left[\frac{1}{2}\left(\overset{\circ}{G}{}^{(0)}_{xx}(0) + \overset{\circ}{G}{}^{(200)}_{xx}(0) \right) + \frac{1}{2f} \right] . \tag{6.11}$$

The importance of the in-band modes of hydrogen has recently been pointed out by LOTTNER and SCHOBER /6.12/. Figure 31a shows the measured energy spectrum from $NbH_{0.05}$. The peak at $\omega \approx 0$ is the quasi-elastic scattering, the peak LA is a Nb--phonon mode. The small peak at 3.8 THz arises from the in-band motion of hydrogen. Model calculations for the local hydrogen spectrum in the band are shown in Fig. 31b. Plotted are the spectrum of pure Nb and the local spectra of hydrogen for vibrations in x and y direction according to a central force nearest neighbour model and a more general model (with additional bending springs between the hydrogen and its nearest neighbours and with changed force constants between the nearest neighbours). The models have been fitted to the observed localized mode frequencies of H in Nb. The spectrum of hydrogen is multiplied by M_{Nb}/M_H in order to be proportional to the thermal vibrations. For both models a pronounced peak for hydrogen motion in x direction appears. (This direction also has the smallest frequency for the localized mode.) The calculated spectra for both models are also plotted in Fig. 31a together with the experimental results. The thermal displacements squared for the hydrogen atom are shown in Fig. 31c. Whereas the contribution from the localized

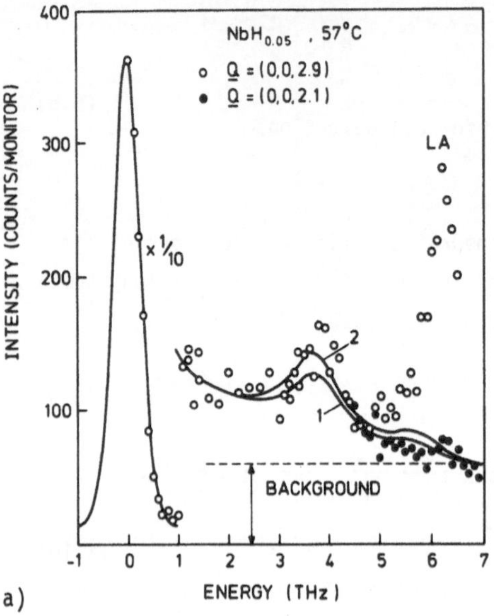

a)

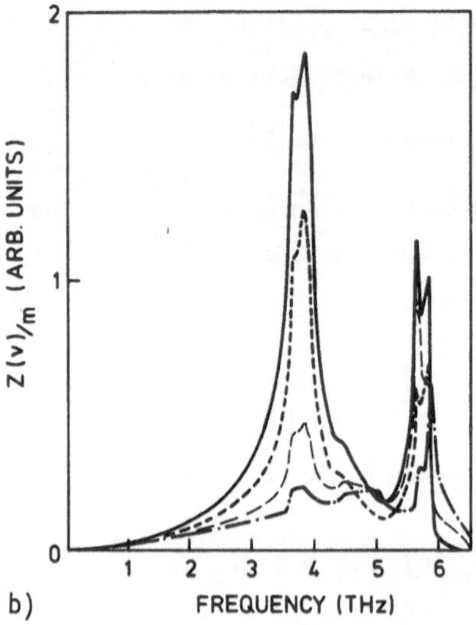

b)

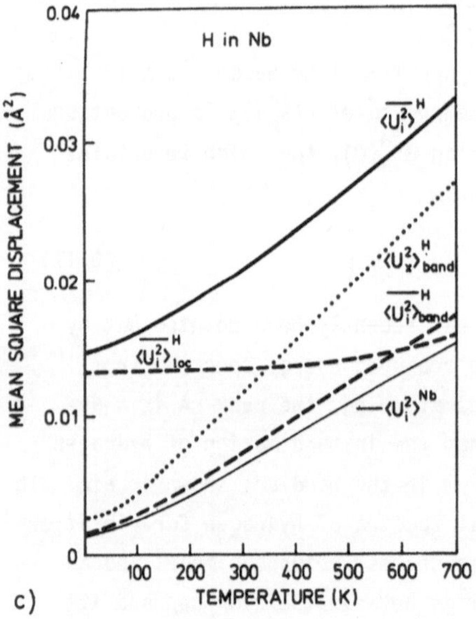

c)

Fig. 31a-c. In-hand motion of H (according to Lottner, Schober and Fitzgerald)

a) Measured energy spectrum of $NbH_{0.05}$ and theoretical results for the central--force nearest neighbour model (curve 1) and the general model (curve 2)

b) Local frequency spectrum $(M_{Nb}/M_H)\cdot z_i^H(\omega)$ for the in-band motion of hydrogen in Nb

—— —— —— $z_x^H(\omega)$ for central-force n.n. model

———————— $z_x^H(\omega)$ for general model

—·—·— $z_z^H(\omega)$ for general model

— — — — $z^{Nb}(\omega)$ for pure Nb

c) Mean square displacements of H in Nb for the general model

$\overline{\langle u_i^2\rangle^H_{\ell oc}}$: contribution of localized modes averaged over all directions

$\langle u_x^2\rangle^H_{x\ band}$: band contribution for x direction

$\overline{\langle u_i^2\rangle^H_{i\ band}}$: band contribution averaged over all directions

$\overline{\langle u_i^2\rangle^H}$: total mean square displacements averaged over all directions

$\langle u_i^2\rangle^{Nb}$: mean square displacements of Nb mode is nearly constant up to high temperatures, the band motion gives for the x direction $\left(\langle u_x^2\rangle^H_{band}\right)$ a strongly increasing contribution. When the contribution of the band is averaged over all directions it nearly coincides with the displacements for pure Nb. It has to be added to $\overline{\langle u_i^2\rangle^H_{\ell oc}}$ in order to give the total displacements $\overline{\langle u_i^2\rangle^H}$.

6.2 Qualitative Explanation of the Dynamics of Self-Interstitials

By studying the dynamics of self-interstitials SCHOLZ and LEHMANN /6.13,14/ found in a computer simulation that the 100-dumbbell vibrates, in addition to several high-frequency localized modes, also with low-frequency resonant modes. This result was quite surprising since in an Einstein model one expects either resonant vibrations or localized vibrations. However we have seen already in the preceeding chapter that the Einstein model for the defect alone can fail since for a very heavy defect with strong force constants we also obtain both localized vibrations (due to the strong force constants) and resonant vibrations (due to the heavy mass). Because for self-interstitials we do not have a big mass, both kinds of modes must be due to the special interaction of the defect with its neighbours, as we will explain in the following.

For a central potential V(R) the coupling constants between two atoms at distance $\underline{R} = (R,0,0)$ are given by

$$\phi_{ij}^{(\underline{R})} = -\partial_{R_i}\partial_{R_j} V(R) = \begin{Bmatrix} f_\parallel & 0 & 0 \\ 0 & f_\perp & 0 \\ 0 & 0 & f_\perp \end{Bmatrix} \quad \text{with} \quad \begin{matrix} f_\parallel = V''(R) \\ f_\perp = V'(R)/R \end{matrix} \quad . \tag{6.12}$$

The "longitudinal" force constant $f_\parallel$ enters for displacements parallel to $\underline{R}$ ("spiral spring"), whereas the "transversal" force constant $f_\perp$ enters for displacements perpendicular to $\underline{R}$ ("leaf" or "bending spring"). For an ideal lattice, the nearest neighbour constant $f_\parallel = f_\parallel^o$ is the dominating force constant, e.g., in fcc crystals it is usually a factor of 10 larger than all other springs, and $f_\perp$ can be neglected in a first approximation.

However, this is not so in the compressed region near an interstitial. A typical interaction potential is shown in Fig. 32. It is a Morse potential

$$V(R) = D\left\{\exp[-2\alpha(R-R_o)] - 2\exp[-\alpha(R-R_o)]\right\} \tag{6.13}$$

with

$$D = 0.18\text{eV}\ , \quad \alpha = 2.23\text{Å}^{-1}, \quad R_o = 2.57\text{Å}\ .$$

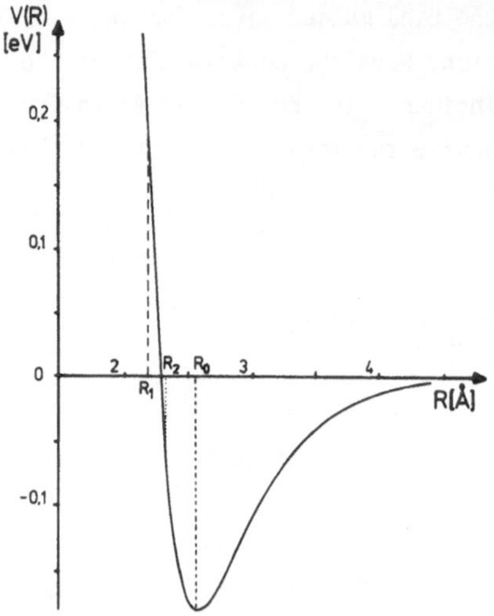

Fig. 32. Morse potential for Cu (according to /6.14/). R_o is about the nearest neighbour distance, R_1 the separation of two dumbbell atoms and R_2 the separation between a dumbbell atom and a nearest neighbour (see Fig. 40)

The parameters D, α and R_o have been fitted to Cu /6.15/. R_o coincides approximately with the nearest neighbour distance $a/\sqrt{2} = 2.55$Å for Cu. Figure 32 shows also the separation R_1 of two atoms forming a 100-dumbbell interstitial (see Fig. 40) and the separation R_2 between a dumbbell atom and a nearest neighbour. The values for R_1 and R_2 are obtained by computer simulation. For such small distances the atoms are strongly repelling each other.

With the Morse potential one obtains, e.g., for the interaction of the two dumbbell atoms $f_\parallel = 8f_\parallel^o$ and $f_\perp = -0.6f_\parallel^o$; and for the interaction between one dumbbell atom and its nearest neighbours: $f_\parallel = 5f_\parallel^o$, $f_\perp = -0.2f_\parallel^o$. Thus, because of the strong interaction, the longitudinal force constant $f_\parallel$ is very large and, because of the repulsion of the atoms, $f_\perp$ is always negative and becomes comparable to $f_\parallel^o$ representing the restoring forces of the ideal lattice. Such a negative bending spring can be visualized as being produced by a compressed spiral spring (see Fig. 33)

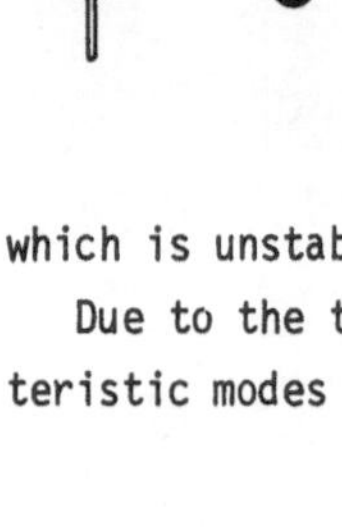

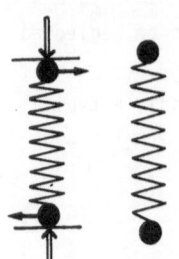

Fig. 33. A compressed spiral spring is instable against displacements perpendicular to the axes, thus representing a negative bending spring

which is unstable against displacements perpendicular to the spring axis.

Due to the two different force constants $f_\parallel$ and $f_\perp$ two different kinds of characteristic modes can occur:

1) Vibrations where neighbouring atoms move in opposite phases so that the strong spiral springs in the interstitial region are stressed which leads to high-frequency localized modes. Then the relatively weak leaf springs are not important.

2) Vibrations where neighbouring atoms are in phase so that the strong spiral springs are only stressed slightly. In the latter case the negative leaf springs become important. They lower the frequency and hence give rise to resonant modes.

As can be seen from Fig. 33 the negative leaf springs introduce a tendency towards instability into the lattice since for some configurations they may even lower the frequency of a resonant mode so much that $\omega_{res}^2 < 0$. Then this interstitial configuration is unstable. Thus the negative bending springs determine the stability or instability of a configuration and lead to resonance modes for the stable configurations. These resonances are very important for an understanding of many physical properties of interstitials /6.14/. For instance, they can explain the experimentally observed large changes of the elastic constants due to interstitials (see Sect. 7.6).

6.3 Octahedral Interstitial

In an fcc lattice the configuration with the highest, i.e., cubic symmetry is the octahedral interstitial, shown in Fig. 34. The high symmetry facilitates the calculation of the dynamical behaviour. We consider the following model with nearest

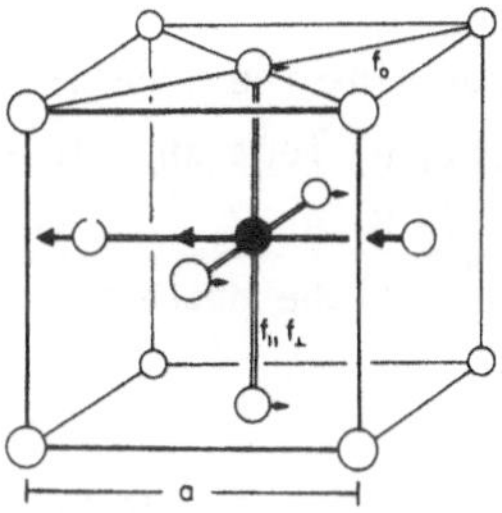

Fig. 34. Octahedral configuration in fcc lattice, showing the longitudinal and transverse force constants $f_\parallel$, f between the interstitial and its nearest neighbours. The arrows indicate the motion of the interstitial and the neighbours in the F_{1u} subspace

neighbour interaction only (Fig. 34): The interstitial - host coupling is described by a longitudinal force constant $f_\parallel$ and a transverse force constant $f_\perp$ whereas the host atoms are coupled only by spiral springs $f_\parallel^o$.

Due to the leaf springs $f_\perp$ this model is not rotationally invariant. This can be seen, if we rotate the crystal as shown in Fig. 35. All longitudinal springs ($f_\parallel$, $f_\parallel^o$) are rotationally invariant and need not be considered. However, due to the rotation the leaf springs $f_\perp$ give rise to forces $\sim f_\perp$ on the neighbouring atoms, which are

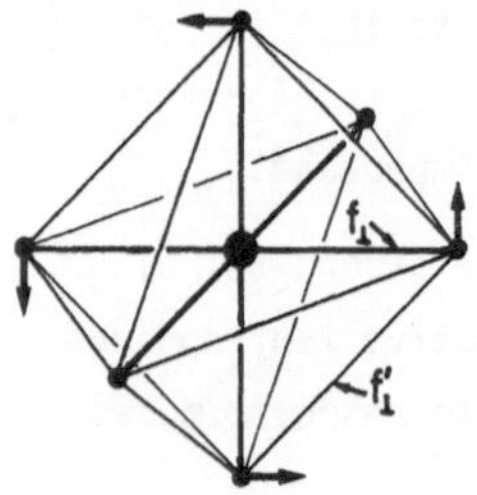

Fig. 35. A rotation of the crystal around the interstitial gives rise to forces due to the leaf springs $f_\perp$. They can be balanced by opposite forces from additional leaf springs $f'_\perp$ between the nearest neighbour, if $f'_\perp = f_\perp/4$

not compensated by other forces. Rotational invariance can be restored by introducing additional bending springs $f'_\perp$ between neighbouring atoms of the octahedron. For $f'_\perp = -f_\perp/4$ no forces arise on any atom. For this model the number of perturbed atoms has not been increased, so that the mathematical effort is essentially the same as before. Since the results, however, show the same qualitative features as the simpler model with $f_\perp$ alone, we will neglect the additional springs $f'_\perp$ in the following.

Only the six atoms at distance a/2 from the interstitial are directly affected by the introduction of the interstitial in the cube center. Thus the calculation of $\hat{G}(\omega)$ (4.9) requires for each frequency ω the inversion of 18-dimensional matrices. We can simplify the calculations by using the cubic symmetry of the point group O_h of the defect /1.2,4.8/. The 18×18 matrices $\hat{G}$, $\overset{\circ}{G}$, φ in (4.9) can be decomposed into their irreducible parts, e.g., $\overset{\circ}{G}$ can be written as

$$\overset{\circ}{G}{}^{\mathfrak{m}\mathfrak{n}}_{ij} = \sum_{\Gamma} \sum_{\gamma,\gamma'=1}^{\sigma_\Gamma} \sum_{\mu=1}^{d_\Gamma} s^{\mathfrak{m}}_i(\Gamma;\gamma;\mu)\, \overset{\circ}{G}{}^{\Gamma}_{\gamma\gamma'}\, s^{\mathfrak{n}}_j(\Gamma;\gamma';\mu) \tag{6.14}$$

where σ_Γ denotes the number of times the total representation contains the irreducible representation Γ with dimension d_Γ. The basis vectors $s^{\mathfrak{m}}_i(\Gamma;\gamma;\mu)$ form an orthonormal, complete set and have been worked out by DETTMANN and LUDWIG /4.8/, d_Γ gives the multiplicity of the distinct eigenvalues of $\overset{\circ}{G}$ and σ_Γ the rank of the matrices in the remaining equations

$$\hat{G}^{\Gamma}_{\gamma\gamma'} = \overset{\circ}{G}{}^{\Gamma}_{\gamma\gamma'} - \sum_{\nu,\nu'=1}^{\sigma_\Gamma} \overset{\circ}{G}{}^{\Gamma}_{\gamma\nu}\, \varphi^{\Gamma}_{\nu\nu'}\, \hat{G}^{\Gamma}_{\nu'\gamma'} \,. \tag{6.15}$$

In our case we obtain the following decomposition of the total representation

$$\Gamma_{\mathrm{octa}} = A_{1g} + E_g + F_{1g} + F_{2g} + 2F_{1u} + F_{2u} \tag{6.16}$$

where A denotes one-dimensional, E two-dimensional, and F three-dimensional representations. The subscripts f and u stand for even and odd symmetry under inversion. The corresponding invariant subspaces are shown in Fig. 36. To the vector E^1_g there are two equivalent vectors $E^{2'}_g$ and $E^{3'}_g$, the displacements of which are confined to

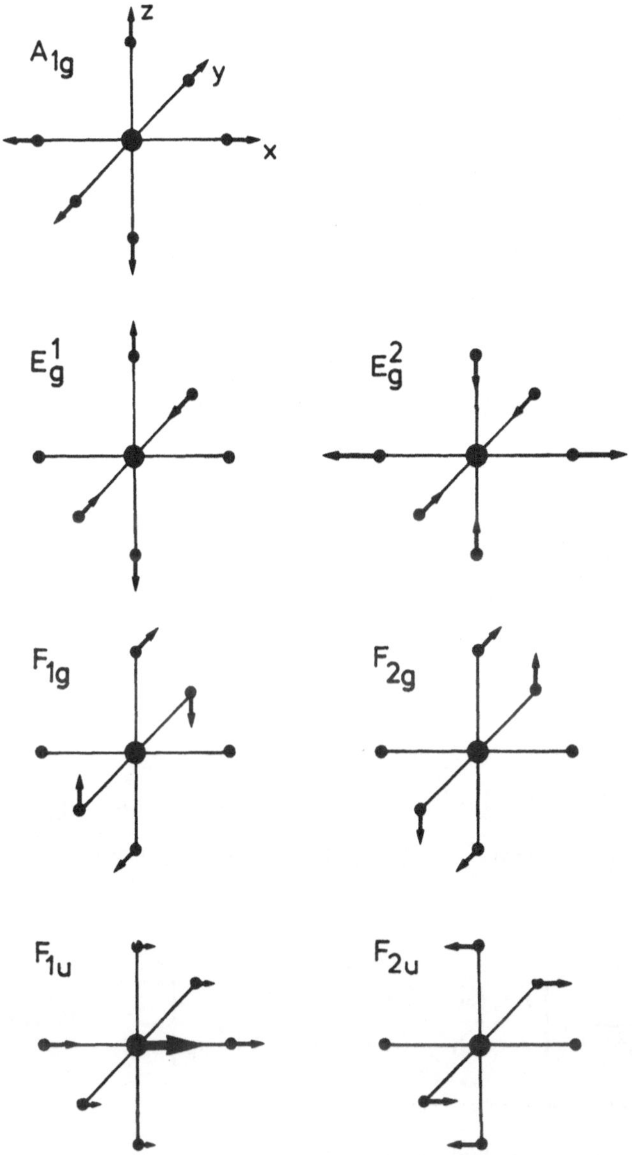

Fig. 36. The irreducible subspaces for the octahedral interstitial

the x-y plane and x-z plane. Since these three vectors are not linearly independent, only the vector E_g^2 is shown which is orthogonal to E_g^1. Of the three basis vectors for the F representations only the one preferring the x direction is shown.

With the exception of F_{1u}, which we discuss below, all representations occur only once. Thus one obtains as the solution of (6.15) for $\hat{G}^\Gamma$

$$\hat{G}^\Gamma = \frac{1}{(\overset{o}{G}{}^\Gamma)^{-1} + \varphi^\Gamma} \qquad \text{for } \Gamma = A_{1g}, E_g, F_{1g}, F_{2g}, F_{2u} \tag{6.17}$$

with

$$
\begin{aligned}
\mathring{G}^{A_{1g}} &= \mathring{G}^{(0)}_{xx} - \mathring{G}^{(200)}_{xx} - 4\mathring{G}^{(110)}_{xy} , & \varphi^{A_{1g}} &= f_{\parallel} \\
\mathring{G}^{E_{g}} &= \mathring{G}^{(0)}_{xx} - \mathring{G}^{(200)}_{xx} + 2\mathring{G}^{(110)}_{xy} , & \varphi^{E_{g}} &= f_{\parallel} \\
\mathring{G}^{F_{1u}} &= \mathring{G}^{(0)}_{xx} - \mathring{G}^{(200)}_{zz} + 2\mathring{G}^{(110)}_{xy} , & \varphi^{F_{1g}} &= f_{\perp} \\
\mathring{G}^{F_{2g}} &= \mathring{G}^{(0)}_{xx} - \mathring{G}^{(200)}_{zz} - 2\mathring{G}^{(110)}_{xy} , & \varphi^{F_{2g}} &= f_{\perp} \\
\mathring{G}^{F_{2u}} &= \mathring{G}^{(0)}_{xx} + \mathring{G}^{(200)}_{zz} - 2\mathring{G}^{(110)}_{zz} , & \varphi^{F_{2u}} &= f_{\perp}
\end{aligned}
\tag{6.18}
$$

For the evaluation of the vibrational behaviour of the interstitial, described by the Green's function $G^{dd}_{ij} = G^{dd}_{xx}\,\delta_{ij}$ or the local vibrational spectrum

$$
z^{d}(\omega) = \frac{2M^{d}\omega}{\pi}\,\mathrm{Im}\{G^{dd}_{xx}(\omega)\}
\tag{6.19}
$$

we have to consider the F_{1u} representation, since the defect moves only in this representation. As can be seen from Figs. 34 - 36, the subspace F_{1u} contains two different displacements for the neighbouring atoms, one for the atoms on the x axis and another one for the atoms in the y-z plane. Thus in order to calculate $\hat{G}_{F_{1u}}$ one has to invert 2×2 matrices. With

$$
\mathring{G}^{F_{1u}} = \begin{Bmatrix} a & b \\ b & c \end{Bmatrix} , \qquad \varphi^{F_{1u}} = \begin{Bmatrix} f_{\parallel} & 0 \\ 0 & f_{\perp} \end{Bmatrix}
$$

$$
a = \mathring{G}^{(0)}_{xx} + \mathring{G}^{(200)}_{xx} , \qquad b = 2\sqrt{2}\,\mathring{G}^{(110)}_{xx} , \qquad c = G^{(0)}_{xx} + \mathring{G}^{(200)}_{zz} + 2\mathring{G}^{(110)}_{zz}
\tag{6.20}
$$

we obtain for $\hat{G}^{F_{1u}}$

$$
\hat{G}^{F_{1u}} = \frac{1}{(1+af_{\parallel})(1+cf_{\perp}) - b^{2}f_{\parallel}f_{\perp}} \begin{Bmatrix} a + (ac-b^{2})f_{\perp} & b \\ b & c + (ac-b^{2})f_{\parallel} \end{Bmatrix} .
\tag{6.21}
$$

Finally we obtain for the Green's function of the defect

$$
G^{dd}_{xx}(\omega) = \left\{ -M^{d}\omega^{2} + \frac{2f_{\parallel} + 4f_{\perp} + f_{\parallel}f_{\perp}(4a + 2c - 4\sqrt{2}b)}{1 + af_{\parallel} + cf_{\perp} + (ac-b^{2})f_{\parallel}f_{\perp}} \right\}^{-1} .
\tag{6.22}
$$

By choosing, as a typical example, a strong longitudinal force constant $f_{\parallel} = 5f^{\circ}_{\parallel}$ between the defect and its neighbours the resulting local spectrum of the defect is shown in Fig. 37 for various values for the leaf spring $f_{\perp}$. The spectrum always contains a localized mode, the frequency of which is almost independent of $f_{\perp}$. Contrary for increasing negative values of $f_{\perp}$, a pronounced resonance mode appears at low frequencies. For $f_{\perp} \cong -0.96f^{\circ}_{\parallel}$ the resonance frequency approaches zero and the configuration becomes unstable. It is seen that the defect hardly vibrates with the

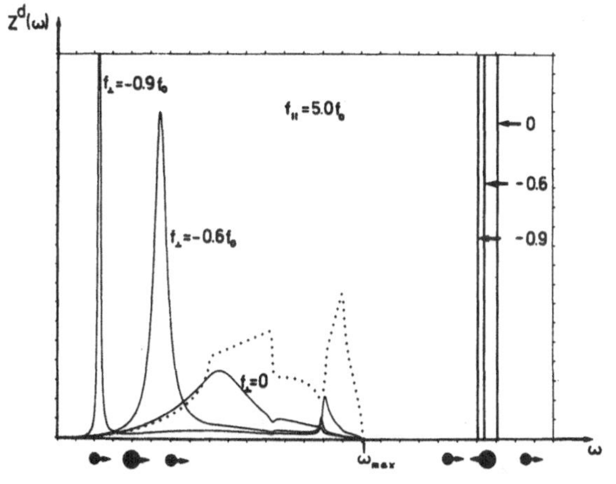

Fig. 37. Local frequency spectrum of the octahedral interstitial for the n.n.model of Fig. 34. The value of the longitudinal force constant $f_{\|}$ is fixed, $f_{\|} = 5f_{\|}^{0}$, whereas $f_{\perp}$ is varied: $f_{\perp} = 0, -0.6f_{\|}^{0}, -0.9f_{\|}^{0}$

eigenfrequencies of the ideal lattice and the local spectrum can be composed in a first approximation of the resonant and localized mode alone. In the localized mode the defect and the nearest neighbours in the direction of the vibration move opposite in phase, thus compressing the strong springs $f_{\|}$, whereas they move in phase in the resonant mode.

The spectrum of a neighbouring atom of the defect is shown in Fig. 38. Contrary to the interstitial, it contains an appreciable intensity at normal lattice frequencies. In addition to the localized F_{1u} mode two other localized modes with A_{1g} and

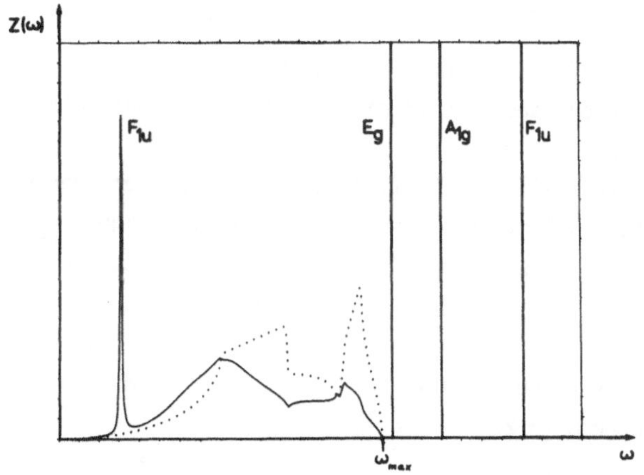

Fig. 38. Local frequency spectrum of a nearest neighbour of the octahedral interstitial

E_g symmetry occur, the displacement patterns of which are sketched in Fig. 36. These are vibrations where the defect itself remains at rest.

For an approximate description of the vibrational behaviour of the interstitial it is sufficient to have approximations for the localized and resonant modes. These can be obtained analytically without using the numerical results of the ideal lattice Green's function. The procedure has been described in Sect. 4.3 for resonant and Sect. 4.5 for localized vibrations.

For the special case $f_{\parallel} = 5f_{\parallel}^{o}$ and $f_{\perp} = -0.85f_{\parallel}^{o}$ we will consider here the Einstein approximation, where only the interstitial moves, and the approximation where also the six nearest neighbours can move. The former approximation gives

$$M^{eff} \cong M^{d} = \overset{o}{M} , \quad f^{eff} \cong 6.6f_{\parallel}^{o} \quad \text{and} \quad \omega_{E} \cong 0.91\omega_{max} \tag{6.23}$$

and yields neither a resonant nor a localized mode. If the motion of the six neighbours is taken into account we obtain the results presented in line 1 of the following table. The damping γ is calculated by (4.64), where the Debye frequency $\omega_{D} \cong 1.03\omega_{max}$ is numerically evaluated from the low-frequency ideal spectrum. Line 2 contains the exact values obtained by the Green's function method. Whereas the agreement is excellent for the localized mode, it is not so good for the resonant mode due to its weaker localization. However, the results can be improved if more atoms are allowed to move /4.12/.

Table 3

M^{eff}	f^{eff}	ω_{res}	γ	$\omega_{\ell oc}$
$1.609\overset{o}{M}$	$0.926f_{\parallel}^{o}$	$0.268\omega_{max}$	$0.134\omega_{res}$	$1.39\omega_{max}$
$1.865\overset{o}{M}$	$0.565f_{\parallel}^{o}$	$0.195\omega_{max}$	$0.057\omega_{res}$	$1.40\omega_{max}$

In order to see what kind of characteristic vibrations can occur for more general values of $f_{\parallel}$ and $f_{\perp}$, we have plotted in Fig. 39 in the $f_{\parallel} - f_{\perp}$ plane the lines $\omega_{res}^{F_{1u}} = 0$ (thick line) and $\omega_{\ell oc}^{F_{1u}} = \omega_{max}$ (thin lines). The line $\omega_{res} = 0$ gives the stability limit of the model: For $f_{\parallel}$, $f_{\perp}$ values below the line the configuration is unstable. A resonance mode occurs if $0 < \omega_{res} \leq \omega_{max}/3$ where the upper limit of $\omega_{max}/3$ (dashed line) has been chosen arbitrarily. For the stability limit ($f^{eff} = 0$ or $G_{xx}^{dd}(0) = \infty$) we obtain from (6.22)

$$2f_{\parallel} + 4f_{\perp} + 1.295 \frac{f_{\parallel} f_{\perp}}{f_{\parallel}^{o}} = 0 . \tag{6.24}$$

An Einstein approximation yields $2f_{\parallel} + 4f_{\perp} = 0$. If in addition also the nearest neighbours are allowed to move, we obtain

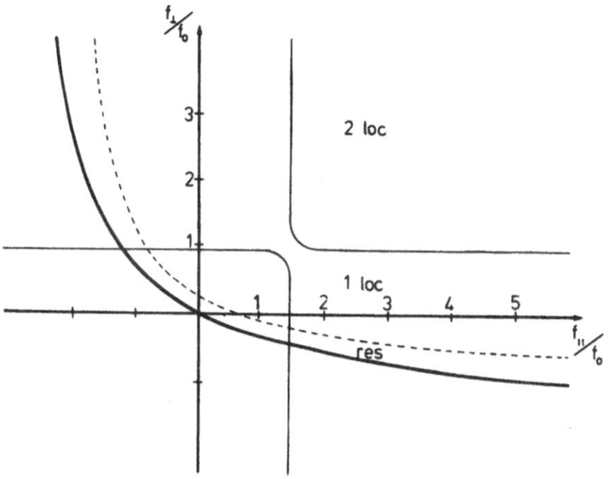

Fig. 39. Conditions for the occurrence of resonant and localized modes of the octahedral interstitial (F_{1u} modes only)

——— $\omega_{res} = 0$ (stability limit)

– – – $\omega_{res} = \frac{1}{3}\,\omega_{max}$

——— $\omega_{\ell oc} = \omega_{max}$

$$2f_{\parallel} + 4f_{\perp} + \frac{8}{7}\,\frac{f_{\parallel}\,f_{\perp}}{f_{\parallel}^{o}} = 0\ . \tag{6.25}$$

This approximation is in good agreement with the exact result (6.24), (thick line). Above the lines $\omega_{\ell oc} = \omega_{max}$ one or two localized modes can occur. The spectra shown in Fig. 37 correspond to the region where both a localized and a resonant mode occur (e.g., $f_{\parallel} = 5f_{\parallel}^{o}$, $f_{\perp} = -0.6f_{\parallel}^{o}$). We would like to emphasize that Fig. 39 refers only to the F_{1u} mode determining the interstitial spectrum. As has been seen already in Fig. 38, localized modes can also occur in other subspaces referring to vibrations where the defect is at rest. However, as has been shown in /4.12/, the highest localized mode as well as the lowest resonant mode, i.e., the one determining the stability, have F_{1u} symmetry.

6.4 100 - Dumbbell Interstitial

Recent X-ray scattering experiments of Ehrhart, Haubold and Schilling show the 100-split interstitial ("dumbbell", see Fig. 40), to be the stable interstitial configuration in Al and Cu. (For a review of these experiments see /6.16 - 18/). This configuration has been also confirmed by mechanical relaxation measurements in Al /6.19/, by magnetic relaxation measurements in Ni /6.20/ and by elastic constant measurements in Al and Cu (see Sect. 7.6). The same configuration was also obtained by numerous computer calculations /6.21,22/.

To get insight into the vibrational behaviour of this 100-dumbbell we consider a simple model with nearest neighbour interaction only. For symmetry reasons we describe the defect by a vacancy at position (0,0,0) and two interstitials at (0,0,±d). The coupling constants are given by a spiral spring $f_{\parallel}$ and a leaf spring $f_{\perp}$ between

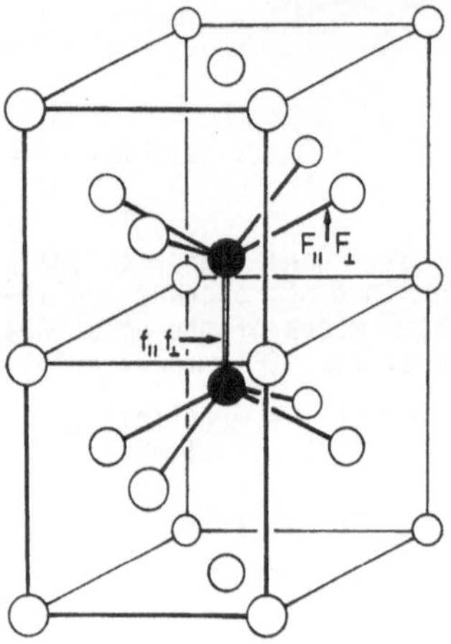

Fig. 40. 100 - dumbbell interstitial in fcc lattice

the dumbbell atoms and by spiral springs $F_{\|}$ and leaf springs $F_{\perp}$ between an interstitial atom and its four nearest neighbours (Fig. 40). To keep the model analytically tractable all other force constants are assumed to be the same as in the ideal lattice, i.e., longer ranged force constant changes are neglected. The vacancy is conveniently described by zero-coupling constants to its neighbours. Since 13 host atoms, i.e., the vacancy and 12 neighbours of the vacancy and the interstitials, are involved in the change of the force constants, we have to invert matrices of order 39×39. Using the tetragonal symmetry of the defect (point group D_{4h}) we can decompose the total representation into irreducible representations as /1.2 , 4.8/.

$$\Gamma_{100\text{-split}} = 3A_{1g} + 2A_{2g} + 2B_{1g} + 3B_{2g} + 4E_g + A_{1u} + 4A_{2u} + 2B_{1u} + 2B_{2u} + 6E_u \ . \qquad (6.26)$$

Thus the highest dimension of the matrices to be inverted is six in the E_u representation.

Whereas for the octahedral position, displacements of the interstitial are only allowed in one representation (F_{1u}), displacements of the dumbbell atoms affect four different representations (A_{1g} , E_g , A_{2u} , E_u). The dumbbell atoms move parallel to their axis in the A-type modes and perpendicular to the axis in the twofold degenerate modes of E type as shown in Fig. 41. Only the even modes (A_{1g} , E_g) stress

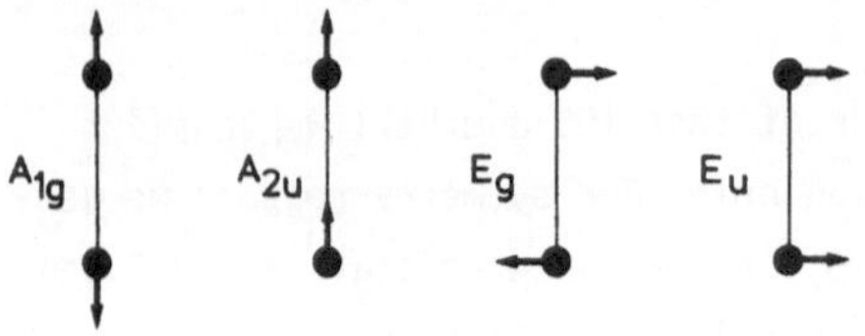

Fig. 41. Irreducible subspaces of the 100 - dumbbell. The E modes are double degenerate

the springs $f_{\parallel}$ or $f_{\perp}$ between the dumbbell atoms. As a consequence the A_{1g} mode leads to a localized mode, whereas the three other modes lead to a resonant and/or localized vibrations depending on whether the motion of the neighbouring atoms is in phase or out of phase with the dumbbell atoms. The four force constants $f_{\parallel}$, $f_{\perp}$, $F_{\parallel}$ and $F_{\perp}$ are fitted to the results of computer simulation by means of two sum rules for $z_x^d(\omega)$ and $z_z^d(\omega)$

$$\langle\omega^2\rangle_x^d = \int_0^\infty d\omega\, \omega^2\, z_x^d(\omega) = \phi_{xx}^{dd}/M \tag{6.27}$$

$$\langle 1/\omega^2\rangle_x^d = \int_0^\infty d\omega \frac{1}{\omega^2}\, z_x^d(\omega) = M\, G_{xx}^{dd}(0)\ . \tag{6.28}$$

The first term $M\langle\omega^2\rangle$ is the Einstein force constant ϕ_{xx}^{dd}, which can be calculated directly from the relaxed configuration and the known interaction potential. The second term $\frac{1}{M}\langle 1/\omega^2\rangle$ is the static response $G_{xx}^{dd}(\omega=0)$ of the defect, which can be calculated by computer simulation by exerting a static force on the defect and allowing all atoms to relax. Since $\langle\omega^2\rangle$ weights the high frequency part of the spectrum $z_x^d(\omega)$, ϕ_{xx}^{dd} virtually determines the localized mode frequencies. On the other hand $\langle 1/\omega^2\rangle$, given by $MG_{xx}^{dd}(0)$, determines the low frequency resonant modes.

Figure 42a,b shows the spectra of a dumbbell atom

$$z^d(\omega) = \frac{1}{3}\sum_{i=1}^{3} z_i^d(\omega)$$

calculated with the fitted force constants for a purely repulsive Born-Mayer potential and for the Morse potential (6.3). In both cases we obtain four different localized modes (A_{1g}, E_u, E_g, A_{2u}) and three different resonant modes (E_g, A_{2u}, E_u). The motions of the dumbbell atoms are indicated by the arrows.

Fig. 43a-c shows the amplitude distribution of the very high-frequency localized mode A_{1g} and the two low-frequency resonant modes of A_{2u} and E_g symmetry. The resonant mode A_{2u} (Fig. 43b) represents a translation of the whole dumbbell along its axis, whereas the mode E_g (Fig. 43c) is a libration of the dumbbell. For both modes the neighbouring atoms move in phase with the dumbbell contrary to the localized mode in Fig. 43a.

Since almost no contribution comes from the eigenfrequencies of the ideal lattice, the spectrum can be described by the resonant and localized modes alone. Therefore, analytical approximations for both kinds of modes, obtained by the technique discussed in Chap. 4, give a good description of the total spectrum. Due to the lower symmetry of the dumbbell (4.19) is no longer a scalar equation but of dimension 6×6 and has to be decomposed into its irreducible parts A_{1g}, A_{2u}, E_g and E_u. Since each representation Γ occurs only once, we obtain from (4.28,29)

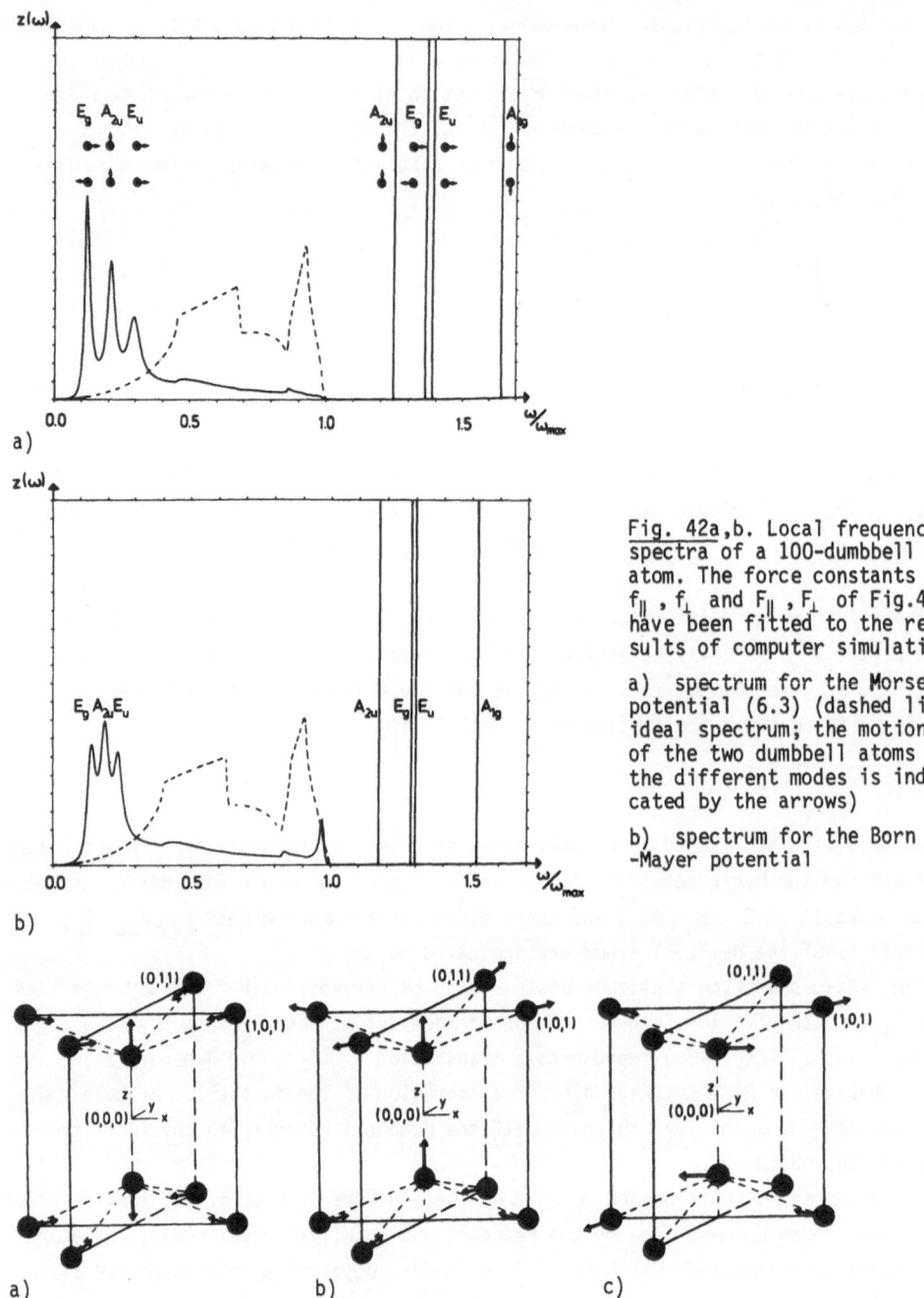

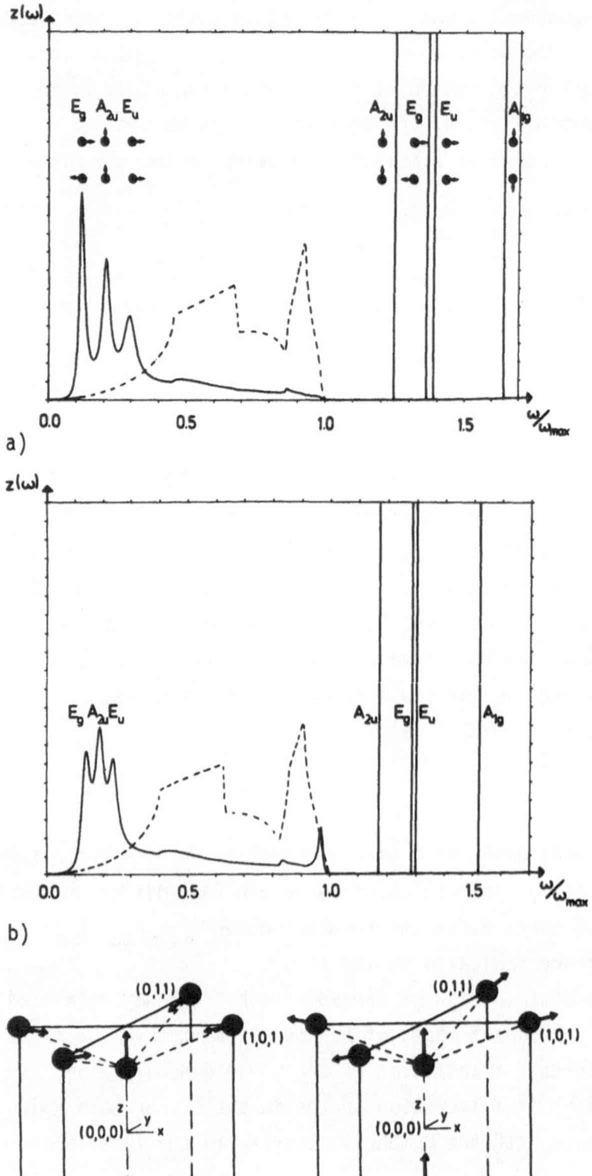

Fig. 42a,b. Local frequency spectra of a 100-dumbbell atom. The force constants $f_{\parallel}$, $f_{\perp}$ and $F_{\parallel}$, $F_{\perp}$ of Fig.40 have been fitted to the results of computer simulation

a) spectrum for the Morse potential (6.3) (dashed line: ideal spectrum; the motion of the two dumbbell atoms in the different modes is indicated by the arrows)

b) spectrum for the Born -Mayer potential

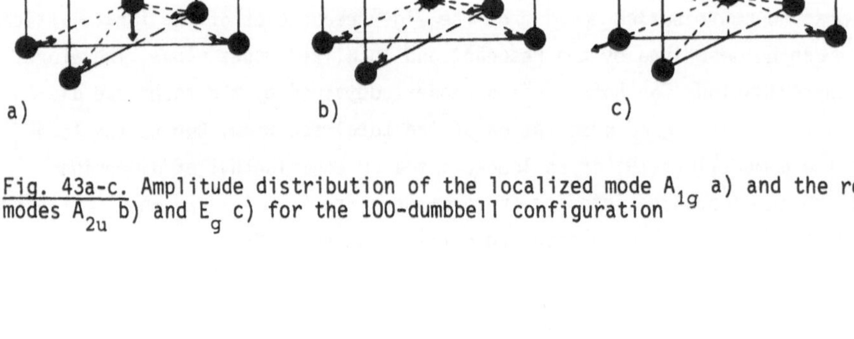

Fig. 43a-c. Amplitude distribution of the localized mode A_{1g} a) and the resonant modes A_{2u} b) and E_g c) for the 100-dumbbell configuration

$$G_{CC}(\omega) = \sum_{\Gamma,\mu} \frac{|\Gamma;\mu\rangle \langle\mu;\Gamma|}{f^{\Gamma}_{eff} - M^{\Gamma}_{eff}\omega^2 - iM^{\Gamma}_{eff}\gamma^{\Gamma}\omega} \tag{6.30}$$

with, e.g., $\gamma^{\Gamma} = \langle\Gamma;\mu|\gamma_{CC}|\Gamma;\mu\rangle$, where μ counts the degeneracy.

The vectors $|\Gamma;\mu\rangle$ represent the eigenvectors in the central subspace of the dumbbell atoms and are sketched in Fig. 41. The resonance frequency for the representation Γ is

$$(\omega^{\Gamma}_{res})^2 = \frac{f^{\Gamma}_{eff}}{M^{\Gamma}_{eff}} \qquad \Gamma = E_g\,, A_{2u}\,, E_u\ . \tag{6.31}$$

The damping γ^{Γ} becomes different for g and u modes

$$\gamma^{\Gamma} = \begin{cases} 2\,\dfrac{3\pi}{2\overset{o}{M}\omega_D^3}\,\dfrac{(f^{\Gamma}_{eff})^2}{M^{\Gamma}_{eff}} & \text{for u modes} \\ \text{const}\cdot\omega^2 & \text{for g modes}\ . \end{cases} \tag{6.32}$$

For u modes (6.32) is a generalization of formula (4.24). On the other hand, for g modes $\gamma^{\Gamma}(\omega = 0)$ vanishes for symmetry reasons, since for even modes ($\underline{u}^{\underline{m}} = -\underline{u}^{-\underline{m}}$) the total restoring force vanishes identically. This means that for g modes the spectrum $z^{\Gamma}(\omega)$ is proportional to ω^4 for $\omega \to 0$ and the resonance curve is more asymmetric than for u modes. This is not in conflict with the general statement of Sect. 4.4 that the local spectrum for an arbitrary atom agrees for $\omega \to 0$ with the ideal spectrum (besides a mass factor $M/\overset{o}{M}$).

Recently SCHOBER and ZELLER /6.23/ calculated the local frequency spectra of di- and tri-interstitials. According to computer simulation /6.23/ the stable di-interstitial consists of two parallel 100-dumbbells on nearest neighbour sites (Fig. 44a). This configuration has also been identified in Al by diffuse X-ray scattering /6.16 - 18/. The spectrum of the di-interstitial is shown in Fig. 44b for the Morse potential. As the one for the single interstitial, it consists of localized and resonant modes. However due to the lower symmetry, all degeneracies are removed leading to many more different localized and resonant modes. The latter partially overlap and appear as broader resonances in the plot of the spectrum. The lowest resonant modes (B_{2g}) is an in-phase libration (E_g) of both dumbbells with a very small resonance frequency.

The thermal displacements $\langle(\underline{s}^{m})^2\rangle$ are related to the local spectrum by (2.53,54), where the local spectrum is essentially weighted by $1/\omega^2$. Thus spring resonances lead to especially large thermal displacements already at low temperatures.

From (2.53) we obtain

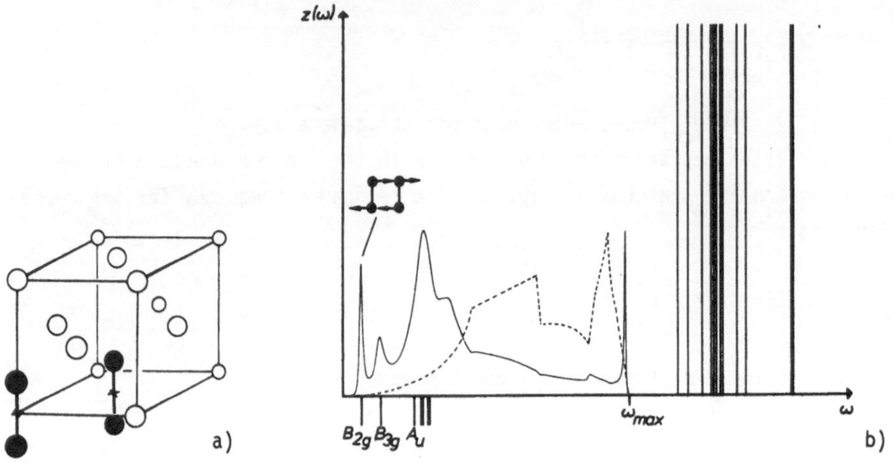

Fig. 44. a) Di-interstitial configuration in fcc lattice
b) Local frequency spectrum of a di-interstitial atom compared with the ideal spectrum (dashed) for the Morse potential (6.3)

$$\langle \underline{s}^2 \rangle \cong \begin{cases} 3\,\dfrac{kT}{M^{eff}\omega_{res}^2} = 3\,\dfrac{kT}{f^{eff}} & \text{for } kT \gtrsim \hbar\omega_{res} \\[2ex] 3\,\dfrac{\hbar}{2M}\left\langle \dfrac{1}{\omega} \right\rangle & \text{for } kT \ll \hbar\omega_{res} \; . \end{cases} \tag{6.33}$$

In Fig. 45 we have plotted $\langle \underline{s}^2 \rangle$ for an atom in the ideal lattice with force constant $f_{\parallel}^{o}$ and for a dumbbell atom with the spectrum of Fig. 42a.
For temperatures $T \geq 40$ K one sees that $\langle \underline{s}^2 \rangle$ for the dumbbell increases strongly and essentially linear with T, whereas the zero-point vibrations are only moderately enhanced. Contrary to $\langle \underline{s}^2 \rangle$ the square of the thermal velocities is, at least at

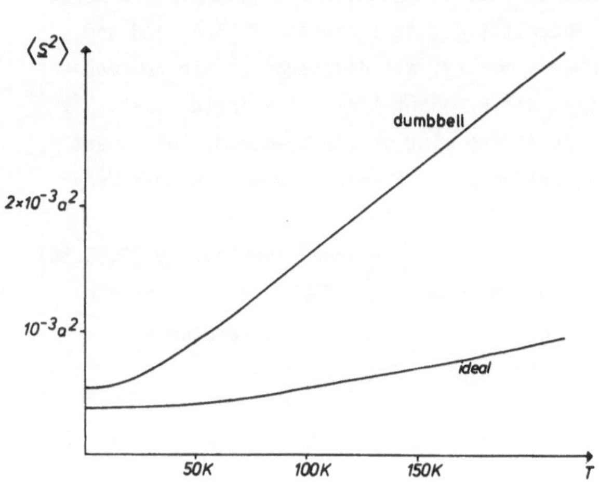

Fig. 45. Thermal displacement squared $\langle s^2 \rangle$ of a dumbbell atom and an atom in the ideal lattice

low temperatures, essentially determined by the localized modes since according to (2.55)

$$\langle \underline{v}^2 \rangle \cong \begin{cases} \dfrac{3\hbar}{2M}\langle\omega\rangle \cong \dfrac{3\hbar}{2M_{eff}^{loc}}\,\omega_{loc} & \text{for } kT \ll \hbar\omega_{loc} \\ 3\,\dfrac{kT}{M} & \text{for } kT \gtrsim \hbar\omega_{loc}\ . \end{cases} \tag{6.34}$$

Mössbauer measurements of VOGL et al. /6.24,25/ at Co^{57} impurities in Al represent the first direct evidence for the resonant and localized modes of interstitials. Normally the Co^{57} atoms are on substitutional sites and show a normal $\langle \underline{s}^2 \rangle$. The measured Debye-Waller factor for these substitutional Co^{57} atoms is shown in Fig. 46. However, after irradiation, migrating interstitials are trapped at the Co^{57} impurities and complicated complexes are formed in which the Co^{57} atoms are no longer

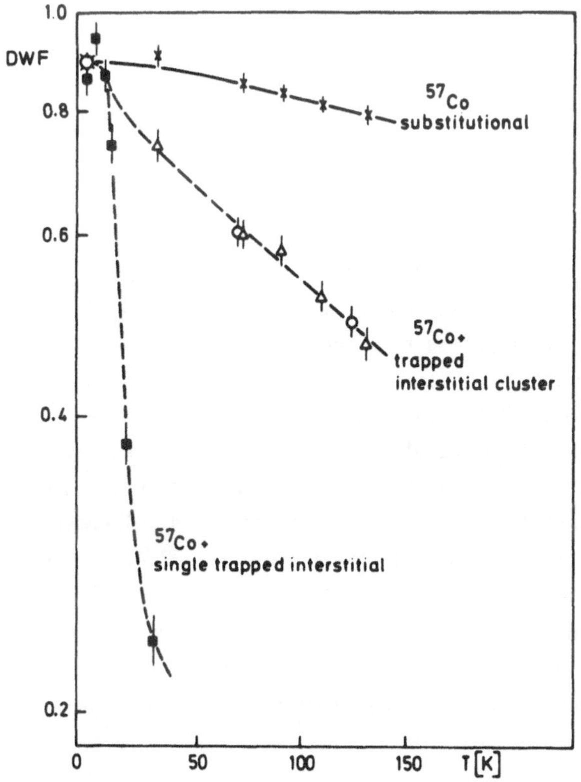

Fig. 46. Debye-Waller factors of Co^{57} (according to VOGL et al. /6.24,25/)
i) on substitutional site
ii) in an interstitial cluster
iii) with single trapped interstitial (reorientation effect)

on substitutional sites, as can be concluded from the appearance of a new and strongly shifted Mössbauer line. According to computer simulation and channeling measurements very probably mixed dumbbells (or similar configurations) are formed which consist of a Co^{57} atom and a host atom /6.22 , 26/. The new Mössbauer line shows an anomalously large Debye-Waller factor (see curve "Co^{57} + trapped interstitial cluster") i.e., large $\langle \underline{s}^2 \rangle$ similar to Fig. 45. On the other hand, the observed larger Doppler shift of the new line indicates the existence of localized modes. The experimental results are complicated due to a dose dependence of the measured Debye-Waller factors. After low-dose electron irradiation only isolated mixed dumbbells (curve "Co^{57} + single trapped interstitial") are formed. Such isolated mixed dumbbells can reorient at very low temperatures, which explains the drastic reduction of the Debye-Waller factor at about 17 K /6.22/. At higher doses, the reorientations are prevented due to the presence of additional trapped interstitials at one and the same Co^{57} atom. The strong reduction of the Debye-Waller factor at about 17 K disappears for higher concentrations (curve "Co^{57} + trapped interstitial cluster") and only the strong reduction due to thermal displacements remains /6.24/.

7. Effects on Phonon Dispersion Curves

After studying the single defect we will in the next two chapters discuss the effects of many randomly distributed defects on the macroscopic properties of the crystal. Many physical properties, like phonon dispersion curves or, e.g., the specific heat, are not determined by the vibrational properties of a single defect, but by the vibrational behaviour of the crystals as a whole, which always contains many defects. The vibrations of such a crystal can be described by a configurationally averaged Green's function, which itself is determined by an effective coupling matrix which is complex and frequency dependent (Sect. 7.1). The imaginary part of this Green's function determines the phonon dispersion curves of the defect crystal (Sect. 7.3) which are normally represented by Lorentzian-shaped ridges in the ω-k plane. Different scattering lengths of the host and the defect, as well as static displacements, lead to a more or less incoherent scattering which is treated in Sect. 7.4. The results of neutron measurements of the phonon dispersion curves and frequency spectra are reviewed in Sect. 7.5. Finally, we discuss the change of the elastic constants due to point defects and their relation to the phonon dispersion curves (Sect. 7.6). In the next chapter the effects of many defects on the thermal properties of the crystal will be discussed.

7.1 The Average Green's Function

If a crystal contains many defects, the coupling parameters and masses are changed in the vicinities of all defects. In a first approximation the changes φ_α of the coupling parameters of the individual defects can be superimposed. This should be allowed for a small concentration of defects as long as overlap effects can be neglected. Then the total perturbation $V(\omega)$ can be split up in contributions V_α due to the individual defects

$$V(\omega) = \varphi - \Delta M\omega^2 = \sum_\alpha V_\alpha = \sum_\alpha (\varphi_\alpha - \Delta M_\alpha \omega^2) \ . \tag{7.1}$$

The Green's function of the defect crystal is then determined by

$$G = \overset{\circ}{G} - \overset{\circ}{G} V G = \overset{\circ}{G} - \overset{\circ}{G} T \overset{\circ}{G} \tag{7.2}$$

where the total T matrix is given in terms of the total perturbation by

$$T = V \frac{1}{1 + \overset{\circ}{G} V} \ . \tag{7.3}$$

This formal solution for G is quite useless for practical purposes, since for the calculation of T one needs the positions of all defects (even for a small defect concentration of $c = N_d/N \sim 10^{-4}$ these are about 3×10^{19} parameters).

However, the macroscopic properties of such a defect crystal should not depend on the exact positions of all N_d defects, but only on the defect type and on the average defect concentration c. Thus one can average the Green's function G over all possible configurations of the defects which are macroscopically equivalent.

The result of the configuration average should be in agreement with a volume average, since any macroscopic crystal practically contains all configurations which are important for the average

$$\langle G^{\underline{m}\underline{n}}_{ij}(\omega)\rangle_{conf} \cong \langle G^{\underline{m}\underline{n}}_{ij}(\omega)\rangle_{vol} = \frac{1}{N} \sum_{\underline{n}'}^{N} G^{\underline{m}-\underline{n}',\underline{n}-\underline{n}'}_{i\ \ j}(\omega) \ . \tag{7.4}$$

By averaging (7.2), we see that $\langle G\rangle$ is determined by the average T matrix, since $\overset{\circ}{G}$ is independent of the defect distribution

$$\langle G\rangle = \overset{\circ}{G} - \overset{\circ}{G} \langle T\rangle \overset{\circ}{G} \ . \tag{7.5}$$

For the caculation of $\langle G\rangle$ it is convenient to introduce an effective change $\Sigma(\omega)$ of the coupling matrix of the ideal lattice

$$\langle G(\omega)\rangle = \frac{1}{\overset{\circ}{\phi} + \Sigma(\omega) - \overset{\circ}{M}\omega^2} = \overset{\circ}{G} - \overset{\circ}{G}\, \Sigma(\omega) \langle G(\omega)\rangle \ . \tag{7.6}$$

Because of the averaging, $\langle G(\omega)\rangle$ and consequently also $\Sigma(\omega)$ have the periodicity and all the symmetries of the "averaged" lattice, i.e., the same symmetries as $\overset{\circ}{\phi}$, if we neglect effects like lattice expansion or superstructure due to the defects. Thus the eigenfunctions of $\overset{\circ}{\phi}+\Sigma$ are plane waves and the problem for the defect crystal has been reduced to the one for an ideal crystal. The remaining problem is only the determination of the "self-energy" $\Sigma(\omega)$.

By comparing (7.5) and (7.6), we see that

$$\langle T\rangle \overset{\circ}{G} = \Sigma \langle G\rangle = \Sigma \left(\overset{\circ}{G} - \overset{\circ}{G} \langle T\rangle \overset{\circ}{G}\right) . \tag{7.7}$$

By solving with respect to Σ, we obtain the result

$$\Sigma(\omega) = \langle T(\omega)\rangle \frac{1}{1 - \overset{\circ}{G}(\omega) \langle T(\omega)\rangle} \quad \text{with } \langle T(\omega)\rangle = \left\langle V \frac{1}{1 + \overset{\circ}{G} V} \right\rangle . \tag{7.8}$$

Similar to $\overset{\circ}{\phi}$, $\Sigma(\omega)$ has the periodicity of the lattice (due to the averaging) and is a symmetrical matrix (since V, $\overset{\circ}{G}$ and T are symmetrical)

$$\Sigma_{ij}^{\underline{m}\underline{n}} = \Sigma_{ij}^{(\underline{m}-\underline{n})} = \Sigma_{ji}^{\underline{n}\underline{m}} . \tag{7.9}$$

However due to its dependence on $\overset{\circ}{G}(\omega)$, it is frequency dependent and complex. The latter is a very characteristic property of the self-energy and describes a damping. Actually there is, of course, no real damping in the lattice, since the total coupling $\phi = \overset{\circ}{\phi} + \varphi$ is real. The apparent damping of the average Green's function is merely due to the production of diffuse, incoherent waves in the lattice being induced by the disorder due to the point defects. As a consequence the intensity of the "average waves" is diminished so that the diffuse waves act as an absorption for $\langle G\rangle$. Thus $\mathrm{Im}\{\Sigma(\omega)\}$ should be negative definite, which follows from (6.6)

$$\mathrm{Im}\{\Sigma(\omega)\} = \mathrm{Im}\left\{\frac{1}{\langle G(\omega)\rangle}\right\} = \frac{1}{2i}\left(\frac{1}{\langle G\rangle} - \frac{1}{\langle G\rangle^*}\right) = -\frac{1}{2i}\frac{1}{\langle G\rangle^*}\left(\langle G\rangle - \langle G\rangle^*\right)\frac{1}{\langle G\rangle} \tag{7.10}$$

since $\mathrm{Im}\{\langle G(\omega)\rangle\}$ is positive definite.

Because of the periodicity, a plane wave representation of $\langle G(\omega)\rangle$ is very useful. By setting

$$\langle G_{ij}^{\underline{m}\underline{n}}(\omega)\rangle = \frac{1}{V_B} \int_{V_B} d\underline{k} \, \langle G_{ij}\rangle (\underline{k},\omega) \, e^{i\underline{k}(\underline{R}^{\underline{m}}-\underline{R}^{\underline{n}})} \tag{7.11}$$

we obtain from (7.6)

$$\sum_{\ell} \left(\overset{\circ}{\phi}_{i\ell}(\underline{k}) + \Sigma_{i\ell}(\underline{k},\omega) - \overset{\circ}{M}\omega^2 \, \delta_{i\ell}\right) \langle G_{\ell j}\rangle (\underline{k},\omega) = \delta_{ij} \tag{7.12}$$

with, e.g., $\Sigma_{i\ell}(\underline{k},\omega) = \sum_{\underline{m}} \Sigma_{i\ell}^{(\underline{m}-\underline{n})}(\omega) \, e^{-i\underline{k}(\underline{R}^{\underline{n}}-\underline{R}^{\underline{m}})}$.

As usual one can define three polarization vectors $\underline{e}^{\sigma}(\underline{k},\omega)$ which for given $\underline{k}$ and ω are eigenvectors of $\overset{\circ}{\phi}+\Sigma$ with eigenvalues $\Lambda_{\sigma}(\underline{k},\omega)$

$$\sum_{\ell}\left(\overset{\circ}{\phi}_{i\ell}(\underline{k}) + \Sigma_{i\ell}(\underline{k},\omega)\right) e_{\ell}^{\sigma}(\underline{k},\omega) = \Lambda_{\sigma}(\underline{k},\omega)\; e_{i}^{\sigma}(\underline{k},\omega) \; . \tag{7.13}$$

Since $\overset{\circ}{\phi}$ and Σ are symmetrical, the Fourier transforms $\overset{\circ}{\phi}_{i\ell}(\underline{k})$ and $\Sigma_{i\ell}(\underline{k},\omega)$ are also symmetrical in i and ℓ, if we have inversion symmetry. Therefore the polarization vectors $\underline{e}^{\sigma}(\underline{k},\omega)$ form for each $\underline{k}$ and ω an orthonormal system. Thus

$$\langle G_{ij}^{\underline{m}\underline{n}}(\omega)\rangle = \frac{1}{V_B}\int_{V_B} d\underline{k} \sum_{\sigma} e_{i}^{\sigma}(\underline{k},\omega)\; e_{i}^{\sigma}(\underline{k},\omega)\; \langle G\rangle_{\sigma}(\underline{k},\omega)\; \exp(i\underline{k}\underline{R}^{\underline{m}-\underline{n}}) \tag{7.14}$$

with

$$\langle G\rangle_{\sigma}(\underline{k},\omega) = \left(\Lambda_{\sigma}(\underline{k},\omega) - \overset{\circ}{M}\omega^2\right)^{-1} \; . \tag{7.15}$$

For the high symmetry direction (100) , (110) and (111) the polarization vectors are determined by symmetry alone. They are independent of $|\underline{k}|$ and ω and the same as in the ideal crystal. Thus we have

$$\Lambda_{\sigma}(\underline{k},\omega) = \left(\overset{\circ}{\underline{e}}{}^{\sigma}, \left(\overset{\circ}{\phi}(\underline{k}) + \Sigma(\underline{k},\omega)\right)\overset{\circ}{\underline{e}}{}^{\sigma}\right) = \overset{\circ}{M}\,\overset{\circ}{\omega}{}_{\sigma}^{2}(\underline{k}) + \Sigma_{\sigma}(\underline{k},\omega) \tag{7.16}$$

and

$$\langle G\rangle_{\sigma}(\underline{k},\omega) = \left(\overset{\circ}{M}\,\overset{\circ}{\omega}{}_{\sigma}^{2}(\underline{k}) + \Sigma_{\sigma}(\underline{k},\omega) - \overset{\circ}{M}\omega^2\right)^{-1} \; . \tag{7.17}$$

Whereas for the calculations of displacement correlations and neutron dispersion curves in defect crystals one needs the average Green's function $\langle G_{ij}^{\underline{m}\underline{n}}(\omega)\rangle$ (see Sect. 7.3), for the frequency spectrum and related thermodynamic properties (see Chap. 8) one needs the average of mass-transformed Green's function

$$\langle \mathcal{G}_{ij}^{\underline{m}\underline{n}}(\omega)\rangle = \langle\sqrt{M^{\underline{m}}}\, G_{ij}^{\underline{m}\underline{n}}(\omega)\sqrt{M^{\underline{n}}}\rangle = \left\langle\left\{\frac{1}{D - (\omega + i\eta)^2}\right\}_{ij}^{\underline{m}\underline{n}}\right\rangle \tag{7.18}$$

where $D = \frac{1}{\sqrt{M}}\,\phi\,\frac{1}{\sqrt{M}}$ is the dynamical matrix. The difference between the two averages is only important for a system with mass disorder, since otherwise $\langle\mathcal{G}_{ij}^{\underline{m}\underline{n}}\rangle = \sqrt{M^{\underline{m}}}\cdot\sqrt{M^{\underline{n}}}\,\langle G_{ij}^{\underline{m}\underline{n}}\rangle$. However for mass disorder $\langle\mathcal{G}\rangle$ and $\langle G\rangle$ are not directly related to each other, but have to be calculated independently. The general theory is essentially the same for both quantities, so that in order to obtain the equations for $\langle\mathcal{G}\rangle$ one may formally replace the coupling constant ϕ in the equation for $\langle G\rangle$ by the dynamical D and all masses M by unit masses M = 1.

In general it is practically impossible to calculate the self-energy $\Sigma(\omega)$ exactly. An approximation for a very small defect concentration is given below. However, there is a simple and exact sum rule which directly relates $\langle\mathcal{G}(\omega)\rangle$ (or $\langle G\rangle$) to averages of the dynamical matrix D. Namely from (2.49) we obtain

$$\int d\omega^2\, \omega^{2n} \frac{1}{\pi}\, \mathrm{Im} \langle \mathcal{G}(\omega) \rangle = \langle D^n \rangle \qquad n = -1,0,1,2,\ldots \quad . \tag{7.19}$$

These averages can either be read in real space or in Fourier space. For instance, for n = 1 we have either

$$\int d\omega^2\, \omega^2 \frac{1}{\pi}\, \mathrm{Im} \langle \mathcal{G}^{\underline{m}\underline{n}}_{ij}(\omega) \rangle = \langle D^{\underline{m}\underline{n}}_{ij} \rangle = \langle \frac{1}{\sqrt{M^{\underline{m}}}}\, \phi^{\underline{m}\underline{n}}_{ij} \frac{1}{\sqrt{M^{\underline{n}}}} \rangle \tag{7.20}$$

or in a reciprocal space with $\langle \mathcal{G} \rangle_\sigma(\underline{k},\omega)$ defined in analogy to (7.14)

$$\int d\omega^2\, \omega^2 \frac{1}{\pi}\, \mathrm{Im} \langle \mathcal{G} \rangle_\sigma(\underline{k},\omega) = \langle D_\sigma(\underline{k}) \rangle = \sum_{\underline{m}\, ij} \exp(-i\underline{k}\underline{R}^{\underline{m}-\underline{n}})\, \overset{\circ}{e}{}^\sigma_i \langle D^{\underline{m}\underline{n}}_{ij} \rangle\, \overset{\circ}{e}{}^\sigma_j \tag{7.21}$$

where we assumed the polarization vectors to be independent of ω.

In order to obtain approximations for $\langle T \rangle$ and Σ in the dilute limit of small concentrations we expand the total T matrix into single t matrices for the individual point defects.

$$t_\alpha = V_\alpha - V_\alpha \overset{\circ}{G} t_\alpha = V_\alpha \frac{1}{1 + \overset{\circ}{G} V_\alpha} = \frac{1}{1 + V_\alpha \overset{\circ}{G}} V_\alpha \ . \tag{7.22}$$

By introducing the sum $V = \sum_\alpha V_\alpha$ into the equation for T

$$T = V - V\overset{\circ}{G} T = \sum_\alpha \left(V_\alpha - V_\alpha \overset{\circ}{G} T \right) , \tag{7.23}$$

we try a similar ansatz for $T = \sum_\alpha T_\alpha$ and obtain

$$\sum_\alpha T_\alpha = \sum_\alpha \left(V_\alpha - V_\alpha \overset{\circ}{G} T_\alpha - V_\alpha \overset{\circ}{G} \sum_{\beta (\neq \alpha)} T_\beta \right) . \tag{7.24}$$

Thus T describes all scattering processes where the last interaction is with defect α. By discarding the sum over α and solving with respect to T, we have an equation for T_α, which can be solved by iteration

$$T_\alpha = t_\alpha - t_\alpha \overset{\circ}{G} \sum_{\beta (\neq \alpha)} T_\beta \cong t_\alpha - t_\alpha \overset{\circ}{G} \sum_{\beta (\neq \alpha)} t_\beta + t_\alpha \overset{\circ}{G} \sum_{\beta (\neq \alpha)} t_\beta \overset{\circ}{G} \sum_{\gamma (\neq \beta)} t_\gamma - + \ldots \quad . \tag{7.25}$$

The final result for T in terms of single scattering matrices is therefore

$$T = \sum_\alpha t_\alpha - \sum_{\substack{\alpha,\beta \\ \alpha \neq \beta}} t_\alpha \overset{\circ}{G} t_\beta + \sum_{\substack{\alpha\beta\gamma \\ \alpha\neq\beta,\, \beta\neq\gamma}} t_\alpha \overset{\circ}{G} t_\beta \overset{\circ}{G} t_\gamma - + \ldots \quad . \tag{7.26}$$

Obviously the summation over two successive and equal indices ($\alpha = \beta$, $\beta = \gamma$, ...) is forbidden since the repeated scattering at the same center has already been summed up by the single t matrices.

By averaging (7.26) over the positions of the defects, we obtain for the first term

$$<t_\alpha> = \frac{1}{N}\sum_{\mu=1}^{N} t^{(\mu)} \;, \qquad \sum_{\alpha=1}^{N_d} <t_\alpha> = c\sum_{\mu=1}^{N} t^{(\mu)} \quad \text{with } c = \frac{N_d}{N} \tag{7.27}$$

where c is the atomic concentration of point defects (total number N_d) and where the sum over μ goes over all N lattice positions which can be occupied by a defect. For the averages of two different defects α and β we obtain a factor c^2 if the defects are randomly distributed. For instance:

$$<t_\alpha \overset{o}{G} t_\beta> = c^2 \sum_{\substack{\mu,\nu \\ \mu\neq\nu}} t^{(\mu)}\overset{o}{G}\, t^{(\nu)} \;. \tag{7.28}$$

However the average of two (or more) t matrices for one and the same defect gives only a factor c.

$$<t_\alpha \,\ldots\, t_\alpha> = c\sum_{\mu} t^{(\mu)} \,\ldots\, t^{(\mu)} \;. \tag{7.29}$$

With these rules we obtain from (7.26)

$$\begin{aligned} <T> = {} & c\sum_{\mu} t^{(\mu)} - c^2 \sum_{\mu,\nu} t^{(\mu)}\overset{o}{G}\, t^{(\nu)} + c^3 \sum_{\mu,\nu,\pi} t^{(\mu)}\overset{o}{G}\, t^{(\nu)}\overset{o}{G}\, t^{(\pi)} - + \ldots \\ & + c^2\sum_{\mu} t^{(\mu)}\overset{o}{G}t^{(\mu)} + c^2(1-c)\sum_{\mu,\nu} t^{(\mu)}\overset{o}{G}t^{(\nu)}\overset{o}{G}t^{(\mu)} - c(1-2c)\sum_{\mu} t^{(\mu)}\overset{o}{G}t^{(\mu)}\overset{o}{G}t^{(\mu)} \\ & - c^3\sum_{\mu,\pi} t^{(\mu)}\overset{o}{G}t^{(\mu)}\overset{o}{G}t^{(\pi)} - c^3\sum_{\mu,\nu} t^{(\mu)}\overset{o}{G}t^{(\nu)}\overset{o}{G}t^{(\nu)} + \ldots \;. \end{aligned} \tag{7.30}$$

By inserting this expression into (7.8) for Σ, we obtain

$$\Sigma = c\sum_{\mu} t^{(\mu)} + c^2\left(\sum_{\mu} t^{(\mu)}\overset{o}{G}\, t^{(\mu)} + \sum_{\substack{\mu,\nu \\ \mu\neq\nu}} t^{(\mu)}\overset{o}{G}\, t^{(\nu)}\overset{o}{G}\, t^{(\mu)} + \ldots\right) + c^3 \ldots \tag{7.31}$$

Thus linearly in c, the self-energy is given by the t matrix of a single defect, averaged over all possible positions μ and multiplied by the total number of defects. This result represents the basis for the following sections

$$\Sigma \cong c\sum_{\mu} t^{(\mu)} = N_d <t> \quad \text{with } <t> = \frac{1}{N}\sum_{\mu=1}^{N} t^{(\mu)} \;. \tag{7.32}$$

By comparing with the dilute limit for $<T>$, we note that this approximation for Σ is a much better approximation than the dilute limit for $<T> \cong c\sum_{\mu} t^{(\mu)}$. Indeed by inserting (7.32) into (7.8), we see that the approximation for Σ corresponds to summing up an infinite series for $<T>$, namely the whole first line in (7.30). This behaviour is connected with the range of the matrices Σ and $<T>$ or $<G>$. $<T>$ and $<G>$

are very long ranged, in so far as they contain terms where two very distant lattice points μ and ν are only connected by one Green's function (e.g., $t^{(\mu)}\overset{\circ}{G}t^{(\nu)}$). Contrary Σ is of shorter range: Distant points are at least connected by two Green's functions (e.g., $t^{(\mu)}\overset{\circ}{G}t^{(\nu)}Gt^{(\mu)}$). Thus due to the longer range, $\langle T\rangle$ and $\langle G\rangle$ are more sensitive to the perturbation than Σ since more positions are sampled and more terms have to be summed up to obtain approximations of the same quality.

For the Fourier component $\Sigma_\sigma(\underline{k},\omega)$ of the self-energy we have in the dilute limit

$$\begin{aligned}\Sigma_\sigma(\underline{k},\omega) &= c\sum_{\underline{m}ij} \exp[-i\underline{k}(\underline{R}^{\underline{m}}-\underline{R}^{\underline{n}})] \sum_{\mu} e_i^\sigma\, t_{i\,j}^{\underline{m}-\underline{\mu},\underline{n}-\underline{\mu}}\, e_j^\sigma \\ &= c\sum_{\substack{\underline{m}\,\underline{n}\\ i\,j}} \exp(-i\underline{k}\underline{R}^{\underline{m}})\, e_i^\sigma\, t_{ij}^{\underline{m}\underline{n}}\, e_j^\sigma \exp(i\underline{k}\underline{R}^{\underline{n}}) = ct_\sigma(\underline{k},\omega)\ ,\end{aligned} \tag{7.33}$$

in the latter formula the t matrix refers to a defect at the origin.

We can also look at the derivation of this result from a different side by directly concentrating on the Fourier transforms $\overset{\circ}{G}(\underline{k},\omega)$ and $\langle G\rangle(\underline{k},\omega)$ of the Green's function $\overset{\circ}{G}$ and $\langle G\rangle$. We will assume $\underline{k}$ to lie in a high symmetry direction so that the polarization vectors $\underline{e}^\sigma(k)$ are for symmetry reasons the same both for $\overset{\circ}{G}$ and $\langle G\rangle$. Firstly we see that $\overset{\circ}{G}_\sigma(\underline{k},\omega)$ has poles at the unperturbed phonons $\omega_\sigma(\underline{k})$

$$\overset{\circ}{G}_\sigma(\underline{k},\omega) = \left(\overset{\circ}{M}\,\overset{\circ}{\omega}{}_\sigma^2(\underline{k}) - \overset{\circ}{M}(\omega+i\eta)^2\right)^{-1} \tag{7.34}$$

whereas for the average Green's function $\langle G\rangle_\sigma(\underline{k},\omega)$ these poles are shifted and "broadened" since $\Sigma_\sigma(\underline{k},\omega)$ has a real and imaginary part

$$\langle G\rangle_\sigma(\underline{k},\omega) = \left(\overset{\circ}{M}\overset{\circ}{\omega}{}_\sigma^2(\underline{k}) + \Sigma_\sigma(\underline{k},\omega) - \overset{\circ}{M}(\omega+i\eta)^2\right)^{-1}\ . \tag{7.35}$$

By looking at the $\langle T\rangle$ matrix expansion

$$\langle G\rangle = \overset{\circ}{G} - \overset{\circ}{G}\langle T\rangle\overset{\circ}{G}$$

with $\langle T\rangle$ given by (7.30), we have to concentrate on such terms, which can lead to a renormalization of the poles of $\overset{\circ}{G}_\sigma(\underline{k},\omega)$. By Fourier transform we get

$$\langle G\rangle_\sigma(\underline{k},\omega) = \overset{\circ}{G}_\sigma(\underline{k},\omega) - \overset{\circ}{G}_\sigma(\underline{k},\omega)\langle T\rangle_\sigma(\underline{k},\omega)\overset{\circ}{G}_\sigma(\underline{k},\omega) \tag{7.36}$$

with

$$\begin{aligned}\langle T\rangle_\sigma(\underline{k},\omega) &= c\,t_\sigma(\underline{k},\omega) - c^2\,t_\sigma(\underline{k})\,\overset{\circ}{G}_\sigma(\underline{k},\omega)\,t_\sigma(\underline{k}) + c^3 t_\sigma\overset{\circ}{G}_\sigma t_\sigma\overset{\circ}{G}\,t_\sigma - + \dots \\ &\quad + c^2\,t_\sigma^{(2)}(\underline{k},\omega) + c^3 \dots\ .\end{aligned} \tag{7.37}$$

Here we have made extensive use of the faltungs theorem: The Fourier tranform of a faltung is the product of the individual Fourier transforms. For instance, the term $\overset{\circ}{G}$ $\langle T\rangle$ $\overset{\circ}{G}$

$$\sum_{\underline{m}'\underline{n}'} \overset{\circ}{G}^{(\underline{m}-\underline{m}')} \langle T\rangle^{(\underline{m}'-\underline{n}')} \overset{\circ}{G}^{(\underline{n}'-\underline{n})} \tag{7.38}$$

is a double faltung, since all three functions depend only on the differences $\underline{m}-\underline{m}'$, $\underline{m}'-\underline{n}'$ and $\underline{n}'-\underline{n}$. Thus its transform is the product $\overset{\circ}{G}(\underline{k})\ \langle T\rangle(\underline{k})\ \overset{\circ}{G}(\underline{k})$. Similar arguments apply to the expression for $\langle T\rangle$. This can be seen more clearly by writing the averages of the single t matrices in the form of (7.27)

$$N_d\langle t\rangle = c \sum_{\mu=1}^{N} t^{(\mu)}$$

where $\langle t\rangle$ is translationally symmetrical, i.e., $\langle t^{\underline{m}\underline{n}}\rangle = \langle t\rangle^{\underline{m}-\underline{n}}$. In this way we obtain from (7.30) for $\langle T\rangle$

$$\begin{aligned}\langle T\rangle = {} & N_d\langle t\rangle - N_d^2\langle t\rangle \overset{\circ}{G}\langle t\rangle + N_d^3\langle t\rangle \overset{\circ}{G}\langle t\rangle \overset{\circ}{G}\langle t\rangle - + \ldots \\ & + N_d c \langle t\overset{\circ}{G}t\rangle + N_d^2(1-c)\ \langle t\ \overset{\circ}{G}\langle t\rangle\ \overset{\circ}{G}t\rangle - N_d(1-2c)\ \langle t\overset{\circ}{G}t\overset{\circ}{G}t\rangle \\ & - N_d^2 c\ \langle t\overset{\circ}{G}t\rangle\ \overset{\circ}{G}\langle t\rangle - N_d^2 c\ \langle t\rangle\ \overset{\circ}{G}\ \langle t\overset{\circ}{G}t\rangle \ \ldots \ .\end{aligned} \tag{7.39}$$

Since all averages are translationally invariant, the faltungs theorem can be applied repeatedly and leads directly to (7.37).

For frequencies ω near $\overset{\circ}{\omega}_\sigma(\underline{k})$, we see that this series (7.36,37) for $\langle G\rangle_\sigma(\underline{k},\omega)$ diverges terribly, since we get terms of the form $(\overset{\circ}{\omega}{}^2-\omega^2)^{-n}$ which are weighted by different factors of the concentration. By summing up the "most divergent" parts in $\langle T\rangle_\sigma(\underline{k},\omega)$

$$c^n\left(\frac{1}{\overset{\circ}{\omega}{}_\sigma^2(\underline{k}) - \omega^2}\right)^{n-1} , \tag{7.40}$$

i.e., all terms $\sim (\overset{\circ}{\omega}{}^2-\omega^2)^n$ weighted by the highest power of the concentration, we obtain the low-concentration result

$$\langle G\rangle_\sigma(k,\omega) = \frac{1}{\overset{\circ}{M}\overset{\circ}{\omega}{}_\sigma^2(k) + ct_\sigma(\underline{k},\omega) - \overset{\circ}{M}\omega^2} . \tag{7.41}$$

For a further study of this subject, especially also for approximations for non-dilute systems, like the average T matrix approximation (ATA) or the coherent potential approximation (CPA), we refer to the recent review article by ELLIOTT et al. /7.1/. The pioneering paper by ELLIOT and TAYLOR /7.2/ should also be mentioned in this context.

7.2 Theory of Thermal Neutron Scattering /1.3, 1.4, 7.3 - 5/

The double differential cross section for the scattering of neutrons into a solid angle Ω with an energy loss $\hbar\omega$ is in the first Born approximation given by

$$\frac{d^2\sigma}{d\Omega\, d\omega} = \frac{k_f}{k_i} \sum_{i,f} p_i(T) \left| <f \left| \sum_{\underline{n}} a_{\underline{n}} \exp(i\underline{K}\underline{r}^{\underline{n}}) \right| i> \right|^2 \delta\left(\omega - \frac{E_i - E_f}{\hbar}\right). \tag{7.42}$$

Here $\underline{k}_i$ and $\underline{k}_f = \underline{k}_i - \underline{K}$ are the initial and final wave vectors of the neutron, respectively, and

$$\hbar\omega = E_i - E_f = \frac{\hbar^2}{2m} (\underline{k}_i^2 - \underline{k}_f^2)$$

is the energy transferred by the neutron to the crystal. $|i>$ and $|f>$ denote the initial and final state of the crystal, $p_i(T)$ the statistical weight of state i, and $a_{\underline{m}}$ is the scattering length of nucleus $\underline{m}$. As has been noted by VAN HOVE /5.6/, the differential cross section can also be written as a Fourier integral over time--dependent correlation functions. The mathematical procedure is the same as for the Mössbauer probability $w_e(\underline{k};\omega)$. Quite analogous to (5.36) one obtains the result

$$\frac{d^2\sigma}{d\Omega\, d\omega} = \frac{k_f}{k_i} \sum_{\underline{mn}} \exp[-i\underline{K}(\underline{R}^{\underline{m}} - \underline{R}^{\underline{n}})]\, a_{\underline{m}} a_{\underline{n}}\, S^{\underline{mn}}(\underline{K},\omega) \tag{7.43}$$

with

$$S^{\underline{mn}}(\underline{K},\omega) = \int \frac{d\omega}{2\pi} \exp(i\omega t) <\exp[-i\underline{K}\underline{s}^{\underline{m}}(t)] \exp[i\underline{K}\underline{s}^{\underline{n}}(0)]>_T .$$

Here we have set $\underline{r}^{\underline{m}}(t) = \underline{R}^{\underline{m}} + \underline{s}^{\underline{m}}(t)$, $\underline{r}^{\underline{n}} = \underline{R}^{\underline{n}} + \underline{s}^{\underline{n}}$. Within the harmonic theory, the correlation functions $S^{\underline{mn}}$ can be further simplified due to

$$\begin{aligned} <\exp[-i\underline{K}\underline{s}^{\underline{m}}(t)] \exp[i\underline{K}\underline{s}^{\underline{n}}(0)]>_T &= \exp[-\tfrac{1}{2} <(\underline{K}\underline{s}^{\underline{m}})^2>_T] \exp[-\tfrac{1}{2} <(\underline{K}\underline{s}^{\underline{n}})^2>_T] \\ &\quad \cdot \exp[<\left(\underline{K}\underline{s}^{\underline{m}}(t)\right)\left(\underline{K}\underline{s}^{\underline{n}}(0)\right)>_T] . \end{aligned} \tag{7.44}$$

The displacement correlations in the exponent have been calculated in Chap. 2 in terms of $\mathrm{Im}\{G(\omega)\}$. For instance, see (2.39):

$$<s_i^{\underline{m}}(t)\, s_j^{\underline{n}}(0)>_T = \int_{-\infty}^{+\infty} \frac{d\omega}{2\pi} \exp(-i\omega t) \frac{2\hbar}{1 - \exp(-\hbar\omega/kT)} \mathrm{Im}\{G_{ij}^{\underline{mn}}(\omega)\} . \tag{7.45}$$

Thus the differential cross section is in the harmonic approximation uniquely determined by the Green's function $G_{ij}^{\underline{mn}}(\omega)$.

By expanding (7.44) in terms of the time-dependent correlations $<\underline{K}\underline{s}^{\underline{m}}(t)\ \underline{K}\underline{s}^{\underline{n}}(0)>$ the first term describes the elastic or zero-phonon scattering

$$\left.\frac{d^2\sigma}{d\Omega\, d\omega}\right|_{\text{elastic}} = \left| \sum_{\underline{m}} a_{\underline{m}} \exp(-M_{\underline{K}}^{\underline{m}}) \exp(i\underline{K}\underline{R}^{\underline{m}}) \right|^2 \delta(\omega) \tag{7.46}$$

where $M_{\underline{K}}^{\underline{m}}$ is the Debye-Waller factor of atom $\underline{m}$, (5.41). All other terms are connected with one or more phonon excitations and represent inelastic scattering processes. Indeed, the elastic scattering $\left(\sim \delta(\omega) \text{ in } (7.43)\right)$ is determined by the correlations for $t \to \infty$, for which $\langle \underline{K}\underline{s}^{\underline{m}}(t)\ \underline{K}\underline{s}^{\underline{n}}(0)\rangle \to 0$. The next term, the one-phonon scattering, can be directly represented by $\mathrm{Im}\{G(\omega)\}$.

$$\begin{aligned}\left.\frac{d^2\sigma}{d\Omega\, d\omega}\right|_{1\text{-phonon}} &= \frac{k_f}{k_i} \sum_{\underline{mn}} \exp[-i\underline{K}(\underline{R}^{\underline{m}}-\underline{R}^{\underline{n}})] a_{\underline{m}} a_{\underline{n}} \exp(-M_{\underline{K}}^{\underline{m}} - M_{\underline{K}}^{\underline{n}}) \\ &\qquad \cdot \int_{-\infty}^{+\infty} \frac{d\omega}{2\pi} \exp(i\omega t) \langle \underline{K}\underline{s}^{\underline{m}}(t)\ \underline{K}\underline{s}^{\underline{n}}(0)\rangle \\ &= \frac{k_f}{k_i} \frac{\hbar}{\pi} \frac{1}{1-\exp(-\hbar\omega/kT)} \sum_{\substack{\underline{mn}\\ ij}} \exp[-i\underline{K}(\underline{R}^{\underline{m}}-\underline{R}^{\underline{n}})] a_{\underline{m}} \exp(-M_{\underline{K}}^{\underline{m}})\, a_{\underline{n}} \exp(-M_{\underline{K}}^{\underline{n}}) \\ &\qquad \cdot K_i K_j \,\mathrm{Im}\{G_{ij}^{\underline{mn}}(\omega)\}\ .\end{aligned} \tag{7.47}$$

In the case of spin or isotopic incoherence we have to average the scattering intensity over the different spin states or isotope distribution on the various sites, which are assumed to be independent.

$$\overline{a_m a_n} = \overline{a_m^2}\, \delta_{mn} + \bar{a}_m \bar{a}_n (1-\delta_{mn}) = \bar{a}_m \bar{a}_n + (\overline{a_m^2} - \bar{a}_m^2)\, \delta_{mn} \tag{7.48}$$

Thus we obtain a "coherent cross section" $d^2\sigma/d\Omega d\omega\big|^{\text{coh}}$ given by (7.43) with a_m and a_n replaced by the coherent scattering lengths $\bar{a}_m$, $\bar{a}_n$. "Coherent" refers to the interference of the waves scattered by different nuclei which enters into (7.43). Analogously we have also a coherent zero-phonon and coherent one-phonon cross section. On the contrary, for the "incoherent scattering", determined by the fluctuations $(a_m - \bar{a}_m)^2$, the interference drops out: Each nucleus scatters independently.

$$\left.\frac{d^2\sigma}{d\Omega\, d\omega}\right|^{\text{incoh}} = \frac{k_f}{k_i} \sum_m \underbrace{\overline{(a_m - \bar{a}_m)^2}}_{=\overline{a_m^2} - \bar{a}_m^2} S^{\underline{mm}}(\underline{K},\omega) \tag{7.49}$$

For an *ideal Bravais lattice* all scattering lengths a_m and Debye-Waller factors $\exp(-M^{\underline{m}})$ are equal. Thus we obtain for the elastic scattering

$$\begin{aligned}\left.\frac{d^2\sigma}{d\Omega\, d\omega}\right|^{\text{coh}}_{\text{elastic}} &= \bar{a}^2 \exp(-2M_{\underline{K}}) \left| \sum_{\underline{m}} \exp(i\underline{K}\underline{R}^{\underline{m}}) \right|^2 \delta(\omega) \\ &= \bar{a}^2 \exp(-2M_{\underline{K}})\, N \frac{(2\pi)^3}{V_c} \sum_{\underline{h}} \delta(\underline{K} - \underline{K}^{\underline{h}})\, \delta(\omega)\ .\end{aligned} \tag{7.50}$$

Thus the coherent elastic intensity consists of different Bragg peaks: The scattering vector $\underline{K}$ has to coincide with a reciprocal lattice vector $\underline{K}^{\underline{h}}$. Further the

$\delta(\omega)$ term postulates $k_f^2 = k_i^2$, meaning that all final wave vectors $\underline{k}_f$ have to lie on the "Ewald sphere" with radius $\underline{k}_i$. In addition we have a structureless incoherent elastic intensity which is practically independent of $\underline{K}$

$$\left.\frac{d^2\sigma}{d\Omega\, d\omega}\right|_{\text{elastic}}^{\text{incoh}} = N(\overline{a^2} - \bar{a}^2)\, \exp(-2M_{\underline{K}})\, \delta(\omega) \;. \tag{7.51}$$

For the coherent one-phonon scattering of a perfect crystal, (7.47) contains the Fourier transform of the Green's function $\mathrm{Im}\{G_{ij}^{\underline{mn}}(\omega)\}$, which can be obtained from (2.19)

$$\begin{aligned}\sum_{\underline{mn}} \exp[-i\underline{K}(\underline{R}^{\underline{m}} - \underline{R}^{\underline{n}})]\; \mathrm{Im}\{G_{ij}^{\underline{mn}}(\omega)\} &= \frac{\pi}{M} \sum_{\sigma} e_i^{\sigma}(\underline{q})\; e_j^{\sigma}(\underline{q})\; \delta\big(\omega^2 - \omega_\sigma^2(\underline{q})\big)\,(\mathrm{sgn}\;\omega) \\ &= \frac{\pi}{M} \sum_{\sigma} \Big[e_i^{\sigma}(\underline{q})\; e_j^{\sigma}(\underline{q})/2\omega_\sigma(\underline{q})\Big] \Big[\delta\big(\omega - \omega_\sigma(\underline{q})\big) - \delta\big(\omega + \omega_\sigma(\underline{q})\big)\Big] \;.\end{aligned} \tag{7.52}$$

Here $\underline{K}$ has been split up in a vector $\underline{q}$ lying in the first Brillouin zone and in a reciprocal lattice vector $\underline{K}^{\underline{h}}$ (the nearest one to $\underline{K}$): $\underline{K} = \underline{K}^{\underline{h}} + \underline{q}$. By inserting this into (7.47) we obtain

$$\begin{aligned}\left.\frac{d^2\sigma}{d\Omega\, d\omega}\right|_{\text{1-phonon}}^{\text{coh}} &= \frac{k_f}{k_i}\frac{\hbar}{2M} N\bar{a}^2 \exp(-2M_{\underline{K}}) \\ &\cdot \sum_{\sigma} \frac{\big(\underline{e}^{\sigma}(\underline{q})\cdot\underline{K}\big)^2}{\omega_\sigma(\underline{q})} \Big\{\Big[n\big(\omega_\sigma(\underline{q})\big) + 1\Big]\, \delta\big(\omega - \omega_\sigma(\underline{q})\big) + n\big(\omega_\sigma(\underline{q})\big)\; \delta\big(\omega - \omega_\sigma(\underline{q})\big)\Big\}\end{aligned} \tag{7.53}$$

with $n(\omega) = \dfrac{1}{\exp(\hbar\omega/kT) - 1}$.

The first term $\sim \delta(\omega - \omega_{q\sigma})$ describes the energy-loss processes of the neutron. Due to the excitation of one phonon with energy $\hbar\omega_{q\sigma}$, the final neutron energy is

$$\frac{\hbar k_f^2}{2m} = \frac{\hbar k_i^2}{2m} - \hbar\omega_{q\sigma} \;.$$

The second term represents the energy-gain processes of the neutron due to the "absorption" of one phonon. Due to the thermal factors $n_{q\sigma} + 1$ and $n_{q\sigma}$, at $T = 0$ only the loss processes remain, whereas at high T both processes have the same factor $kT/\hbar\omega$.

The most characteristic property of (7.53) is that for given ω we only get intensity for those k_f values for which

$$\hbar\omega = \frac{\hbar^2}{2m}(k_i^2 - k_f^2) = \hbar\omega_\sigma(\underline{q}) = \hbar\omega_\sigma(\underline{k}_f - \underline{k}_i) \;.$$

For each σ, this describes a surface in $\underline{k}_f$ space. Thus for each value of $\underline{k}_f$ we will in general get three different intensity peaks corresponding to the three different phonon polarizations. All multiphonon processes no longer have such a strong selection rule and form a more or less continuous background intensity against which the one-phonon scattering is easy to discriminate.

The incoherent contribution to the one-phonon scattering is essentially given by the spectrum of the ideal lattice

$$\mathrm{Im}\{G_{ij}^{mn}(\omega)\} = \frac{\pi}{2\omega M} z(|\omega|)\ \delta_{ij} \quad \text{for cubic crystals.} \tag{7.54}$$

Thus we obtain

$$\begin{aligned} \left.\frac{d^2\sigma}{d\Omega\, d\omega}\right|_{1\text{-phonon}}^{\text{incoh}} &= \frac{k_f}{k_i}\frac{\hbar}{2M} N(\overline{a^2} - \bar{a}^2)\exp(-2M_{\underline{K}})\frac{K^2}{\omega}\frac{1}{1-\exp(-\hbar\omega/kT)} z(|\omega|) \\ &= N(\overline{a^2} - \bar{a}^2)\exp(-2M_{\underline{K}})\frac{m}{M}\frac{k_f}{k_i}\frac{(\underline{k}_i - \underline{k}_f)^2}{k_i^2 - k_f^2}\frac{z(|\omega|)}{1-\exp(-\hbar\omega/kT)} . \end{aligned} \tag{7.55}$$

As a function of $\underline{K}$, this scattering is rather structureless which is due to the incoherent superposition of the scattering intensities of the individual nuclei. However, it has the advantage to allow for strong incoherent scattering $\overline{a^2} \gg \bar{a}^2$ a direct determination of $z(\omega)$.

The scattering theory of a *defect lattice* is much more complicated. Whereas the inelastic one-phonon scattering will be discussed in detail in the next sections, we will here sketch the main features of the *elastic scattering*. For more details the reader is referred to the book by KRIVOGLAZ /7.5/, the review paper by DEDERICHS /7.6/, or the forthcoming review article by TRINKAUS in this series. Compared to the Bragg scattering (7.50) for the ideal lattice, several effects occur:

1) Due to the lattice expansion induced by the defect, the Bragg peaks are shifted to the positions corresponding to the reciprocal lattice vectors of the expanded lattice.

2) Because of the static displacements due to the defect, the Bragg intensities are reduced by an additional, temperature-independent static Debye - Waller factor $\exp(-2L_{\underline{K}})$.

3) Due to the lack of translation symmetry, an additional diffuse intensity appears between the Bragg peaks. This diffuse scattering is especially strong for defects with strong displacement fields and has proved to be one of the best tools for studying the structure of point defects in metals. The intensity is especially high near the Bragg reflections ("Huang scattering").

4) For defects with especially long-ranging displacement fields, e.g., for dislocations (but not for point defects or defect clusters), the static Debye - Waller factor vanishes, $\exp(-2L_{\underline{K}}) = 0$, and therefore also the Bragg intensity. This case is usually referred to as line broadening since the diffuse scattering usually forms new, but broadened lines resembling the Bragg peaks. Physically the dis-

appearance of the Bragg peaks is due to the loss of long-range order; under these conditions an average periodic lattice no longer exists.

7.3 Change of Phonon Dispersion Curves

As one expects, also the *inelastic scattering* for a *defect lattice* gets very complicated. We will only consider the one-phonon scattering. Whereas the elastic scattering is mostly influenced by the static structure of the defects, e.g., by the static displacements, the inelastic one-phonon scattering is mostly influenced by the "dynamical structure" of the defect, i.e., the defect mass and the changed force constants. The influence of the static displacements on the one-phonon scattering is in general small and will be discussed in the next section. For the moment we will neglect the static displacements.

The coherent one-phonon cross section is given by the configurational average of (7.47)

$$\left.\frac{d^2\sigma}{d\Omega\, d\omega}\right|_{\text{1-phonon}} = \frac{k_f}{k_i}\frac{\hbar}{\pi}\frac{1}{1-\exp(-\hbar\omega/kT)} \cdot \sum_{\substack{\underline{m}\underline{n}\\ ij}} \exp[-i\underline{K}(\underline{R}^{\underline{m}}-\underline{R}^{\underline{n}})]\, \langle a_{\underline{m}} \exp(-M^{\underline{m}}_{\underline{K}})\, K_i\, \mathrm{Im}\{G^{\underline{m}\underline{n}}_{ij}(\omega)\}\, K_j\, a_{\underline{n}} \exp(-M^{\underline{n}}_{\underline{K}})\rangle . \tag{7.56}$$

Since the thermal displacements are usually small, $\exp(-M) \approx 1$ is a good approximation for the Debye-Waller factor. For the moment we assume all scattering lengths to be equal ($a_{\underline{m}} = \bar{a}$). (The general case will be treated in the next section.) Then the cross section is directly given by the average Green's function $\langle G(\omega)\rangle$ or more precisely by the imaginary part of $\langle G\rangle_\sigma(\underline{k},\omega)$ (7.17)

$$\left.\frac{d^2\sigma}{d\Omega\, d\omega}\right|_{\text{1-phonon}} = \frac{k_f}{k_i}\frac{\hbar}{\pi}\frac{1}{1-\exp(-\hbar\omega/kT)}\, N\, \bar{a}^2 \sum_{\underline{m},ij} \exp\left(-i\underline{K}(\underline{R}^{\underline{m}}-\underline{R}^{\underline{n}})\right) \cdot K_i\, \mathrm{Im}\{\langle G^{\underline{m}\underline{n}}_{ij}(\omega)\rangle\}\, K_j \tag{7.57}$$

$$= \frac{k_f}{k_i}\frac{\hbar}{\pi}\frac{1}{1-\exp(-\hbar\omega/kT)}\, N\, \bar{a}^2 \sum_\sigma |\underline{K}\cdot\underline{e}^\sigma(\underline{q},\omega)|^2\, \mathrm{Im}\{\langle G\rangle_\sigma(\underline{q},\omega)\}$$

$$= \frac{k_f}{k_i}\frac{\hbar}{\pi}\frac{1}{1-\exp(-\hbar\omega/kT)}\, N\, \bar{a}^2 \sum_\sigma |\underline{K}\cdot\underline{e}^\sigma(\underline{q},\omega)|^2 \cdot \frac{-\Sigma''_\sigma(\underline{q},\omega)}{\left[\mathring{M}\left(\mathring{\omega}^2_\sigma(\underline{q})-\omega^2\right)+\Sigma'_\sigma(\underline{q},\omega)\right]^2+\left(\Sigma''_\sigma(\underline{q},\omega)\right)^2} \tag{7.58}$$

where we have split up the self-energy into its real part Σ' and its imaginary part Σ'' which is negative; see (7.10).

$$\Sigma_\sigma(\underline{q},\omega) = \Sigma'_\sigma(\underline{q},\omega) + i\,\Sigma''_\sigma(\underline{q},\omega)\ , \quad \Sigma''_\sigma < 0 \tag{7.59}$$

Since in the ideal crystal we get only intensity if $\omega^2 = \overset{\circ}{\omega}{}^2_\sigma(\underline{q})$, we expect also for the defect crystal the major intensity in the vicinity of these ideal phonon lines. Thus we replace the ω- and q-dependent self-energy $\Sigma_\sigma(\underline{q},\omega)$ by the purely q-dependent $\Sigma_\sigma\left(\underline{q}, \overset{\circ}{\omega}_\sigma(\underline{q})\right) = \Sigma_\sigma(\underline{q})$. In this approximation the phonons of the ideal crystal given by

$$\sim \delta\left(\overset{\circ}{\omega}{}^2_\sigma(\underline{q}) - \omega^2\right) \quad \text{or} \quad \omega = \overset{\circ}{\omega}_\sigma(\underline{q})$$

are replaced by Lorentzian lines

$$\frac{-\Sigma''_\sigma(\underline{q})}{\left(\overset{\circ}{M}\overset{\circ}{\omega}{}^2_\sigma(\underline{q}) + \Sigma'_\sigma(\underline{q}) - \overset{\circ}{M}\omega^2\right)^2 + \left(\Sigma''_\sigma(\underline{q})\right)^2} \tag{7.60}$$

being broadened by $\Sigma''_\sigma(\underline{q})$ and shifted relative to the ideal phonons by $\Sigma'_\sigma(\underline{q})$. By introducing a new complex dispersion relation $\omega_\sigma(\underline{q}) = \omega'_\sigma(\underline{q}) + i\omega''_\sigma(\underline{q})$, we can write

$$\Delta\left(\overset{\circ}{M}\omega^2_\sigma(\underline{q})\right) = \Sigma_\sigma(\underline{q}) \ . \tag{7.61}$$

For the real part $\omega'_\sigma(\underline{q})$ and imaginary part $\omega''_\sigma(\underline{q})$ we obtain to first order in c as *phonon shift*:

$$\Delta\omega'_\sigma(\underline{q}) = \frac{1}{2\overset{\circ}{M}\overset{\circ}{\omega}_\sigma(\underline{q})}\,\Sigma'_\sigma(\underline{q}) = \frac{c}{2\overset{\circ}{M}\overset{\circ}{\omega}_\sigma(\underline{q})}\,t'_\sigma\left(\underline{q},\omega_\sigma(\underline{q})\right) \tag{7.62}$$

and as *phonon width*:

$$\Delta\omega''_\sigma(\underline{q}) = \frac{1}{2\overset{\circ}{M}\overset{\circ}{\omega}_\sigma(\underline{q})}\,\Sigma''_\sigma(\underline{q}) = \frac{c}{2\overset{\circ}{M}\overset{\circ}{\omega}_\sigma(\underline{q})}\,t''_\sigma\left(\underline{q},\omega_\sigma(\underline{q})\right) \ . \tag{7.63}$$

Let us in particular discuss the case of a *resonance vibration*. The single t matrix

$$t_\alpha = V_\alpha - V_\alpha G_\alpha V_\alpha \quad \text{with} \quad G_\alpha = \overset{\circ}{G}\,\frac{1}{1 + V_\alpha \overset{\circ}{G}} \tag{7.64}$$

can show a resonance behaviour since we may write the defect Green's function G_α as

$$G_\alpha(\omega) \cong \frac{-\,|R\rangle\langle R|}{\overset{\circ}{M}(\omega^2 - \omega^2_{res} + i\gamma\omega)} \ . \tag{7.65}$$

Here $|R\rangle$ represents a normalized resonance state being identical with $\sqrt{\overset{\circ}{M}/M_{eff}}\,|u(0)\rangle$ of (4.55). Then $\langle G_\sigma(\underline{q},\omega)\rangle$ is given by

$$\langle G_\sigma(\underline{q},\omega)\rangle = \left(-\overset{\circ}{M}\omega^2 + \overset{\circ}{M}\overset{\circ}{\omega}{}^2_\sigma(\underline{q}) + c\,\frac{|\langle q\sigma|\,V\,|R\rangle|^2}{M(\omega^2 - \omega^2_{res} + i\gamma\omega)}\right)^{-1} \ . \tag{7.66}$$

The shift $\Delta\omega'$ and the width $\Delta\omega''$ are obtained by replacing ω by $\omega_\sigma(\underline{q})$ in the t matrix

$$\Delta\omega'_\sigma(\underline{q}) \cong c\,\frac{|\langle q\sigma|\,V\,|R\rangle|^2}{2\overset{\circ}{M}{}^2\omega_\sigma(\underline{q})}\;\frac{\omega^2_\sigma(\underline{q}) - \omega^2_{res}}{\left(\omega^2_\sigma(\underline{q}) - \omega^2_{res}\right)^2 + \gamma^2\omega^2_\sigma(\underline{q})} \tag{7.67}$$

$$\Delta\omega''_\sigma(\underline{q}) \cong c\, \frac{|\langle q\sigma|\, V\, |R\rangle|^2}{2\overset{\circ}{M}{}^2\omega_\sigma(\underline{q})}\ \frac{\gamma\omega_\sigma(\underline{q})}{\left(\omega_\sigma^2(\underline{q}) - \omega_{res}^2\right)^2 + \gamma^2\omega_\sigma^2(\underline{q})} \tag{7.68}$$

This behaviour is schematically plotted in Fig. 47a. For $\omega < \omega_{res}$ $(\omega > \omega_{res})$ the new phonon line is shifted below (above) the ideal line, while the damping is largest at $\omega = \omega_{res}$.

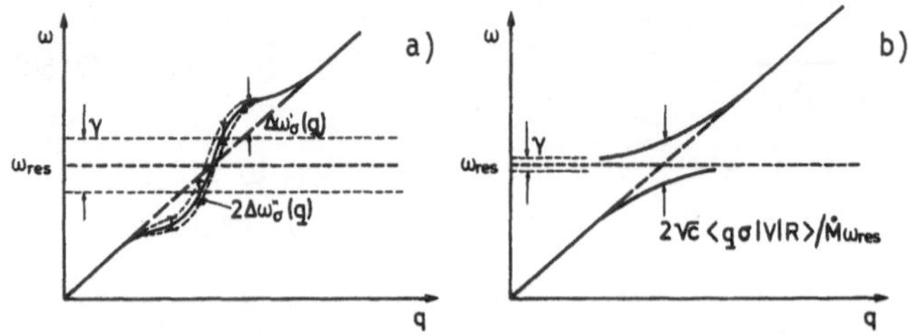

Fig. 47a,b. Effects of resonance vibrations on the phonon dispersion curves
a) a mere "distortion" of the phonon line being shifted by $\Delta\omega'_\sigma(\underline{q})$ given by (7.67) and broadened by $\Delta\omega''_\sigma(\underline{q})$ or (7.68) (valid if $2\Delta\omega''_\sigma(\underline{q}) \ll \gamma$)
b) a hybridization of the ideal phonon line with the resonance mode resulting in a split of the dispersion curves (valid $2\Delta\omega''_\sigma(\underline{q}) \gg \gamma$)

In some cases we can also obtain a quite different resonance behaviour. To see this we have to take the $\underline{q}$- *and* ω-dependence of the self-energy $\Sigma_\sigma(\underline{q},\omega)$, or of the t matrix $t_\sigma(q,\omega)$, respectively, more serious. First, we have as condition for the validity of the single t matrix approximation

$$\overset{\circ}{M}\overset{\circ}{\omega}{}^2_\sigma(\underline{q}) \gg c t_\sigma(\underline{q},\omega) \tag{7.69}$$

or from (7.66) for $\omega = \omega_{res}$

$$\overset{\circ}{M}\overset{\circ}{\omega}{}^2_\sigma(\underline{q}) \gg c\, \frac{|\langle q\sigma|\, V\, |R\rangle|^2}{\overset{\circ}{M}\gamma\omega_{res}}\,. \tag{7.70}$$

On the other hand, we can replace ω by $\overset{\circ}{\omega}_\sigma(\underline{q})$ in the single t matrix

$$t(\omega) \cong -\frac{V\,|R\rangle\,\langle R|\,V}{\overset{\circ}{M}(\omega^2 - \omega_{res}^2 + i\gamma\omega)} \tag{7.71}$$,

if (for $\overset{\circ}{\omega}_\sigma(\underline{q}) = \omega_{res}$):

$$\omega^2 - \overset{\circ}{\omega}{}^2_\sigma(\underline{q}) \ll \gamma\omega \tag{7.72}$$

leading to the above results (7.67,68) for $\Delta\omega'_\sigma(\underline{q})$ and $\Delta\omega''_\sigma(\underline{q})$. By inserting $\omega^2 \approx \omega_\sigma^2(\underline{q}) + (1/\overset{\circ}{M})\cdot ct_\sigma(\underline{q},\omega)$, this condition is equivalent to (for $\overset{\circ}{\omega}_\sigma(\underline{q}) \approx \omega_{res}$)

$$2\Delta\omega''_\sigma(\underline{q}) = c\,\frac{\left|\langle q\sigma|\,V\,|R\rangle\right|^2}{\overset{\circ}{M}{}^2\gamma\omega^2_{res}} \ll \gamma\ , \tag{7.73}$$

i.e., the damping of the renormalized phonons has to be much smaller than the damping γ of the resonance vibration.

Clearly this condition can only be met, if γ is not too small. If however $\gamma \ll 2\Delta\omega''_\sigma(\underline{q})$, but still large enough that the single t matrix approximation (7.66) holds, in other words, if

$$\overset{\circ}{M}\overset{\circ}{\omega}{}^2_\sigma(\underline{q}) \gg c\,\frac{\left|\langle q\sigma|\,V\,|R\rangle\right|^2}{\overset{\circ}{M}\gamma\omega_{res}} \gg \overset{\circ}{M}\gamma\omega_{res} \tag{7.74}$$

then the phonon dispersion is changed dramatically. In the t matrix (7.66) we can now neglect the damping term $\gamma\omega$ against $\omega^2 - \omega^2_{res}$, so that the new phonon lines are determined by

$$\overset{\circ}{M}\left(\omega^2 - \overset{\circ}{\omega}{}^2_\sigma(\underline{q})\right) = c\,\frac{\left|\langle q\sigma|\,V\,|R\rangle\right|^2}{\overset{\circ}{M}(\omega^2 - \omega^2_{res})}\ . \tag{7.75}$$

This describes a *hybridization* of the "ideal phonon line" $\omega^2 = \overset{\circ}{\omega}{}^2_\sigma(\underline{q})$ with the "resonance line" $\omega^2 = \omega^2_{res}$ and results in a splitting of the phonon lines, as schematically shown in Fig. 47b. At $\overset{\circ}{\omega}_\sigma(\underline{q}) = \omega_{res}$ the separation $\delta\omega$ of the two lines is

$$\delta\omega = \frac{\sqrt{c}}{\overset{\circ}{M}\omega_{res}}\left|\langle q\sigma|\,V\,|R\rangle\right|\ . \tag{7.76}$$

Thus even a small concentration of point defects can lead to drastic changes of the phonons, provided the defects exhibit resonance vibrations with sufficiently small damping.

For the case of *localized vibrations* the single t matrix approximation breaks down. The condition (7.69) no longer holds since $t_\sigma(\underline{q}\omega)$ has a pole at the localized mode frequency $\omega = \omega_{\ell oc}$. This situation can only be dealt with by a self-consistent theory like the coherent potential approximation (CPA) /7.1/ which is outside the scope of our article. However qualitatively we can understand the behaviour by invoking an anharmonic damping γ_{an} of the localized mode, so that the t matrix becomes finite at the localized mode frequency $\omega_{\ell oc}$ and (7.69) still holds for small concentrations. For $\omega \approx \omega_{\ell oc}$ we can then expand $\langle G\rangle_\sigma(\underline{q},\omega)$ linear in c

$$\langle G\rangle_\sigma(\underline{q},\omega) - \overset{\circ}{G}_\sigma(\underline{q},\omega) \approx \frac{-1}{(\overset{\circ}{M}\overset{\circ}{\omega}{}^2_\sigma(\underline{q}) - \overset{\circ}{M}\omega^2_{\ell oc})^2}\, ct_\sigma(\underline{q},\omega) \tag{7.77}$$

since $\omega^2_{\ell oc} - \overset{\circ}{\omega}{}^2_\sigma(\underline{q})$ is always positive and assumed to be larger than $ct_\sigma(\underline{q},\omega)$. For frequencies near $\omega_{\ell oc}$ we now make a similar expansion for $t_\sigma(\underline{q},\omega)$ as in the case of a resonance vibration (7.66), so that

$$t_\sigma(\underline{q},\omega) \approx -\frac{\left|\langle q\sigma\,|\,V(\omega_{\ell oc})\,|\,\ell\rangle\right|^2}{\overset{\circ}{M}(\omega^2 - \omega^2_{\ell oc} + i\gamma_{an}\omega_{\ell oc})} \tag{7.78}$$

where $|\ell\rangle$ refers to the localized state. Since

$$\{\overset{\circ}{\phi} - \overset{\circ}{M}\omega_{\ell oc}^2 + V(\omega_{\ell oc})\} \, |\ell\rangle = 0 \tag{7.79}$$

we have

$$\langle q\sigma \,|\, V(\omega_{\ell oc}) \,|\, \ell\rangle = \langle q\sigma \,|\, (\overset{\circ}{M}\omega_{\ell oc}^2 - \overset{\circ}{\phi}) \,|\, \ell\rangle = \overset{\circ}{M}\left(\omega_{\ell oc}^2 - \overset{\circ}{\omega}{}_{\sigma}^2(\underline{q})\right) \langle q\sigma|\ell\rangle \tag{7.80}$$

so that finally

$$\langle G\rangle_{\sigma}(\underline{q},\omega) - \overset{\circ}{G}_{\sigma}(\underline{q},\omega) \approx c \, \frac{\left|\langle q\sigma|\ell\rangle\right|^2}{\overset{\circ}{M}(\omega^2 - \omega_{\ell oc}^2 + i\gamma_{an}\omega_{\ell oc})} \; . \tag{7.81}$$

Thus we obtain a new phonon line $\omega \approx \omega_{\ell oc}$ which is practically free of dispersion. Its exact shape is determined by the anharmonic damping γ_{an} or the "statistical" damping of the CPA. More important is that the intensity of this line is directly proportional to the concentration c and the square of the matrix element $\langle q\sigma|\ell\rangle$.

Let us discuss now the *isotopic defect* as the most simple example. With

$$V_{\alpha}\,{}^{\underline{mn}}_{ij} = - \Delta M\omega^2 \, \delta^{\underline{m}\underline{\alpha}} \, \delta^{\underline{n}\underline{\alpha}} \, \delta_{ij} \tag{7.82}$$

we obtain immediately from (7.33)

$$\Sigma_{\sigma}(\underline{q},\omega) = ct_{\sigma}(\underline{q},\omega) = -c \, \frac{\Delta M\omega^2}{1 - \overset{\circ}{G}{}^{(0)}_{xx}(\omega) \, \Delta M\omega^2} \; . \tag{7.83}$$

The result is independent of the wave vector $\underline{q}$ and the polarization σ.

As an example we have chosen $M = 4\overset{\circ}{M}$ corresponding to a substitutional Ag defect in Al. Figure 48a,b shows equal intensity contours of the "local" spectrum $z_{\sigma}(\underline{q},\omega)$

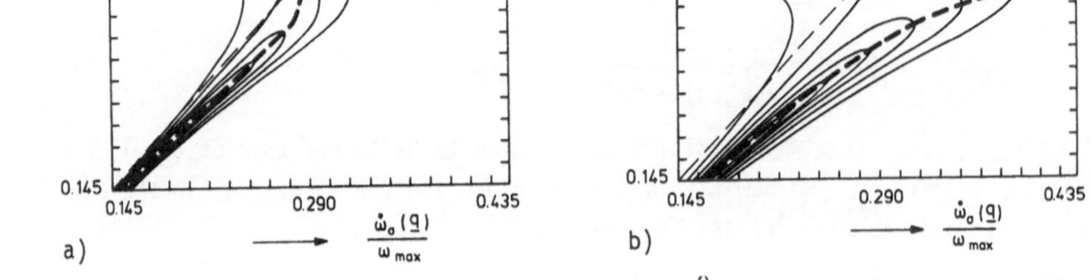

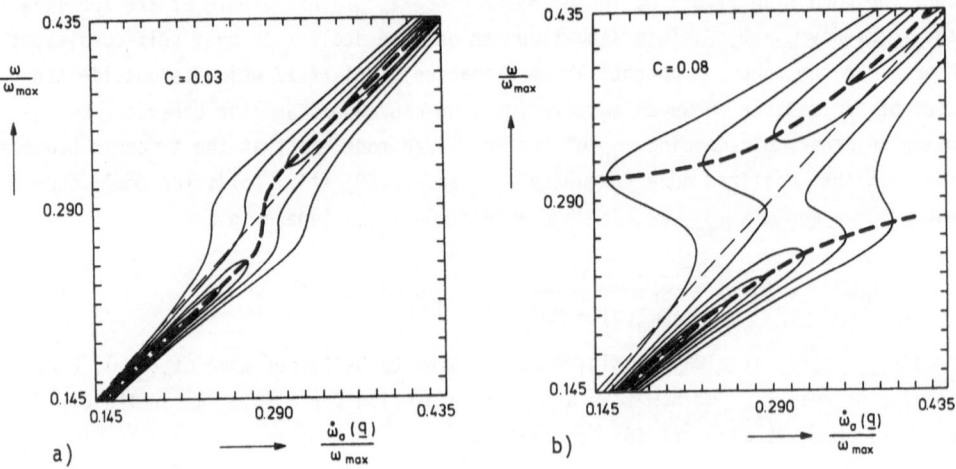

Fig. 48a,b. Effect of a heavy isotopic defect ($M = 4\overset{\circ}{M}$, corresponding to Ag in Al) on the phonon dispersion. Shown are equal intensity contours, for which the intensity on neighbouring lines differs by a factor of 2.
a) c = 0.03: distortion of the phonon line as schematically shown in Fig. 47a
b) c = 0.08: hybridization (compare Fig. 47b)

$$z_\sigma(\underline{q},\omega) = \frac{2\overset{\circ}{M}\omega}{\pi}\,\mathrm{Im}\{\langle G\rangle_\sigma(\underline{q},\omega)\} \tag{7.84}$$

for an isotopic defect as a function of ω/ω_{max} and $\overset{\circ}{\omega}_\sigma(\underline{q})/\omega_{max}$. Figure 48a refers to a concentration c = 0.03, leading to a mere distortion and broadening of the phonons as described by (7.67,68). Figure 48b corresponds to a concentration c = 0.08, where the tendency towards a hybridization (7.75) is clearly visible. In fact the calculations are based on the average t matrix approximation (ATA) which takes also higher order terms in c into account and which is more accurate than the single t matrix approximation, especially for c = 0.08. Both approximations yield qualitatively the same result: a mere distortion for small c and a hybridization for larger concentrations.

The effects of point defects on the phonon dispersion curves have been treated by LAKATOS and KRUMHANSL /7.7/, MANNHEIM /7.8/, ZHERNOV and AUGUST /7.9/, COHEN /7.10/, and others. The hybridization effect due to resonance modes has been discussed by SCHOBER et al. /7.11/, and independently by WOOD and MOSTOLLER /7.12/ (see also the paper of ZHERNOV and KALUGINA /7.13/).

Note added in proof: As we have learned in the meantime the hybridisation effect has been already discussed earlier by IVANOV /7.55/.

7.4 Effects of Different Scattering Lengths and Static Displacements

7.4.1 Effects of Different Scattering Lengths

In general the scattering length a_d of the defect is different from the scattering length a_h of the host atoms. Then the cross section is not directly given by $\langle G^{mn}\rangle$. Consider, e.g., a substitutional alloy with A and B atoms and scattering lengths a_A and a_B. Then we have to calculate

$$m \neq n: \quad \langle a^m a^n G^{mn}(\omega)\rangle = c_A^2 a_A^2 \langle G^{mn}_{AA}\rangle + c_A c_B a_A a_B \{\langle G^{mn}_{AB}\rangle + \langle G^{mn}_{BA}\rangle\} + c_B^2 a_B^2 \langle G^{mn}_{BB}\rangle \tag{7.85a}$$

$$m = n: \quad \langle a^m a^m G^{mm}(\omega)\rangle = c_A a_A^2 \langle G^{mm}_{AA}\rangle + c_B a_B^2 \langle G^{mm}_{BB}\rangle \;. \tag{7.85b}$$

Here the Green's functions $\langle G_{AA}\rangle$, $\langle G_{BB}\rangle$, $\langle G_{AB}\rangle$ are conditional averages. For instance, $\langle G^{mn}_{AB}\rangle$ means the average Green's function under the subsidiary condition that one has an A atom at m and a B atom at n. Evidently

$$\text{for } m \neq n: \quad \langle G^{mn}\rangle = c_A^2 \langle G^{mn}_{AA}\rangle + c_A c_B(\langle G^{mn}_{AB}\rangle + \langle G^{mn}_{BA}\rangle) + c_B^2 \langle G^{mn}_{BB}\rangle \tag{7.86a}$$

$$\text{for } m = n: \quad \langle G^{mm}\rangle = c_A \langle G^{mm}_{AA}\rangle + c_B \langle G^{mm}_{BB}\rangle \;. \tag{7.86b}$$

One way of obtaining the cross section is then to write down directly equations for the averages $\langle G^{mn}_{AB}\rangle$, as has been done for the isotopic defects by /1.10/. There exists however also a general prescription to evaluate the average $\langle a^m a^n G^{mn}\rangle$ by the same procedure as the one for $\langle G^{mn}\rangle$. Namely one can introduce renormalized force constants $\tilde{\phi}$ and masses $\tilde{M}$ as follows

$$\langle a^m G^{mn}_{ij} a^n\rangle = \langle a^m \left\{\frac{1}{\phi - M(\omega + i\eta)^2}\right\}^{mn}_{ij} a^n\rangle = a^2 \langle\left\{\frac{1}{\tilde{\phi} - \tilde{M}(\omega + i\eta)^2}\right\}^{mn}_{ij}\rangle = a^2 \langle\tilde{G}^{mn}_{ij}\rangle \quad (7.87)$$

where

$$\tilde{\phi}^{mn}_{ij} = \frac{a}{a^m}\phi^{mn}_{ij}\frac{a}{a^n}\,, \quad \tilde{M}^m = \left(\frac{a}{a^m}\right)^2 M^m\,.$$

Thus the problem has been reduced to calculate the average of the Green's function $\tilde{G}$ characterized by the force constants $\tilde{\phi}$ and the masses $\tilde{M}$. The general procedure to calculate $\langle\tilde{G}\rangle$ is then the same as the one for $\langle G\rangle$: Write the coupling parameters and masses as

$$\tilde{\phi} = \mathring{\phi} + \Delta\tilde{\phi}\,, \quad \tilde{M} = \mathring{M} + \Delta\tilde{M}\,, \quad \tilde{V} = \Delta\tilde{\phi} - \Delta\tilde{M}\omega^2 \quad (7.88)$$

which defines a perturbation $\tilde{V}$, from which a T matrix $\tilde{T}$ can be calculated. The self-energy $\tilde{\Sigma}$ is then given by

$$\langle\tilde{G}\rangle = \frac{1}{\mathring{\phi} + \tilde{\Sigma} - \mathring{M}\omega^2}\,, \quad \tilde{\Sigma} = \langle\tilde{T}\rangle\frac{1}{1 - \mathring{G}\langle\tilde{T}\rangle}\,, \quad \tilde{T} = \tilde{V}\frac{1}{1 + \mathring{G}\tilde{V}}\,. \quad (7.89)$$

By restricting ourselves again to the dilute limit ($c = c_B \ll 1$), the self-energy is given by the single t matrix

$$\tilde{\Sigma} \cong c\sum_\mu \tilde{t}^{(\mu)} \quad \text{with} \quad \tilde{t}^{(\mu)} = \tilde{V}^{(\mu)}\frac{1}{1 + G\tilde{V}^{(\mu)}} \quad (7.90)$$

where $\tilde{V}^{(\mu)}$ is the renormalized perturbation for a defect at position μ, i.e.,

$$\tilde{V}^{(\mu)} = \frac{a_h}{a^{(\mu)}}V^{(\mu)}\frac{a_h}{a^{(\mu)}}\,, \quad \{a^{(\mu)}\}^{mn} = a_h\delta_{mn} + (a_d - a_h)\,\delta_{m\mu}\,\delta_{n\mu}\,. \quad (7.91)$$

By some tricky manipulations, the renormalized t matrix $\tilde{t}^{(\mu)}$ can be directly expressed in terms of $t^{(\mu)}$. Starting with the Green's functions $G^{(\mu)}$ and $\tilde{G}^{(\mu)}$ for a single defect

$$G^{(\mu)} = \mathring{G} - \mathring{G}t^{(\mu)}\mathring{G}\,, \quad \tilde{G}^{(\mu)} = \mathring{G} - \mathring{G}\tilde{t}^{(\mu)}\mathring{G} \quad (7.92)$$

and using the relation between $G^{(\mu)}$ and $\tilde{G}^{(\mu)}$

$$\tilde{G}^{(\mu)} = \frac{a^{(\mu)}}{a_h}G^{(\mu)}\frac{a^{(\mu)}}{a_h} = \left(1 + \frac{\Delta a^{(\mu)}}{a_h}\right)(\mathring{G} - \mathring{G}t^{(\mu)}\mathring{G})\left(1 + \frac{\Delta a^{(\mu)}}{a_h}\right) \quad (7.93)$$

one obtains by comparison

$$\tilde{t}^{(\mu)} = t^{(\mu)} + (\overset{o}{\phi} - \overset{o}{M}\omega^2)\,\frac{\Delta a^{(\mu)}}{a_h}\,(1 - \overset{o}{G}t^{(\mu)}) + (1 - t^{(\mu)}\overset{o}{G})\,\frac{\Delta a^{(\mu)}}{a_h}\,(\overset{o}{\phi} - \overset{o}{M}\omega^2)$$
$$+ (\overset{o}{\phi} - \overset{o}{M}\omega^2)\,\frac{\Delta a^{(\mu)}}{a_h}\,\overset{o}{G}(1 - t^{(\mu)}\overset{o}{G})\,\frac{\Delta a^{(\mu)}}{a_h}\,(\overset{o}{\phi} - \overset{o}{M}\omega^2)\;. \qquad (7.94)$$

After Fourier transformation, the $\underline{k}\omega$-dependent Green's function is given by

$$\langle\tilde{G}_\sigma(\underline{k}\omega)\rangle = \frac{1}{\overset{o}{M}\left(\overset{o}{\omega}{}^2_\sigma(\underline{k}) - \omega^2\right) + c\tilde{t}_\sigma(\underline{k}\omega)} \approx \frac{1}{\overset{o}{M}\left(\overset{o}{\omega}{}^2_\sigma(\underline{k}) - \omega^2\right) + ct_\sigma(\underline{k}\omega)} \qquad (7.95)$$

$$\text{for } \omega \approx \omega_\sigma(\underline{k})\;.$$

Since all terms in (7.94) except $t^{(\mu)}$ contain a factor $\overset{o}{\phi} - \overset{o}{M}\omega^2$, $\tilde{t}_\sigma(\underline{k}\omega)$ contains factors $\overset{o}{\omega}{}^2_\sigma(\underline{k}) - \omega^2$, so that in the vicinity of the phonon lines $\omega^2 \approx \overset{o}{\omega}{}^2_\sigma(\underline{k})$ the renormalized t matrix $\tilde{t}_\sigma(\underline{k}\omega)$ can be replaced by the original one $t_\sigma(\underline{k}\omega)$. Thus the different scattering lengths do not influence the phonon lines, but only lead to a more or less incoherent or diffuse scattering.

7.4.2 Incoherent Scattering

Let us discuss this incoherent scattering in more detail by going back to the conditional Green's functions $\langle G_{AA}\rangle$, $\langle G_{AB}\rangle$ etc. In the dilute limit ($c = c_B \ll 1$) we obtain for the average $\langle a^m G^{mn} a^n\rangle$ up to terms linear in c

$$m \neq n:\quad \langle a^m G^{mn} a^n\rangle = a_h^2\,\langle G^{mn}\rangle + ca_h(a_d - a_h)\{\langle G^{mn}_{hd}\rangle + \langle G^{mn}_{dh}\rangle\} \qquad (7.96a)$$

$$m = n:\quad \langle a^m G^{mm} a^m\rangle = a_h^2\,\langle G^{mm}\rangle + c(a_d^2 - a_h^2)\,\langle G^{mm}_{dd}\rangle\;. \qquad (7.96b)$$

Here a_h, a_d stands for the scattering lengths of the host and defect atoms. $\langle G^{mn}\rangle$ is the average Green's function determining the phonon dispersion curves (last section). $\langle G^{mm}_{dd}(\omega)\rangle$ is the impurity Green's function $G^{dd}(\omega)$, $\langle G^{mn}_{hd}\rangle$ and $\langle G^{mn}_{dh}\rangle$ the interference terms between the impurity and host atoms. By inserting this expression for the averages into the one-phonon cross section, we obtain

$$\left.\frac{d^2\sigma}{d\Omega\,d\omega}\right|_{1\text{-phonon}} = \frac{k_f}{k_i}\,\frac{\hbar}{\pi}\,\frac{N}{1 - \exp(-\hbar\omega/kT)}\left[a_h^2\sum_\sigma\left|\underline{K}\cdot\underline{e}^\sigma(\underline{q}\omega)\right|^2\,\mathrm{Im}\{\langle G\rangle_\sigma(\underline{q}\omega)\}\right.$$
$$+ ca_h(a_d - a_h)\sum_{ij} K_i\left(\Gamma^{hd}_{ij}(\underline{q}\omega) + \Gamma^{dh}_{ij}(\underline{q}\omega)\right)K_j \qquad (7.97)$$
$$\left. + c(a_d^2 - a_h^2)\sum_{ij} K_i\,\mathrm{Im}\{G^{dd}_{ij}(\omega)\}\,K_j\right]\;.$$

In addition to the first term, representing the result of the last section, we obtain due to the different scattering lengths an interference term determined by the mixed Green's function

$$\Gamma_{ij}^{hd}(\underline{q},\omega) = \sum_{m,ij} \exp[-i\underline{q}(\underline{R}^m - \underline{R}^n)] \; \mathrm{Im}\{G_{\substack{hd\\ij}}^{mn}(\omega)\} \tag{7.98}$$

and a term proportional to $a_d^2 - a_h^2$, being determined by the local Green's function of the impurity, respectively its local frequency spectrum.

Evidently the last term is very important, if $a_d^2 \gg a_h^2$. Thus in this limit one directly measures the local frequency spectrum of the defect, giving rise to an incoherent scattering intensity since this term does not show dispersion.

There are two situations, for which the scattering is completely incoherent:

i) For a host similar to V, which due to isotopic incoherence scatters completely incoherent ($a_h \cong 0$). In this case all nondiagonal elements $m \neq n$ vanish, since $\overline{a^m \cdot a^n} = \overline{a^m} \cdot \overline{a^n} = 0$.

ii) For polycrystalline samples one essentially averages over all directions of $\underline{K}$, which in (7.47) practically wipes out the nondiagonal elements $m \neq n$ (7.54).

In both cases only the diagonal elements remain leading to an incoherent scattering

$$\frac{d^2\sigma}{d\Omega\, d\omega} \cong \frac{k_f}{k_i} \frac{\hbar}{\pi} \frac{N}{1 - \exp(-\hbar\omega/kT)} \left[\sum_{ij} a_h^2 K_i K_j \; \langle \mathrm{Im}\{G_{ij}^{mm}(\omega)\}\rangle + c(a_d^2 - a_h^2) K_i K_j \; \mathrm{Im}\{G_{ij}^{dd}(\omega)\} \right] . \tag{7.99}$$

Now the average Green's function $\mathrm{Im}\langle G_{xx}^{mm}(\omega)\rangle$ can be related to the average density of states as defined in Chap. 8

$$\langle z_x^m(\omega)\rangle = \frac{2\omega}{\pi} \langle M^m \; \mathrm{Im}\{G_{xx}^{mm}(\omega)\}\rangle = \frac{2\omega\mathring{M}}{\pi} \left(\mathrm{Im}\langle G_{xx}^{mm}(\omega)\rangle + c\, \frac{M^d - \mathring{M}}{\mathring{M}} \, \mathrm{Im}\{G_{xx}^{dd}(\omega)\} \right) . \tag{7.100}$$

Since the last term essentially represents the local spectrum of the defect $z_x^d(\omega)$, we obtain

$$\mathrm{Im}\langle G_{xx}^{mm}(\omega)\rangle = \frac{\pi}{2\omega\mathring{M}} \left(\langle z_x^m(\omega)\rangle - c\left(1 - \frac{\mathring{M}}{M_d}\right) z_x^d(\omega) \right) . \tag{7.101}$$

By setting for small c

$$\langle z_x^m(\omega)\rangle = \mathring{z}_x(\omega) + c \cdot \Delta z_x(\omega) \tag{7.102}$$

where $\Delta z_x(\omega)$ is the change of the frequency spectrum $\mathring{z}_x(\omega)$, we obtain for the change of the incoherent scattering due to $N_d = cN$ defects

$$\Delta \frac{d^2\sigma}{d\Omega\, d\omega} = \frac{k_f}{k_i} \frac{\hbar K^2}{2\omega\mathring{M}} \frac{a_h^2}{1 - \exp(-\hbar\omega/kT)} \, N_d \left\{ \Delta z_x(\omega) + \left[\left(\frac{a_d}{a_h}\right)^2 \frac{\mathring{M}}{M_d} - 1 \right] z_x^d(\omega) \right\} . \tag{7.103}$$

Note that the local spectrum of the defect does not enter with the same weight as that of the host atoms, but is multiplied by $(a_d/a_h)^2 \cdot \mathring{M}/M_d$.

As an example, let us discuss the case of resonant and localized vibrations. As we will show in Chap. 8 $\Delta z_x(\omega)$ can be approximated near the localized or resonant mode frequency by a normalized δ or Lorentz function

$$\Delta z_x(\omega) \approx \begin{cases} \delta(\omega - \omega_{\ell oc}) & \text{for } \omega \approx \omega_{\ell oc} \\ \dfrac{2}{\pi} \dfrac{\gamma\omega^2}{(\omega^2 - \omega_{res}^2)^2 + \gamma^2\omega^2} & \text{for } \omega \approx \omega_{res} \ . \end{cases} \tag{7.104}$$

According to the results of Chap. 4, for the local frequency spectrum $z_x^d(\omega)$ the δ or Lorentz function has to be multiplied by $M_d/M_{eff}^{\ell oc}$ or M_d/M_{eff}^{res} so that

$$\Delta \frac{d^2\sigma}{d\Omega\, d\omega} = \frac{k_f}{k_i} \frac{\hbar K^2}{2\omega \overset{\circ}{M}} \frac{a_h^2}{1 - \exp(-\hbar\omega/kT)} N_d\, f(\omega) \left[\left(1 - \frac{M_d}{M_{eff}}\right) + \left(\frac{a_d}{a_h}\right)^2 \frac{\overset{\circ}{M}}{M_{eff}}\right] \tag{7.105}$$

with

$$f(\omega) = \begin{cases} \dfrac{2}{\pi} \dfrac{\gamma\omega^2}{(\omega^2 - \omega_{res}^2)^2 + \gamma\omega} & \text{for resonant modes} \\ \delta(\omega - \omega_{\ell oc}) & \text{for localized modes} \ . \end{cases}$$

For substitutional defects usually $M_{eff} \approx M_d$. Thus the changed density of states is multiplied by a factor $(a_d/a_h)^2\, \overset{\circ}{M}/M_d$. Even for large M^d this factor can be appreciable, if $a_d \gg a_h$. On the contrary, there is only a very small effect, if the scattering lengths are equal (note that in the specific heat directly $f(\omega)$ enters, yielding always a large effect for resonant modes).

7.4.3 Effects of Static Displacements

So far we have totally neglected the static displacements induced by the defect. They lead to a positional disorder which also influences the inelastic scattering. In (7.47) the positions $\underline{R}^m$ can be split up into the positions $\underline{\overset{\circ}{R}}^m$ in the ideal lattice and the displacements $\underline{s}^m$ from these positions which are fluctuating quantities. Therefore the one-phonon scattering is proportional to (assuming equal scattering lengths $\bar{a}$)

$$\left.\frac{d^2\sigma}{d\Omega\, d\omega}\right|_{1\text{-phonon}} \sim \bar{a}^2 \sum_{\substack{mn\\ij}} \exp[-i\underline{K}(\underline{\overset{\circ}{R}}^m - \underline{\overset{\circ}{R}}^n)] \times \langle \exp[-i\underline{K}\underline{s}^m]\, K_i\, \mathrm{Im}\{G_{ij}^{mn}(\omega)\}\, K_j\, \exp(i\underline{K}\underline{s}^n)\rangle \ . \tag{7.106}$$

The problem is now much more involved, since the potential disorder is coupled to the mass- and force-constant disorder. In order to see qualitatively, which new effects are introduced by the static displacements, we neglect the mass- and force-constant disorder totally and replace $\mathrm{Im}\{G_{ij}^{mn}(\omega)\}$ by $\mathrm{Im}\{\overset{\circ}{G}{}_{ij}^{mn}(\omega)\}$. Then only the configurational average $\langle \exp[i\underline{K}(\underline{s}^m - \underline{s}^n)]\rangle$ remains. In analogy to the elastic scattering due to static displacements /7.5,6/ this average is split up into an uncorrelated term (modifying the "Bragg scattering") and a correlated term (giving rise to "diffuse scattering")

$$\langle e^{-i\underline{K}\underline{s}^m} e^{i\underline{K}\underline{s}^n}\rangle = \langle e^{-i\underline{K}\underline{s}^m}\rangle \langle e^{i\underline{K}\underline{s}^n}\rangle + \left(\langle e^{-i\underline{K}\underline{s}^m} e^{i\underline{K}\underline{s}^n}\rangle - \langle e^{-i\underline{K}\underline{s}^m}\rangle \langle e^{i\underline{K}\underline{s}^n}\rangle\right). \quad (7.107)$$

The first term will be written as follows

$$\langle e^{-i\underline{K}\underline{s}^m}\rangle = e^{-i\underline{K}\langle\underline{s}^m\rangle} \langle e^{-i\underline{K}(\underline{s}^m - \langle\underline{s}^m\rangle)}\rangle = e^{-i\underline{K}\langle\underline{s}^m\rangle} e^{-L_{\underline{K}}} \quad (7.108)$$

where the average displacement $\langle\underline{s}^m\rangle$ describing an expansion of the lattice has been taken out. The remaining factor $\exp(-L_{\underline{K}})$ is the analog to the thermal Debye-Waller factor $\exp(-M_{\underline{K}})$ and independent of m. By inserting this expression into (7.106) and replacing $G(\omega)$ by $\overset{\circ}{G}(\omega)$, we obtain phonon lines $\sim \delta\left(\omega^2 - \omega_\sigma^2(\underline{q})\right)$ which are diminished in intensity by the Debye-Waller factor $\exp(-L_{\underline{K}} - L_{-\underline{K}})$ and shifted in $\underline{q}$ space due to the lattice expansion.

Thus the new phonons $\omega_\sigma(\underline{q})$ are given in terms of the unshifted ones by the expression

$$\omega_\sigma^2(\underline{q}) = \overset{\circ}{\omega}{}_\sigma^2(\underline{q} + \Delta\underline{q}) \quad \text{with} \quad \Delta\underline{q} = \frac{\Delta a}{a}\,\underline{q} = c\,\frac{1}{3}\,\frac{\Delta V}{V_c}\,\underline{q} \quad (7.109)$$

or

$$\Delta\omega_\sigma(\underline{q}) = \frac{\partial\overset{\circ}{\omega}_\sigma(\underline{q})}{\partial\underline{q}} \cdot \underline{q}\; c\,\frac{\Delta V}{3V_c} \quad (7.110)$$

where $\Delta a/a$ is the relative macroscopic lattice parameter change and ΔV the volume change of a single defect.

The correlated term in (7.106), when introduced into (7.107), leads to a faltung between the original phonons $\sim \delta\left(\omega^2 - \omega_\sigma^2(\underline{q})\right)$ and the correlation function $C_{\underline{K}}(\underline{q})$ giving an intensity contribution

$$\sim \int d\underline{q}\; \delta\left(\omega^2 - \omega_\sigma^2(\underline{q})\right)\, C_{\underline{K}}(\underline{K} - \underline{q}) \quad (7.111)$$

with

$$C_{\underline{K}}(\underline{q}) = \sum_{mn} e^{-i\underline{q}(\underline{R}^m - \underline{R}^n)} \left(\langle e^{-i\underline{K}(\underline{s}^m - \underline{s}^n)}\rangle - \langle e^{-i\underline{K}\underline{s}^m}\rangle \langle e^{i\underline{K}\underline{s}^n}\rangle\right).$$

Thus the sharp phonon lines $\delta\left(\omega - \omega_\sigma(\underline{q})\right)$ are smeared out by the correlation function $C_{\underline{K}}(\underline{q})$ and become diffuse.

One sees that the effect of static displacements on the inelastic scattering is very similar to their effects on the elastic scattering: The phonon lines (corresponding to the Bragg peaks) become shifted and diminished in intensity, and an additional nondispersive scattering (corresponding to the diffuse scattering) appears. However, note that in most cases these effects are rather small due to the smallness of the static displacements so that the mass- and force-constant disorder represent the more important effect.

Static displacements also lead to changes of the coupling parameters. Here we have to distinguish between two different displacements of the point defect. i) The displacements $\underline{s}^\infty$ which would occur in an infinite crystal. $\underline{s}^\infty(\underline{r})$ varies asymptotically as $1/r^2$ and the dilatation is asymptotically given by

$$\partial_{\underline{r}}\, \underline{s}^\infty(\underline{r}) \approx \frac{1}{r^3}\, f(\Omega) \quad \text{with} \quad \int d\Omega\, f(\Omega) = 0 \tag{7.112}$$

so that the asymptotic displacement field $\underline{s}^\infty(\underline{r})$ is dilatation free when averaged over the angle (the lattice dilatation is concentrated in the immediate vicinity of the defect). ii) In addition to $\underline{s}^\infty$ a very small image field $\underline{s}^{Im}(\underline{r})$ exists, which allows the surface of the finite crystal to be force free. It is slowly varying over the crystal dimension and leads to a dilatation which is practically uniformly spread over the whole crystal (especially when we consider that we always have many statistically distributed defects, the total image field practically represents a uniform expansion). The total volume change ΔV is the sum of the corresponding volume changes ΔV_∞ and ΔV^{Im}

$$\Delta V = \Delta V_\infty + \Delta V^{Im} \cong 3\, \frac{1-\nu}{1+\nu}\, \Delta V_\infty\,, \qquad \nu = \text{Poisson number}\,. \tag{7.113}$$

This volume change ΔV is, e.g., directly measured by lattice parameter measurements, since for cubic crystals the change Δa of the lattice constant a is

$$\frac{\Delta a}{a} = c\, \frac{\Delta V}{3V_c}\,.$$

Here one measures the average expansion over large distances (of the order of an extinction length, i.e., about 10^4 lattice constants). On the contrary, the coupling parameters are determined by the change of the distance between near neighbours, which locally changes considerably. Very near the defect, where the total displacements are large and given by $\underline{s}^\infty(\underline{r})$, we have appreciable changes of the nearest neighbour distances which lead to the local force-constant changes $\varphi^{(\mu)}$ near the defect μ. However in the defect free region, the lattice is only dilated due to $\underline{s}^{Im}(\underline{r})$, ΔV^{Im}, respectively. Thus in addition to the local changes $\varphi^{(\mu)}$ we have a homogeneous softening of the force constants due to the image expansion

$$\left(\Delta\phi_{ij}^{mn}\right)^{Im} = \left(\frac{\partial}{\partial V}\, \overset{\circ}{\phi}{}_{ij}^{mn}\right) N_d\, \Delta V^{Im} = \left(\frac{\partial}{\partial V}\, \overset{\circ}{\phi}{}_{ij}^{mn}\right) N_d\, \frac{2(1-2\nu)}{3(1-\nu)}\, \Delta V\,. \tag{7.114}$$

In addition we have the change $\Delta\omega_\sigma(q)$ of (7.110) due to the lattice expansion, i.e., due to the total volume change $\Delta V = \Delta V_\infty + \Delta V^{Im}$ entering in the positions $\underline{R}^m$. One part of this expression, the one proportional to ΔV^{Im}, can be lumped together with the above term by introducing the volume dependence of the unperturbed phonon, so that the sum of both effects gives

$$\Delta\omega_\sigma(\underline{q}) = \frac{\partial\overset{\circ}{\omega}_\sigma}{\partial V} N_d \Delta V^{Im} + \frac{\partial\overset{\circ}{\omega}_\sigma}{\partial \underline{q}} \cdot \underline{q} \frac{N_d \Delta V_\infty}{3V} . \tag{7.115}$$

Instead of $\partial\overset{\circ}{\omega}_\sigma/\partial V$ one may also introduce Grüneisen parameters $\gamma_\sigma(\underline{q})$

$$\gamma_\sigma(\underline{q}) = \frac{V_c}{\overset{\circ}{\omega}_\sigma(\underline{q})} \frac{\partial\overset{\circ}{\omega}_\sigma(\underline{q})}{\partial V} . \tag{7.116}$$

Erroneously it has been assumed in the past that the change of coupling parameters ϕ_{ij}^{mn} due to lattice expansion is proportional to ΔV as the lattice parameter change is, instead of being proportional to ΔV^{Im} (which, e.g., is about $(1/3)\Delta V$ for $\nu \approx 1/3$).

The effects of different scattering lengths and static displacements have not yet been treated in all details. Equation (7.105) for the dilute limit is due to KAGAN and ZHERNOV /7.14/. The effects of static displacements have only been discussed by KRIVOGLAZ /7.5/.

7.5 Results of Neutron Scattering Experiments

7.5.1 Effects due to Resonance Modes

MØLLER and MACKINTOSH /7.15/ measured the coherent inelastic scattering of the Cr-3at% W alloy (mass ratio: $M_W/M_{Cr} = 3.5$). The results for the transverse 100 and 110 direction are shown in Fig. 49. They clearly show a resonance behaviour as schematically shown in Fig. 47a. However due to lattice expansion leading to a homogeneous softening of the coupling parameters, the phonon lines for the alloy do not cross the ideal phonon lines of Cr at higher frequencies. CUNNINGHAM et al. /7.16/ studied the same alloy for a number of smaller concentrations. An analytical analysis by

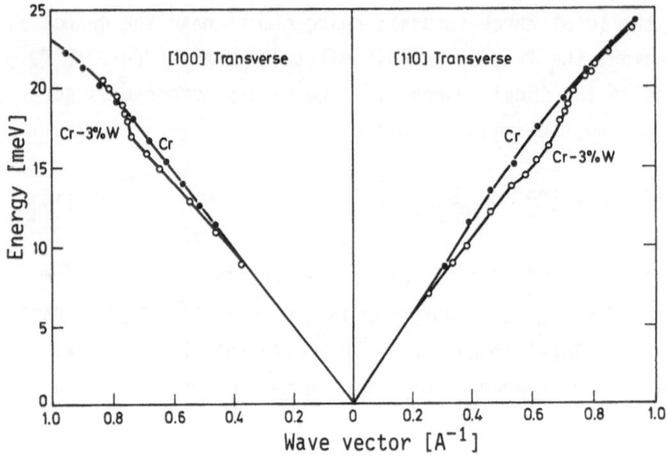

Fig. 49. Phonon dispersion in Cr (•) and Cr-3at% W (∘); according to /7.15/

KERSHARWANI and AGRAWAL /7.17/ indicated roughly a 20% strengthening of the nearest neighbour W-Cr force constants over those of the pure Cr.

SVENSON and BROCKHOUSE /7.18/, and SVENSON and KAMITAKAHARA /7,19/ studied a Cu alloy containing 3% Au (mass ratio M_{Au}/M_{Cu} = 3.1). A theoretical analysis of the system has been worked out by BRUNO and TAYLOR /7.20/. Figure 50 shows the measured frequency shifts $\Delta\omega$ for the T_1 (110) branch together with a theoretical curve corresponding to a fitted n.n. Au-Cu force constant strengthened by about 26%.

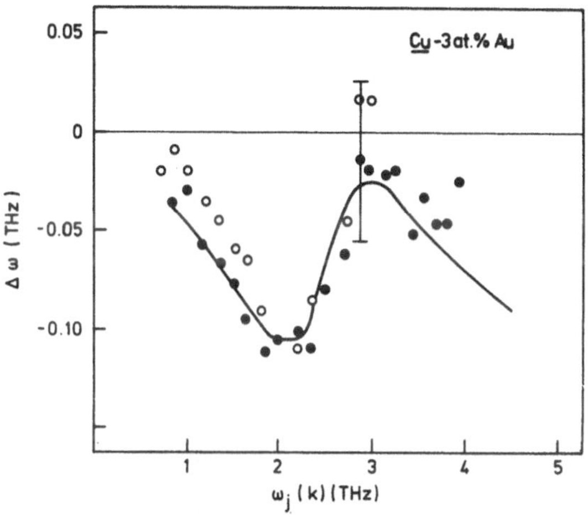

Fig. 50. Phonon shift $\Delta\omega$ for the T_1 (110) branch of Cu-3at% Au. The open und solid circles are the experimental values of SVENSON and KAMITAKAHARA /7.19/. The line represents a theoretical fit of BRUNO and TAYLOR /7.20/ with a n.n. Au-Cu force constant about 26% larger than that of Cu-Cu

MOSTOLLER et al. /7.21/ measured the phonons in the hcp rare earth alloy $Y_{0.9}Tb_{0.1}$ and compared the result with the mass defect CPA theory. The system shows a resonance mode at a relatively large frequency. Good agreement with the mass defect theory is obtained.

The first indication of the hybridization effect shown in Fig. 47b is due to WALTON et al., who measured the system KCl doped with CN^- /7.22/. This system has been studied extensively by thermal conductivity /7.23/ and ultrasonic measurements /7.24/. Especially the (110) phonons couple strongly to molecular transitions between the ground state and hindered rotational modes of CN^-. The whole effect is strongly temperature dependent indicating that the internal motions are strongly anharmonic. The system may be treated approximately as a two-level system, for which the t matrix has to be multiplied by a thermal population factor tanh $(\hbar\omega_{libr}/kT)$ (see /7.25/) so that the effect vanishes for higher temperatures. The measurements of WALTON et al. show just this (Fig. 51): Whereas for T = 10 K a clear splitting of the phonon dispersion curves occurs, for higher temperatures the measured values approach the ideal crystal results (dashed line).

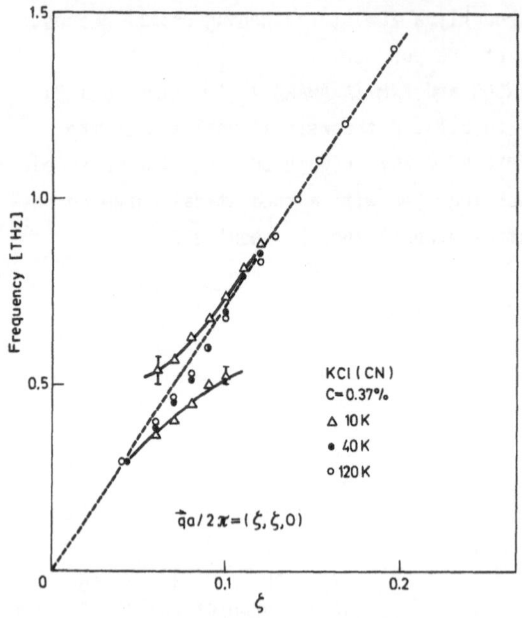

Fig. 51. Dispersion relation of phonons in $KCl_{1-c}(CN)_c$ for c = 0.37% at different temperatures. The dotted line represents pure KCl. At T = 10 K, hybridization is observed (according to /7.22/)

The more simple system Al-3.5% Ag and Al-8% Ag with a mass ratio $M_{Ag}/M_{Al} = 4$ has been studied recently by ZINKEN et al. /7.26/. Due to the large mass ratio and an additional softening of the local force constants by about 10% the expected splitting of the phonon line should occur, if c > 3%. Figure 52 shows the results for the transverse 001 mode. The tendency for splitting is clearly seen, especially for c = 8%. The arrows mark the position of the resonance frequency. A similar splitting of the phonon lines has been observed recently by SMITH et al. /7.27/ in the system $Mo_{0.85}Re_{0.15}$. (See also the article by NICKLOW /1.14/ for a recent review).

KUNITOMI et al. /7.28/ measured the phonon dispersion in the disordered $Ni_{1-x}Pt_x$ system over the whole range of decomposition. In the dilute limit of Pt in the Ni host one expects a resonance mode due to the large mass ratio ($M_{Pt}/M_{Ni} = 3.3$). Figure 53 shows the dispersion in 100 direction for a 5% concentration of Pt atoms. Due to the resonance mode at 3.3 THz a hybridization and splitting of the longitudinal and transversal phonons is observed. From mass-defect theory (CPA) the resonance frequency is predicted at 3.1 THz, indicating a slight increase of the Pt-Ni force constant over the Ni one, which is plausible due to the larger size of the Pt atom.

Similar resonance effects should also occur for the low-frequency libration mode of dumbbell interstitials in fcc, as has been discussed by WOOD et al. /7.29/, and SCHOBER et al. /7.30/. An experimental observation of these effects would be interesting, since it would give us additional information about the symmetry and force constants of the selfinterstitial. However, due to the very small defect concentra-

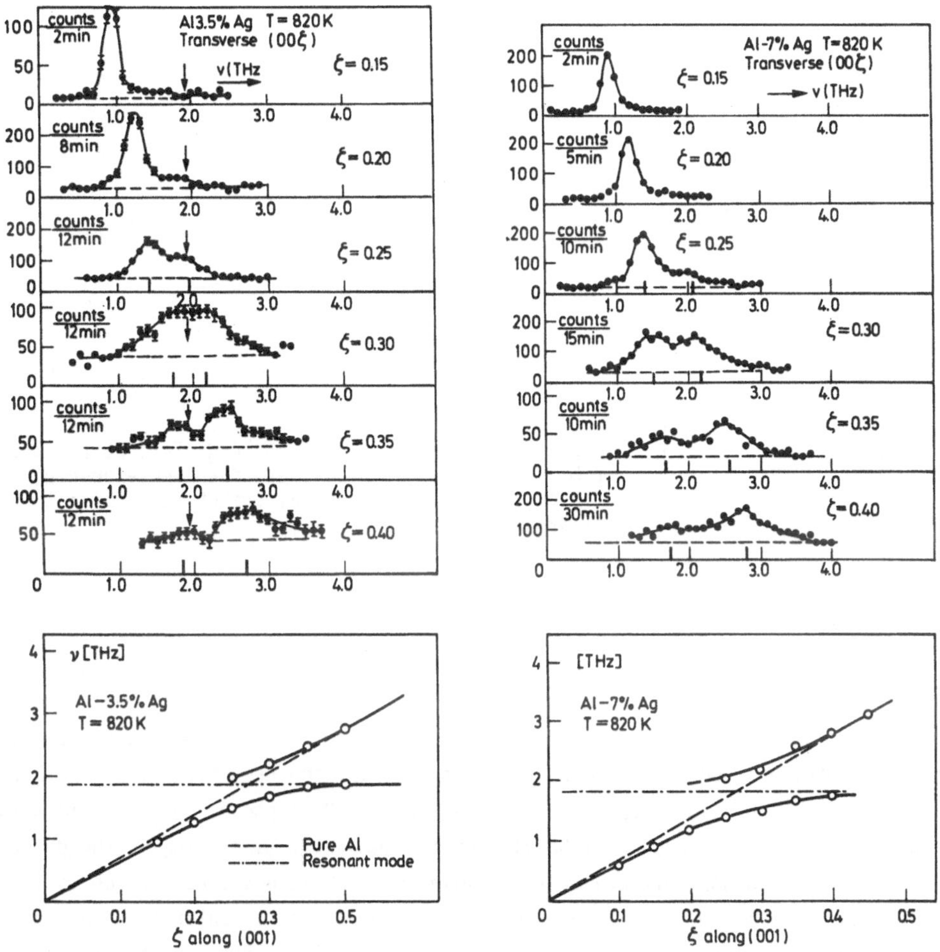

Fig. 52. Constant q scans and phonon dispersion at 820 K for Al-3.5at% Ag and Al-7at% Ag for the transverse (001)-branch (according to /7.30/)

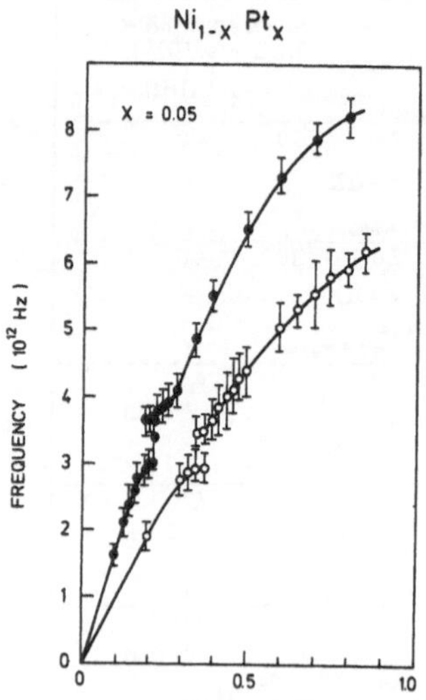

Fig. 53. Dispersion relation in Ni-5at% Pt for the (001)-direction; open circles: transverse branch, closed circles: longitudinal branch (according to /7.32/)

tions obtainable by irradiation such experiments are extremely difficult. Two experimental groups /7.31,32/ have worked on these problems. NICKLOW et al. measured the phonon dispersion in Cu crystals irradiated at 4.2 K with thermal neutrons to produce Frenkel pair concentrations of 40 ppm /7.31a/, and recently 120 ppm /7.31b/. Figure 54a shows the line shape of the transversal T (00ζ) phonon which should most strongly couple to the resonant libration of the dumbbell. Surprisingly, the phonon peak shapes show unusual structure, the origin of which is not clear. Since it is not removed after a 72 K anneal and even a 300 K anneal, it is unlikely to be connected with single interstitials. In addition to this effect the central line is slightly shifted and the resulting peak shift, plotted in Fig. 54b, is consistent with theoretical expectations /7.29,30/. BÜNING et al. /7.32/ measured the phonon dispersion of the T(00ζ) branch of Al irradiated by fast neutrons to produce a much higher Frenkel pair concentration of 810 ppm. Figure 54c shows the resulting shift of the phonon versus q together with a theoretical prediction. Apparently only a general softening is measured but not a resonance structure as, e.g., in Fig. 54b. While the interpretations of both experiments are not quite clear, the discrepancies can most likely be attributed to the different kind of defect damage produced by thermal and fast neutron irradiation. Whereas by thermal n irradiation predominantly isolated Frenkel pairs are formed, by fast n irradiation a high percentage of clustered interstitials is formed, so that the shift shown in Fig. 54c is presumably due to a mixture of small interstitial agglomerates of different sizes. Note, for instance, that already for di-interstitials we obtain quite a number of resonance

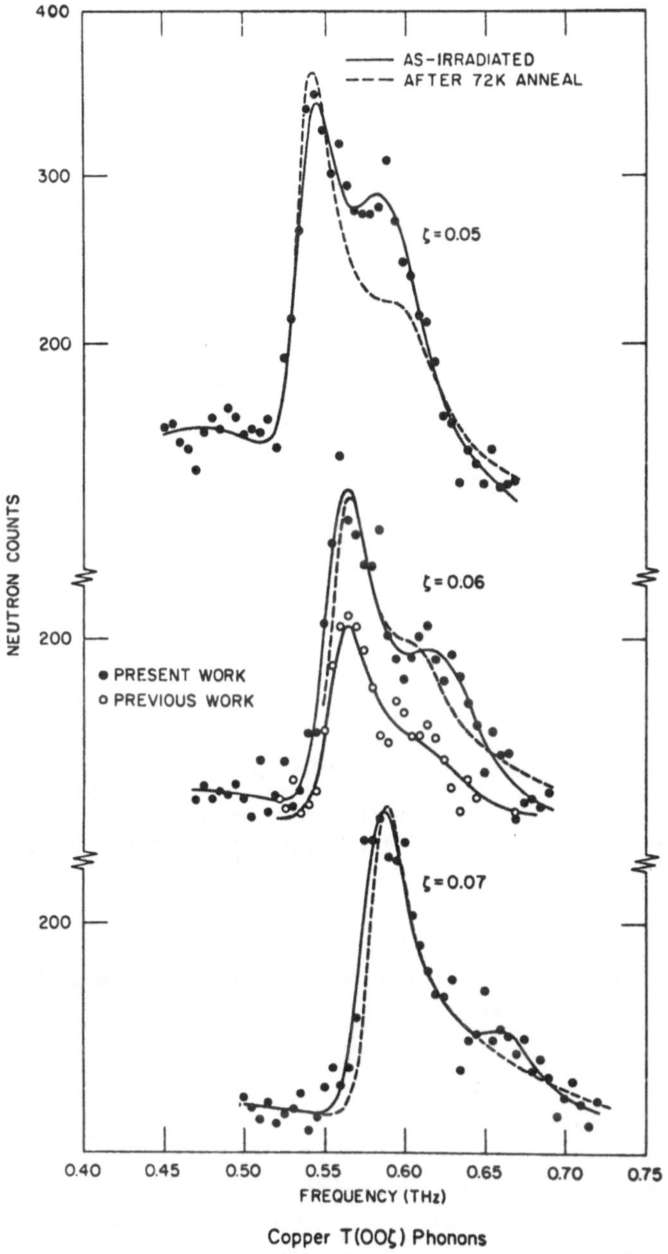

Fig. 54. a) Frequency distributions of neutrons scattered from irradiated Cu at 4 K and after 72 K anneal for q = (00ζ) 2π/a (after /7.31b/)

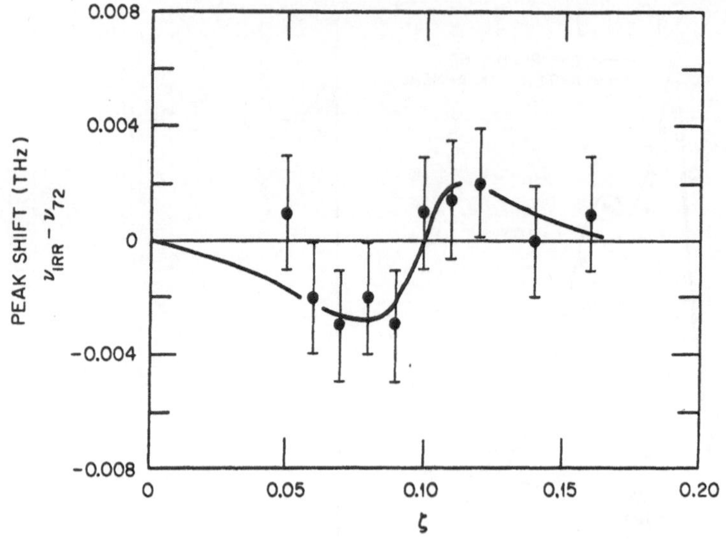

b)

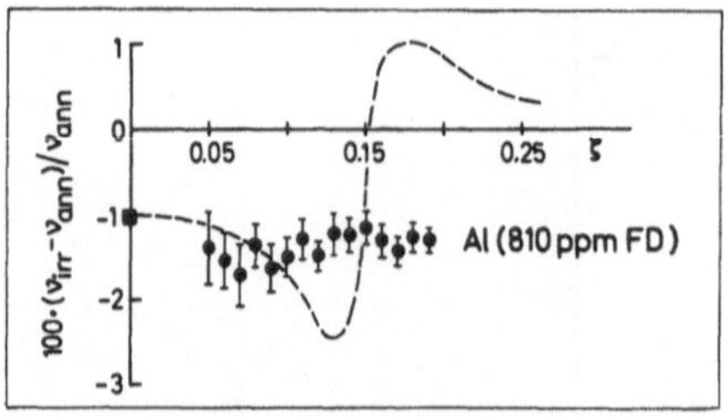

c)

Fig. 54. b) The frequency shifts of phonons in irradiated Cu relative to the values measured after a 72 K anneal (after /7.31b/)

c) The relative frequency shifts of phonons in fast neutron-irradiated Al for q = (00ζ) 2π/a. The dashed line represents a theoretical curve which has been fitted to the experimentally observed changes of the elastic constant c_{44} (▪) /7.50/ (after /7.32/)

modes coupling to elastic deformations (see Fig. 44b) so that for a mixture of different clusters the typical resonance structure is expected to disappear due to the superposition of different resonance curves.

7.5.2 Observation of Localized Modes

NICKLOW et al. /7.33/ studied localized vibrations in CuAl by coherent neutron scattering (mass ratio M_{Al}/M_{Cu} = 0.43). Fig. 55 shows one of their results: Constant q scans as obtained for pure Cu shows only the expected peak from the coherent neutron scattering by the L(ζζζ) mode, and this zone boundary mode has the highest frequency ω_{max} observed in Cu. The results for the Cu-4at% Al alloy are considerably different: The frequency of the L(ζζζ) mode has decreased significantly. At the same time its intensity is diminished. An additional peak has appeared at a frequency slightly above the band edge due to coherent scattering at the Al-induced local mode. For the Cu-10at% Al alloy the intensity of the local mode has increased roughly

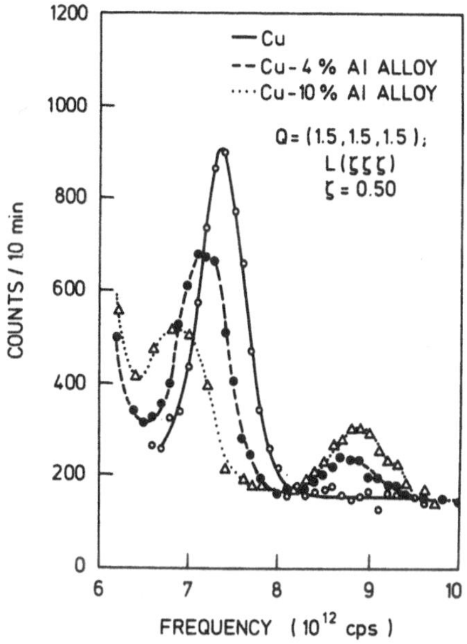

Fig. 55. Constant q scans for Cu and two Cu-Al alloys showing peaks from scattering by in-band modes (high peaks) and localized modes (smaller peaks). The q components are in units of $2\pi/a$, where a is the lattice parameter (according to /7.33/)

proportional to the concentration, while the frequency remains stable. On the contrary, the frequency of the $L(\zeta\zeta\zeta)$ band mode has decreased further. As a function of the wave vector $\underline{q}$ the intensity of the localized mode is maximal near $\underline{q} = \frac{2\pi}{a}(1.5\,,1.5\,,1.5)$, i.e., at the zone boundary where $\omega^2_{\ell oc} - \omega^2_{\sigma}(\underline{q})$ is smallest [compare (7.81)]. Figure 56 shows this intensity variation of the local-mode scattering as a function of $\underline{q}$ in the (111) direction. The strong $\underline{q}$ dependence directly shows that the scattering is coherent. The agreement with the calculated curve (mass-defect CPA) is very good.

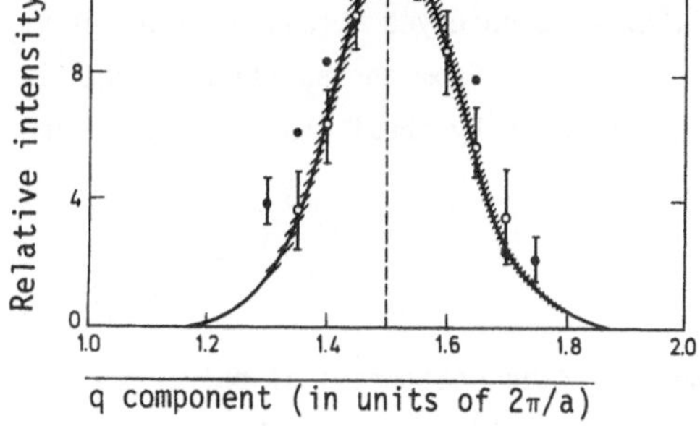

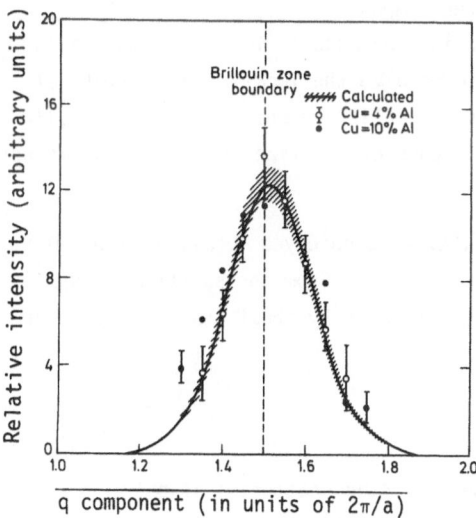

Fig. 56. Comparison of the q dependence of the measured *relative* intensity of the local-mode scattering in Cu-Al with that calculated from mass-defect theory (according to /7.33/)

WAKABAYASHI et al. /7.34/ studied impurity modes in Ge-9.2% Si. WAKABAYASHI /7.35/ compared the results with the mass-defect CPA-theory. Figure 57 shows the experimental and theoretical line shapes of the scattered neutron groups for several momentum tranfers. The large peak is an in-band mode, the higher smaller peak the localized mode due to Si. The theoretical results of the CPA (full line) agree very well with

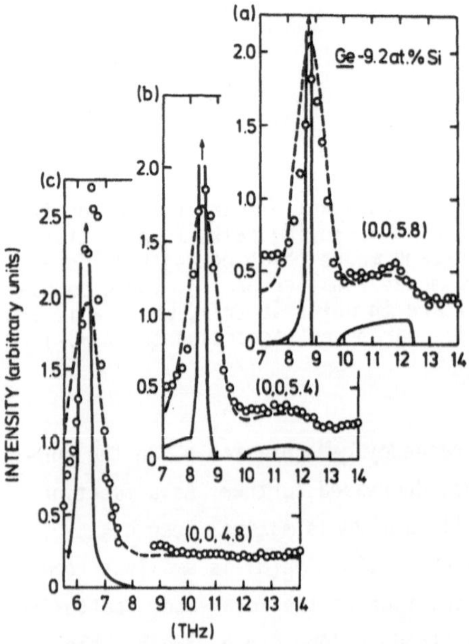

Fig. 57. Experimental and theoretical line shapes of the scattered neutron groups of Ge-9.2at% Si for the momentum transfer q = (0,0,5.8), (0,0,5.4) and (0,0,4.8). The solid {dotted} lines represent calculations (mass-defect CPA) without {with} instrumental resolution corrections (according to /7.35/)

the experimental results, when the instrumental broadening is taken into account. Apparently, force-constant changes are not very important.

ALS-NIELSON /7.36/ observed localized and in-band mode in Ta-12% Nb. Due to instrumental broadening, it was not possible to resolve the localized mode peak, although the data demonstrate the existence of a localized mode. SMITH et al. /7.37/ found by coherent neutron scattering the existence of localized translational as well as torsional modes of NH_4 in KCl.

Note added in proof: NATKANIEC et al. /7.56/ have observed localized modes in MgLi and CuBe for impurity concentrations of 5 and 10%. From the position of the local mode frequencies a decrease of the n.n. force constants by about 45% (for MgLi) and 55% (for CuBe) is estimated.

7.5.3 Incoherent Scattering Experiments

CHERNOPLEKOV and ZEMLYANOV /7.38/ studied the inelastic scattering of polycrystalline Mg-2.8% Pb alloys by the time of flight method (mass ratio M_{Pb}/M_{Mg} = 8.6). They ob-

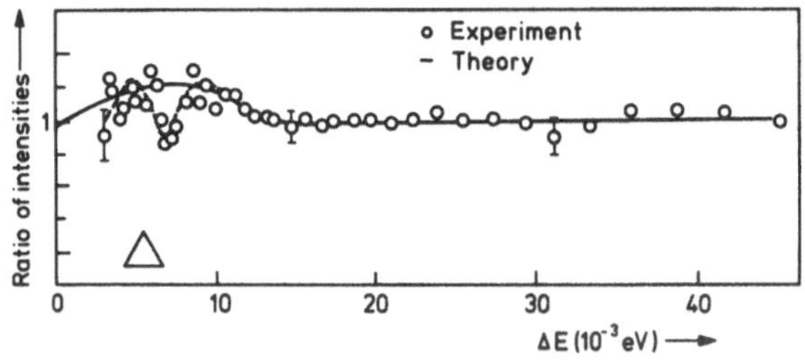

Fig. 58. Experimental ratio of the inelastic neutron-scattering cross sections of the Ti-5% U alloy and of pure Ti. The triangle indicates the experimental resolution. The theoretical curve is calculated using an isotropic Debye-approximation for the ideal spectrum (according to /7.39/)

served a resonance peak at low frequencies in agreement with mass-defect theory. In a similar measurement CHERNOPLEKOV et al. /7.39/ studied the system Ti-5% U (mass ratio 5). Figure 58 shows some of the experimental results: The ratio of the intensity of the inelastic scattering from Ti-5% U to the inelastic scattering of pure Ti. A split peak is observed centered at about 7meV, which is due to the splitting of the resonance mode in the hexagonal Ti lattice.

MOSER et al. /7.40/ first studied dilute V alloys (V-Be, V-Pt and V-Ni). They clearly observed a localized mode for V-Be, but for V-Pt a resonance peak was not clearly observed. SYRYKH et al. /7.41/, and CHERNOPLEKOV et al. /7.42/ made a study of the scattering from V-Ta and V-W alloys. Here only very small irregularities in the spectrum at small frequencies are observed which have been explained by the near coincidence of the cross sections of Ta and W with V. A much bigger effect is found in V-U alloys due to the large U-scattering amplitude /7.43/. Figure 59a,b shows the measured and calculated intensity of V-3% U. A peak due to the U-resonance mode is clearly seen. For the theoretical curve the U-V force constant has been decreased by a factor of 2 compared to the V-V one.

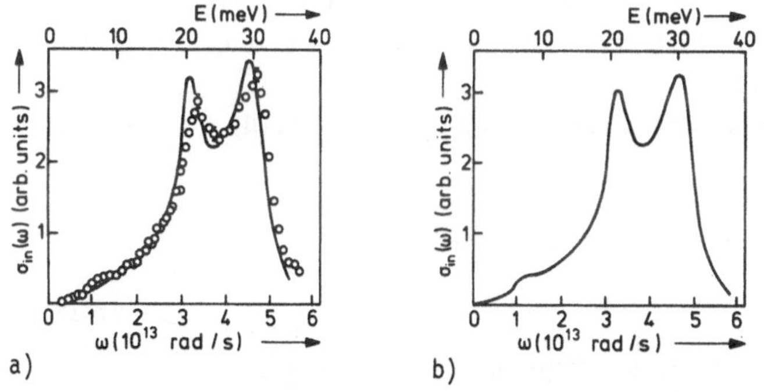

Fig. 59. a) Incoherent cross section $\sigma_{in}(\omega)$ for V-3at% U (o) and pure U
b) Calculated cross section $\sigma_{in}(\omega)$ for V-3at% U with allowance for a V-U force constant decreased by 50% compared with that for V-V.
(according to /7.43/)

7.6 Change of Elastic Constants

The elastic properties of the disordered crystal are described by the static Green's function $\langle G(\omega=0)\rangle$ which itself is determined by the effective static coupling constant $\overset{\circ}{\phi}+\Sigma(\omega=0)$

$$\langle G(0)\rangle = \left\langle \frac{1}{\phi}\right\rangle = \frac{1}{\overset{\circ}{\phi}+\Sigma(0)} \, . \tag{7.117}$$

In a plane wave representation $\langle G(0)\rangle$ may be written as

$$\langle G^{mn}_{ij}(0)\rangle = \sum_{\sigma} \int_{V_B} \frac{d\underline{k}}{V_B} \frac{e^{\sigma}_i(\underline{k})\, e^{\sigma}_j(\underline{k})}{\phi_\sigma(\underline{k},0)} \exp[i\underline{k}(\underline{R}^m - \underline{R}^n)]$$
$$\text{with } \phi_\sigma(\underline{k},0) = \overset{\circ}{M}\omega^2_\sigma(\underline{k}) + \Sigma_\sigma(\underline{k},0) \, . \tag{7.118}$$

For large distances $|\underline{R}^m - \underline{R}^n| \to \infty$, the major contribution to the integral comes from the $1/k^2$ singularity of $1/\phi_\sigma(\underline{k},0)$ at the origin, which is determined by the elasticity tensor $C_{ijk\ell}$

$$\phi_\sigma(\underline{k},0) \cong V_c \sum_{ijk\ell} e^{\sigma}_i(\hat{\underline{k}})\, C_{ijk\ell} k_j k_k\, e^{\sigma}_\ell(\hat{\underline{k}}) \quad \text{for } \underline{k} \to 0 \, . \tag{7.119}$$

Thus asymptotically the lattice Green's function $\langle G^{mn}_{ij}(0)\rangle$ approaches the continuum Green's function $\langle G_{ij}(\underline{R}^m - \underline{R}^n)\rangle$

$$\langle G_{ij}(\underline{R}^m - \underline{R}^n)\rangle = \int_{-\infty}^{\infty} \frac{d\underline{k}}{(2\pi)^3} \exp[i\underline{k}(\underline{R}^m - \underline{R}^n)] \left\{\underline{\underline{\tau}}^{-1}(\underline{k})\right\}_{ij}$$
$$\text{with } \tau_{i\ell}(\underline{k}) = \sum_{jk} C_{ijk\ell} k_j k_k \, . \tag{7.120}$$

Due to the $1/k^2$ singularity, $G_{ij}(\underline{R}^m - \underline{R}^n)$ varies asymptotically as $1/|\underline{R}^m - \underline{R}^n|$ times a complicated angular function, which can only be calculated analytically for isotropic and for hexagonal crystals.

Since in the dilute limit the static self-energy is given by $\Sigma(0) = N_d\langle t(0)\rangle$ (7.32) we have for $\Sigma_\sigma(\underline{k},0)$

$$\Sigma_\sigma(\underline{k},0) = c \sum_{\substack{mn\\ik}} \exp(i\underline{k}\underline{R}^m)\, e^{\sigma}_i(\underline{k})\, t^{mn}_{ik}(0)\, e^{\sigma}_k(\underline{k}) \exp(-i\underline{k}\underline{R}^n)$$
$$\cong c \sum_{ijk\ell} e^{\sigma}_i(\hat{k})\, e^{\sigma}_k(\hat{k})\, k_j k_\ell \sum_{mn} R^m_j\, t^{mn}_{ik}(0)\, R^n_\ell \qquad \text{for } \underline{k} \to 0. \tag{7.121}$$

Here we have expanded the exponentials: $\exp(i\underline{k}\underline{R}^m) \cong 1 + i\underline{k}\underline{R}^m$. The first terms vanish due to the translational invariance of the coupling parameters $\left(\sum_m t^{mn}_{ik}(0) = 0 = \sum_n t^{mn}_{ik}(0)\right)$.

By comparing with (7.119) we see that the change $\Delta C_{ijk\ell}$ of the elasticity tensor is given by

$$\Delta C_{ijk\ell} = \frac{c}{V_c} \sum_{mn} R_j^m \, t_{ik}^{mn}(0) \, R_\ell^n \, . \tag{7.122}$$

Note that the elastic data are independent of the atomic masses. Therefore mass resonances do not influence the elastic behaviour. On the contrary, spring resonances, signalizing a near instability of the defect, can lead to a strong softening of the elastic constants, but only if the resonance modes have the proper symmetry. By using the resonance representation (7.71) of the t matrix

$$t(0) = \varphi - \varphi \frac{|\alpha\rangle\langle\alpha|}{\mathring{M}\omega_{res}^2} \varphi \cong - \varphi \frac{|\alpha\rangle\langle\alpha|}{\mathring{M}\omega_{res}^2} \varphi \tag{7.123}$$

we see that such a resonance $|\alpha\rangle$ leads to large and negative changes of the elastic moduli, since t(0) and $\Delta C_{ijk\ell}$ are negative definite in the approximation (7.123). However for a defect with inversion symmetry only those resonance modes $|\alpha\rangle$ are effective and can couple to elastic deformations $\sum_j \varepsilon_{ij} R_j^m$, which have even symmetry

$$\alpha_i^{\underline{m}} = - \alpha_i^{-\underline{m}} , \tag{7.124}$$

i.e., the same symmetry as elastic deformations have ($R_j^{\underline{m}} = -R_j^{\underline{m}}$). Otherwise the matrix element $\langle \varepsilon R | \varphi | \alpha \rangle$ vanishes due to $\varphi_{ij}^{\underline{mn}} = \varphi_{ij}^{-\underline{m}-\underline{n}}$.

The connection of the elastic constants with the phonon dispersion curves can be seen by looking at the small frequency and long wavelength limit of the dispersion. By expanding $\Sigma(\underline{k},\omega)$ for small $\underline{k}$ and small ω, we obtain

$$\Sigma_{ik}(\underline{k},\omega) \cong -(\langle M\rangle - \mathring{M})\omega^2 \, \delta_{ik} + c \sum_{\substack{mn \\ j\ell}} k_j R_j^m \, t_{ik}^{mn}(0) \, k_\ell R_\ell^n \, . \tag{7.125}$$

By inserting into (7.62), we obtain for the phonon shift

$$\Delta\omega_\sigma^2(\underline{k}) = \frac{1}{\mathring{M}} \left[\sum_{\substack{ij \\ k\ell}} V_c \, \Delta C_{ijk\ell} \, k_j k_\ell \, e_i^\sigma(\hat{\underline{k}}) \, e_j^\sigma(\hat{\underline{k}}) - c(M_d - \mathring{M}) \, \mathring{\omega}^2(\underline{k}) \right] . \tag{7.126}$$

The elastic constants of a crystal can also be defined via the *potential energy of a deformed crystal*. If ε_{ij} is the homogeneous strain in an ideal crystal, the elastic energy U is

$$U = \frac{V}{2} \sum_{\substack{ij \\ k\ell}} C_{ijk\ell} \, \varepsilon_{ij} \, \varepsilon_{k\ell} \tag{7.127}$$

where V is the crystal volume. Also a crystal with many, statistically distributed defects can be homogeneously deformed on a macroscopic scale, even though microscopically the deformations are inhomogeneous. Therefore, in a defect crystal, a homogeneous deformation $\overset{o}{s}{}^m_i = \sum_j \varepsilon_{ij} R^m_j$ of the crystal surface determines the average, homogeneous strain of the defect crystal. Thus we can also define the elastic constants for a defect crystal by means of the strain dependence of the deformation energy U.

For a crystal with a small concentration $c = N_d/N$ of defects, the energy contribution of the different defects is additive

$$U = U_o + N_d u \;, \quad u = U_1 - U_o \tag{7.128}$$

so that it is sufficient to calculate the difference u between the energies of a crystal with a single defect (U_1) and of the ideal crystal (U_o).

In order to calcuate U_1, we have to know the displacements s^m_i of the crystal with one defect under a uniform displacement $\overset{o}{s}{}^m_i = \varepsilon_{ij} R^m_j$ of the surface atoms. Since the atoms in the bulk are force free, we have for interior atoms

$$\sum_{nj} \phi^{mn}_{ij} s^n_j = 0 \quad \text{or} \quad \sum_{nj} \overset{o}{\phi}{}^{mn}_{ij} s^n_j = -\sum_{nj} \varphi^{mn}_{ij} s^n_j \;. \tag{7.129}$$

By introducing the ideal Green's function $\overset{o}{G}{}_s{}^{mn}_{ij}$ for a fixed surface

$$\overset{oo}{\phi G_s} = 1 \quad \text{for interior atoms,} \qquad \overset{o}{G}{}_s{}^{mn}_{ij} = 0 \quad \text{for m and/or n at the surface,}$$

the displacements can be written as

$$\underline{s} = \underline{\overset{o}{s}} - \overset{o}{G}_s \varphi \underline{s} = \underline{\overset{o}{s}} - \overset{o}{G}_s t \underline{\overset{o}{s}} \quad \text{with} \quad t = \varphi \frac{1}{1 + \overset{o}{G}_s \varphi} \;; \quad \underline{\overset{o}{s}} = \varepsilon \underline{R} \tag{7.130}$$

which directly shows that in addition to the homogeneous field $\underline{\overset{o}{s}}$ one obtains displacements $-\overset{o}{G}_s \varphi \underline{s}$ localized near the defects. They can be thought of as arising due to induced "Kanzaki-forces"

$$\delta \underline{f}(\varepsilon) = -\varphi \underline{s} = -t \varepsilon \underline{R} \;. \tag{7.131}$$

The energy U_1 is then

$$U_1 = \frac{1}{2} (\underline{s}, \phi \underline{s}) = \frac{1}{2} (\underline{\overset{o}{s}}, \phi \underline{s}) \tag{7.132}$$

since $\phi \underline{s} = 0$ only for surface atoms, for which $\underline{s}$ coincides with $\underline{\overset{o}{s}} = \varepsilon \underline{R}$. Introducing $\phi = \overset{o}{\phi} + \varphi$, we obtain

$$U_1 = \frac{1}{2} (\underline{\overset{o}{s}}, \overset{o}{\phi} \underline{\overset{o}{s}}) + \frac{1}{2} (\underline{\overset{o}{s}}, \varphi \underline{s}) \tag{7.133}$$

since again $\overset{o}{\phi} \underline{\overset{o}{s}} = 0$ only at the surface, where $\underline{s} = \underline{s}_o$. Thus u is given by

$$u = \frac{1}{2}(\overset{\circ}{\underline{s}}, \varphi \underline{s}) = \frac{1}{2}(\overset{\circ}{\underline{s}}, t\overset{\circ}{\underline{s}}) = \frac{1}{2} \sum_{\substack{mn \\ ijk\ell}} \varepsilon_{ij} R_j^m t_{ik}^{mn}(0) \varepsilon_{k\ell} R_\ell^n . \tag{7.134}$$

By comparing with (7.127) we see that the change of the elasticity tensor is given by

$$\Delta C_{ijk\ell} = \frac{c}{V_C} \sum_{mn} R_j^m t_{ik}^{mn}(0) R_\ell^n . \tag{7.135}$$

This is in agreement with (7.122), if in the t matrix the Green's function $\overset{\circ}{G}_s$ for the finite crystal is replaced by the one for the infinite crystal, which is practically always allowed.

The derivation of $\Delta C_{ijk\ell}$ from the energy is especially interesting since it allows one to calculate the elastic data by a *variational technique*. Since ϕ is positive definite, one has for arbitrary $\hat{\underline{s}}$

$$\frac{1}{2}\left(\underline{s} - \hat{\underline{s}}, \phi(\underline{s} - \hat{\underline{s}})\right) \geqslant 0 . \tag{7.136}$$

If now at the surface $\hat{\underline{s}}$ coincides with the exact solution $\underline{s}$, i.e., if

$$\hat{\underline{s}} = \varepsilon \underline{R} + \delta \hat{\underline{s}} \quad \text{with} \quad \delta \hat{\underline{s}} = 0 \text{ for surface atoms,} \tag{7.137}$$

we obtain from above

$$\frac{1}{2}(\hat{\underline{s}}, \phi \hat{\underline{s}}) \geqslant \frac{1}{2}(\underline{s}, \phi \underline{s}) . \tag{7.138}$$

The mixed terms are identical to $(\underline{s}, \phi \underline{s})$,

$$(\hat{\underline{s}}, \phi \underline{s}) = (\underline{s}, \phi \underline{s}) = (\underline{s}, \phi \hat{\underline{s}}) \tag{7.139}$$

since $\phi \underline{s} \neq 0$ only at the surface, where $\hat{\underline{s}} = \varepsilon \underline{R} = \underline{s}$. Thus (7.138) gives an upper bound to the true deformation energy for all $\delta \hat{\underline{s}}$, which vanish at the surface. An example is the first Born approximation, $\hat{\underline{s}} = \varepsilon \underline{R}$ or $\delta \hat{\underline{s}} = 0$

$$\frac{1}{2}(\varepsilon \underline{R}, \phi \varepsilon \underline{R}) \geqslant \frac{1}{2}(\underline{s}, \phi \underline{s}) . \tag{7.140}$$

This is plausible, since the additional relaxations $\delta \hat{\underline{s}}$ should lower the energy. Other and better bounds can be obtained, by allowing only relaxations $\delta \hat{\underline{s}} \neq 0$ in a restricted region near the defect. For instance, in an Einstein-type approximation only the defect is allowed to relax, or only the defect and its nearest neighbours, or only the atoms in the "defect subspace" where $\varphi \neq 0$, etc. By denoting the corresponding Einstein-type Green's function by $\mathscr{G}_E$, i.e.,

$$\begin{aligned} &\overset{\circ}{\phi} \overset{\circ}{\mathscr{G}}_E = 1 && \text{for all atoms in the "relaxed" subspace} \\ &\overset{\circ}{\mathscr{G}}{}_{E\,ij}^{\;mn} = 0 && \text{for all m and/or n outside} \end{aligned} \tag{7.141}$$

we have

$$\hat{\underline{s}} = \varepsilon\underline{R} - \overset{\circ}{\mathscr{G}}_E\varphi\hat{\underline{s}} = \varepsilon\underline{R} - \overset{\circ}{\mathscr{G}}_E t_E \varepsilon\underline{R} \quad \text{with } t_E = \varphi\frac{1}{1+\mathscr{G}_E\varphi} . \tag{7.142}$$

For the energy (7.133) one obtains by substracting the zero order term

$$\frac{1}{2}(\varepsilon\underline{R}\,,\, t_E\varepsilon\underline{R}) \geq \frac{1}{2}(\varepsilon\underline{R}\,,\, t\varepsilon\underline{R}) = u \tag{7.143}$$

It is easy to prove, that by continuously enlarging the relaxed subspace, the corresponding upper bound continuously decreases, until it approaches the exact value.

In this way, certain combinations of the elastic constants, the elastic moduli $c^{(\rho)}$, can be calculated by variation. The moduli $c^{(\rho)}$ may be defined as the eigenvalues of the elasticity tensor $C_{ijk\ell}$ with suitable elastic strains as eigenvectors

$$\sum_{k\ell} C_{ijk\ell}\,\varepsilon_{k\ell}^{(\rho)} = c^{(\rho)}\,\varepsilon_{ij}^{(\rho)} \;,\quad \rho = 1,2,\ldots,6 \;,\quad \sum_{ij}\varepsilon_{ij}^{(\rho)}\,\varepsilon_{ij}^{(\rho')} = \delta_{\rho\rho'} \;. \tag{7.144}$$

By introducing the eigenstrains $\varepsilon_{ij}^{(\rho)}$, the energy can be written as a sum of 6 positive contributions

$$U = \frac{V}{2}\sum_{ijk\ell} C_{ijk\ell}\,\varepsilon_{ij}\,\varepsilon_{k\ell} = \frac{V}{2}\sum_{\rho=1}^{6} c^{(\rho)}\,\varepsilon_\rho^2 \quad \text{with } \varepsilon_\rho = \sum_{ij}\varepsilon_{ij}^{(\rho)}\,\varepsilon_{ij} \;. \tag{7.145}$$

For instance, for a cubic crystal we have three different $c^{(\rho)}$'s: the compression modulus $3K = c_{11} + 2c_{12}$, the 100-shear modulus $2c_{44}$ and the 110-shear modulus $c_{11} - c_{12}$. For the corresponding eigenstrains $\varepsilon_{ij}^{(\rho)}$ we refer to /2.4/.

By using the normalized eigenstrains $\varepsilon^{(\rho)}$ the change of the modulus $c^{(\rho)}$ follows from (7.135) as

$$\Delta c^{(\rho)} = \frac{c}{V_C}(\varepsilon^{(\rho)}\underline{R}\,,\, t\varepsilon^{(\rho)}\underline{R}) \leq \frac{c}{V_C}(\varepsilon^{(\rho)}\underline{R}\,,\, t_E\varepsilon^{(\rho)}\underline{R}) \;. \tag{7.146}$$

So far we have neglected in our discussion the effects of the *long-ranging* permanent *displacements* around the point defects. They affect the elastic constants in two ways. First, the volume V, entering in (7.127) is changed by $N_d\Delta V$, where ΔV is the volume change of the single defect. Also the positions $\underline{R}^m$ in the ideal lattice are changed by $\Delta\underline{R}^m = c(\Delta V/3V_C)\underline{R}^m$ due to the homogeneous lattice expansion. Secondly, and more important, the coupling parameters far away from the defects are changed, but only due to the homogeneous expansion induced by the image forces alone. This term can be expressed by a volume or pressure derivative of the elastic constants. Together one obtains

$$\Delta C_{ijk\ell} = -\,c\frac{\Delta V}{3V_C}\overset{\circ}{C}_{ijk\ell} - c\frac{\Delta V}{V_C}^{\mathrm{Im}}\frac{c_{11}+2c_{12}}{3}\frac{\partial\overset{\circ}{C}_{ijk\ell}}{\partial p} + \frac{c}{V_C}\sum_{mn} R_j^m\, t_{ik}^{mn}(0)\, R_\ell^n \;. \tag{7.147}$$

This result is quite analogous to the corresponding result (7.115) for the phonon dispersion curves.

As a simple example we will consider a substitutional defect in fcc with nearest neighbour interaction and longitudinal force constants f and $\overset{\circ}{f}$ for the impurity - host and host - host interaction, respectively. The results have been worked out by BREUER /2.4/. A homogeneous compression only couples to the A_{1g} subspace, yielding a compression modulus

$$\Delta(c_{11}+2c_{12}) = c\,\frac{8}{a}\,(f-\overset{\circ}{f})\left[1 + 0.24\,\frac{f-\overset{\circ}{f}}{\overset{\circ}{f}}\right]^{-1} . \tag{7.148}$$

A 110-shear couples to the E_g subspace. The corresponding modulus $(c_{11}-c_{12})$ is changed by

$$\Delta(c_{11}-c_{12}) = c\,\frac{2}{a}\,(f-\overset{\circ}{f})\left[1 + 0.38\,\frac{f-\overset{\circ}{f}}{\overset{\circ}{f}}\right]^{-1} . \tag{7.149}$$

Finally a 100-shear couples to the F_{2g} subspace. The change of the 100-shear modulus follows as

$$\Delta(2c_{44}) = c\,\frac{4}{a}\,(f-\overset{\circ}{f})\left[1 + 0.33\,\frac{f-\overset{\circ}{f}}{\overset{\circ}{f}}\right]^{-1} . \tag{7.150}$$

For the Born approximation, where $t \approx \varphi$, the denominators are replaced by 1.

In all three cases the change of the elastic moduli is essentially proportional to the change of the force constants. Note that for $f > 0$ no resonances can occur for which the denominator is small, since one needs negative force constants in order to get an instability (mass resonances do not influence the elastic behaviour).

Especially interesting are two limiting cases. First, if the impurity force constant gets infinitely large, the elastic moduli nevertheless approach finite values which are determined by

$$\Delta(c_{11}+2c_{12}) \cong c\;33.3\,\frac{\overset{\circ}{f}}{a}\,,\quad \Delta(c_{11}-c_{12}) \cong c\;5.26\,\frac{\overset{\circ}{f}}{a}\,,\quad \Delta(2c_{44}) \cong c\;12.1\,\frac{\overset{\circ}{f}}{a} \tag{7.151}$$

In this limit the Born approximation, yielding infinite values, is completely wrong. As a second case we consider a vacancy, which can be described by setting $f = 0$. This is allowed since the three pure translational degrees of freedom of the atom at the vacancy site are of odd symmetry and do not couple to homogeneous deformations. For this model we obtain a weakening of all elastic constants

$$\Delta(c_{11}+2c_{12}) = -c\;10.5\,\frac{\overset{\circ}{f}}{a}\,,\quad \Delta(c_{11}-c_{12}) = -c\;3.23\,\frac{\overset{\circ}{f}}{a}\,,\quad \Delta(2c_{44}) = -c\;6.06\,\frac{\overset{\circ}{f}}{a} \tag{7.152}$$

For the same model the elastic constants of the ideal crystal are given by

$$c_{11} + 2c_{12} = \frac{4\overset{\circ}{f}}{a}\,,\quad c_{11} - c_{12} = \frac{\overset{\circ}{f}}{a}\,,\quad 2c_{44} = \frac{2\overset{\circ}{f}}{a}\,. \tag{7.153}$$

Therefore the relative changes per atomic concentration are $\dfrac{\Delta c^{(\rho)}}{c\,c^{(\rho)}} \approx -3$ for all three moduli.

$$\frac{\Delta(c_{11}+2c_{12})}{c(c_{11}+2c_{12})} = -2.64 \ , \quad \frac{\Delta(c_{11}-c_{12})}{c(c_{11}-c_{12})} = -3.20 \ , \quad \frac{\Delta c_{44}}{cc_{44}} = -3.00 \ . \qquad (7.154)$$

There exist extensive *experimental studies* about the *elastic data* of dilute alloys. For instance, PURWINS et al. /7.44/ measured the elastic data of single crystalline Ag alloys. KÖSTER /7.45/ reported older measurements together with new results for Cu, Ag, and Au alloys. LENKKERI and LÄHTEENKORVA /7.46/ investigated the V-Cr system, HUBBELL and BROTZEN /7.47/ the Nb-Mo system. As yet, there is practically no theoretical effort to explain these data or to extract from them information about the dynamics of these systems.

As an instructive example for these measurements, Fig. 60a-c shows the Young modulus of the Ag and Au alloys, normalized to the atomic concentration c, as a

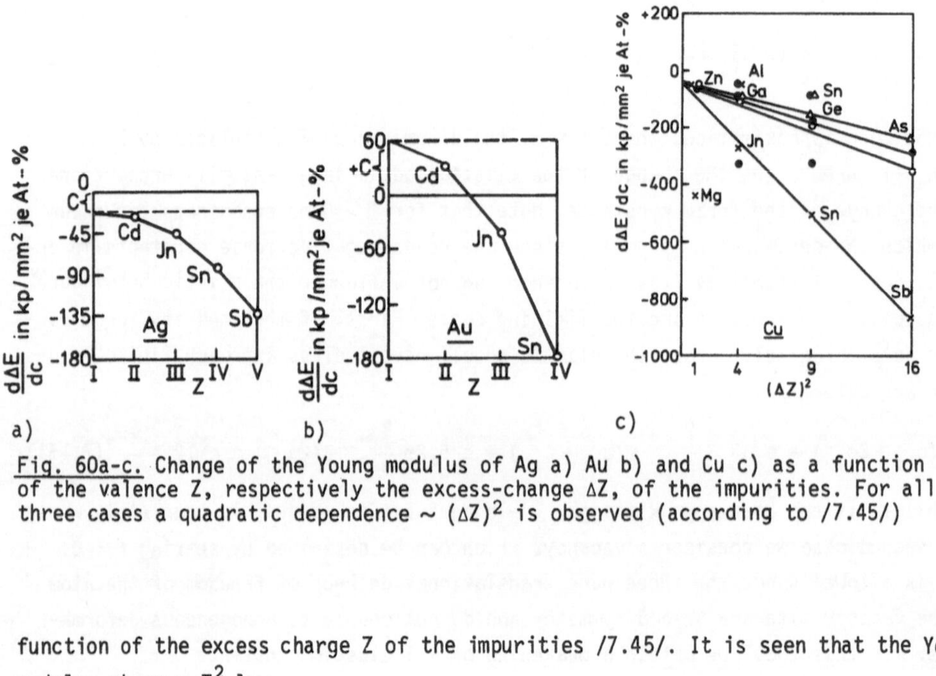

Fig. 60a-c. Change of the Young modulus of Ag a) Au b) and Cu c) as a function of the valence Z, respectively the excess-change ΔZ, of the impurities. For all three cases a quadratic dependence ~ $(\Delta Z)^2$ is observed (according to /7.45/)

function of the excess charge Z of the impurities /7.45/. It is seen that the Young modulus obeys a Z^2-law

$$E = c(C - \alpha Z^2) \qquad (7.155)$$

where the constants C and α are different for each alloy system. Figure 60c shows the corresponding results for Cu alloys, this time plotted against $(\Delta Z)^2$. The Z dependence can be explained by the theory of FUCHS /7.48/. Unforunately, no pseudopotential calculations have been reported until now.

In the last years the effects of *interstitials and vacancies* on the elastic data have been studied in detail. REHN et al. /7.49/ measured the change of the elastic constants of Cu irradiated by thermal neutrons at low temperatures. ROBROCK and SCHILLING /7.50/ measured the change of the Al elastic constants after electron irradiation. The results for the relative changes, normalized by the concentration, are summerized in the following table (relative change in % per atomic % defects).

Table 4

	$\frac{\Delta c_{44}}{c \cdot c_{44}}$	$\frac{\Delta(c_{11}-c_{12})}{c(c_{11}-c_{12})}$	$\frac{\Delta(c_{11}+2c_{12})}{c(c_{11}+2c_{12})}$
Experiment			
Frenkel pair Cu /7.49/	-31	-15	0
Frenkel pair Al /7.50/	-27	-15	—
Theory (Morse pot. Cu)			
100-dumbbell	-36.5	-10.5	- 8.9
vacancy	- 3.1	- 4.3	- 2.7
Frenkel pair	-39.6	-14.8	-11.6

From the table it is clear that

1) The changes of the shear moduli are negative and very large (the corresponding changes for alloys are typically of the order ±1). On the contrary, the compression modulus is changed only little.
2) The effect on the shear moduli is anisotropic, being appreciably larger for $2c_{44}$ (100-shear) than for $c_{11}-c_{12}$ (110-shear).

The magnitude and anisotropy of these values can be traced back to the dynamics of the 100-dumbbell (see Chap. 6). Of the three resonance modes E_g, A_{2u}, E_u only the libration mode E_g with even symmetry can couple to elastic deformations. However, for symmetry reasons this mode can only couple to 100-shear deformations (Fig. 61a) which determine the shear modulus c_{44}, but not to a homogeneous compression (Fig. 61b)

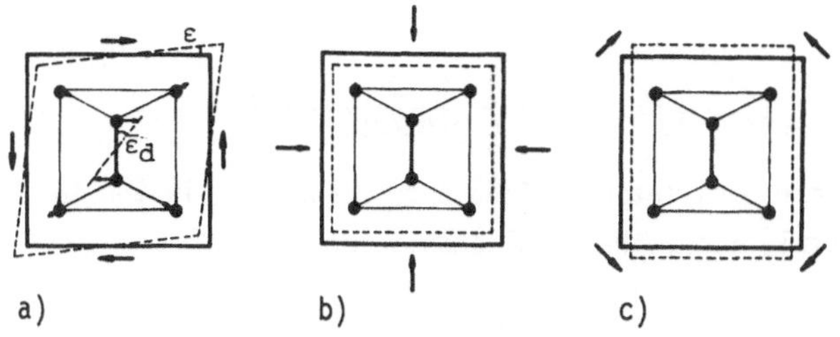

Fig. 61a-c. Homogeneous deformation of a crystal with a 100-dumbbell
a) 100-shear (modulus c_{44})
b) Compression [modulus $(c_{11}+2c_{12})/3$]
c) 110-shear [modulus $(c_{11}-c_{12})/2$]

determining the compression modulus $c_{11}+2c_{12}$, nor to 110-shear deformations (Fig. 61c) determining the shear modulus $c_{11}-c_{12}$.

For the 100-dumbbell the changes $\Delta c_{\alpha\beta}$ have been calculated by computer simulation /7.51/. Due to the lower tetragonal symmetry, one has to average over all three equivalent 100-directions. The results for the Morse potential (6.13) are summarized in the table above. Results for other potentials are given in /6.22/. All potentials give qualitatively the same result: a strong negative change of $\Delta(2c_{44})$ and a moderate negative change of $\Delta(c_{11}-c_{12})$. Contrary to the experimental result, for the Morse potential the change of the compression modulus is too large. Listed are also the theoretical results for a vacancy in Cu, which are about the values obtained by the nearest neighbour model. The results for the Frenkel pair, i.e., vacancy plus interstitial, should be compared with the experimental values. The theoretical results give the correct order of magnitude and the right anisotropy for the changes of the shear moduli. Since the latter is due to the symmetry (E_g) of the resonant libration mode of the 100-dumbbell, this is further evidence for the existence of the 100-dumbbell in Cu and Al and in agreement with recent X-ray and elastic relaxation results /6.1,16-20/.

The theoretical description of the change of the elastic constants is due to LUDWIG /7.52/ - who started from the deformation energy - and ELLIOTT et al /7.53/ (t matrix formula). The present formulation follows our work in /7.51/.

8. Thermodynamic Properties

8.1 Free Energy of Defect Crystal /1.5,8/

In the harmonic approximation, the internal energy E(T,V) as well as the free energy F(T,V) (see below) are given as the sum of an elastic part depending only on crystal volume and a vibrational part depending only on temperature

$$E(T,V) = E_{\mathrm{elast}}(V) + E_{\mathrm{vibr}}(T)\ , \tag{8.1}$$

E_{vibr} is the sum of single oscillator comtributions

$$E_{\mathrm{vibr}}(T) = \int d\omega\ Z(\omega)\ \varepsilon(\omega,T) \qquad \text{with}\quad \varepsilon(\omega,T) = \hbar\omega\left(n(\omega)+\frac{1}{2}\right) = \hbar\omega\left(\frac{1}{2}+\frac{1}{\exp(\hbar\omega/kT)-1}\right)\ . \tag{8.2}$$

Here $\varepsilon(\omega,T)$ is the energy of a single oscillator of frequency ω and $Z(\omega)$ is the total frequency spectrum of the crystal, i.e., the total density of states

$$Z(\omega) = \sum_{\alpha=1}^{3N} \delta(\omega-\omega_\alpha) = 2\omega\ \mathrm{Tr}\ \delta(\omega^2 - D) \quad \text{with}\quad D = \frac{1}{\sqrt{M}}\phi\frac{1}{\sqrt{M}}\ . \tag{8.3}$$

The last equation follows since the ω_α^2's are the eigenvalues of the dynamical matrix D.

From the internal energy we can calculate the free energy F(T,V) and all other thermodynamic quantities. For instance, from the thermodynamic relation

$$F = E - TS = E + T \left.\frac{\partial F}{\partial T}\right|_V \tag{8.4}$$

we can evaluate directly F(T,V). The result for F(T,V) and for the entropy $S = -\partial F/\partial T|_V$ is

$$F(T,V) = E_{elast}(V) + \int d\omega\, Z(\omega)\, \varphi(\omega,T) \quad \text{with } \varphi = \frac{\hbar\omega}{2} + kT\, \ln\left(1 - \exp(-\hbar\omega/kT)\right) \tag{8.5}$$

$$S(T) = \int d\omega\, Z(\omega)\, \sigma(\omega,T) \quad \text{with } \sigma = k\left[\frac{\hbar\omega/kT}{\exp(\hbar\omega/kT) - 1} - \ln\left(1 - \exp(-\hbar\omega/kT)\right)\right]. \tag{8.6}$$

In (8.3), the trace can also be performed in a real space representation. Then the total spectrum $Z(\omega)$ can be written in terms of the local spectra $z_i^m(\omega)$ for each atom m and direction i as introduced in Chap. 2

$$Z(\omega) = 2\omega \sum_{mi} \langle mi \mid \delta(\omega^2 - D) \mid mi\rangle = \sum_{mi} z_i^m(\omega) \tag{8.7}$$

$$\text{with} \quad z_i^m(\omega) = \frac{2\omega}{\pi}\, \mathrm{Im}\{\mathcal{G}_{ii}^{mm}(\omega)\} = \frac{2\omega M^m}{\pi}\, \mathrm{Im}\{G_{ii}^{mm}(\omega)\} \text{ and } \mathcal{G}(\omega) = \left[D - (\omega + i\eta)^2\right]^{-1}.$$

For a *disordered crystal* the thermodynamic properties are therefore determined by the configurational averaged Green's function $\langle\mathcal{G}\rangle$

$$\langle\mathcal{G}_{ij}^{mn}(\omega)\rangle = \langle \sqrt{M^m}\, G_{ij}^{mn}(\omega)\, \sqrt{M^n}\rangle \tag{8.8}$$

instead of the average $\langle G_{ij}^{mn}\rangle$ needed for, e.g., phonon dispersion curves and correlation functions.

For a small concentration of point defects we need only to consider changes of E(T,V) or F(T,V) linear in c

$$E(T,V) = E_o(T,V) + N_d\, e(T,V) \tag{8.9}$$

$$\text{with } N_d = c\, N$$

$$F(T,V) = F_o(T,V) + N_d\, f(T,V) \tag{8.10}$$

where e(T,V) is the formation energy, f(T,V) = e - Ts the formation free energy and s the formation entropy. Thus for the temperature dependence of these quantities we need the change $\Delta Z(\omega)$ of the total spectrum linear in c. From the exact equation

$$Z(\omega) = \frac{2\omega}{\pi}\, \mathrm{Tr}\; \mathrm{Im}\{\mathcal{G}(\omega)\} = 2\omega\, \mathrm{Tr}\; \mathrm{Im}\{\mathring{\mathcal{G}}(\omega) - \mathring{\mathcal{G}}(\omega)\, T(\omega)\, \mathring{\mathcal{G}}(\omega)\} \tag{8.11}$$

we obtain by approximating the total T matrix by the sum of the individual t matrices ($T \cong \sum_\alpha t_\alpha$)

$$\Delta Z(\omega) = N_d \, \Delta z(\omega) = -N_d \frac{2\omega}{\pi} \, \mathrm{Im} \, \mathrm{Tr}\{\overset{\circ}{\mathcal{G}}(\omega) \, t(\omega) \, \overset{\circ}{\mathcal{G}}(\omega)\} \quad . \tag{8.12}$$

Thus the vibrational contribution to the formation energies e and f are given by

$$e_{vibr}(T) = \int d\omega \, \Delta z(\omega) \, \varepsilon(\omega,T) \; , \quad f_{vibr}(T) = \int d\omega \, \Delta z(\omega) \, \varphi(\omega,T) \tag{8.13}$$

where $\Delta z(\omega)$ is the change of the spectrum $Z(\omega)$ due to a single defect.

From an experimental point of view, two thermodynamic quantities are of special interest since they are directly accessible to experimental observations:

1) The change $\Delta c_L(T)$ of the *lattice specific heat* due to point defects, which due to $c(T) = T(\partial S/\partial T)|_V$ can be directly calculated from the entropy

$$\Delta c_L(T) = N_d \, k \int d\omega \, \Delta z(\omega) \, (\hbar\omega/kT)^2 \exp(\hbar\omega/kT) \left(\exp(\hbar\omega/kT) - 1\right)^{-2} \; . \tag{8.14}$$

Theoretical and experimental results for the specific heat are discussed in Sect. 8.3.

2) *Formation energy e and formation entropy s:* In thermal equilibrium the concentration c of point defects is for given T and V given by

$$c = \exp[-f(T,V)/kT] = \exp(s/k) \exp(-e/kT) \; , \quad \text{since } f = e - Ts. \tag{8.15}$$

Most interesting is the case of high temperatures, where $\varepsilon(\omega,T) \approx kT$. For the formation energy e we obtain then

$$e_{vibr}(T) = kT \int d\omega \, \Delta z(\omega) = 0 \quad \text{since} \quad \int d\omega \, \Delta z(\omega) = 0 \tag{8.16}$$

as will be proved below (8.45). Thus the formation energy is in the classical limit independent of T. For the formation entropy s(T) and the free energy we obtain analogously

$$s(T) = -\frac{1}{T} f_{vibr}(T) = k \int d\omega \, \left(1 + \ln \frac{kT}{\hbar\omega}\right) \Delta z(\omega) = -k \int d\omega \, \ln(\omega) \, \Delta z(\omega) \tag{8.17}$$

where we have again used the sum rule (8.45). The theoretical problems connected with the calculation of the entropy s(T) will be discussed in Sect. 8.4.

The formulation of the free energy in terms of the changed density of state goes back to LIFSHITZ /8.1/. See also the article by LIFSHITZ and KOSEVICH /1.8/ and the book by MARADUDIN et al. /1.5/.

8.2 Properties of the Changed Density of States $\Delta z(\omega)$

According to (8.12) the change $\Delta Z(\omega)$ of the total spectrum is proportional to the number N_d of defects times the change $\Delta z(\omega)$ due to a single defect which itself is given by

$$\begin{aligned}\Delta z(\omega) &= \sum_{\alpha} \left[\delta(\omega - \omega_\alpha^1) - \delta(\omega - \omega_\alpha^o) \right] = \frac{2\omega}{\pi} \operatorname{Tr} \operatorname{Im}\{\overset{1}{\mathcal{G}}(\omega) - \overset{o}{\mathcal{G}}(\omega)\} \\ &= -\frac{2\omega}{\pi} \operatorname{Tr} \operatorname{Im}\{\overset{o}{\mathcal{G}}(\omega)\, t(\omega)\, \overset{o}{\mathcal{G}}(\omega)\} = -\frac{2\omega}{\pi} \operatorname{Tr} \operatorname{Im}\{t\overset{o}{\mathcal{G}}\overset{o}{\mathcal{G}}\}\end{aligned} \tag{8.18}$$

where ω_α^1 are the eigenfrequencies of a crystal with a single defect and ω_α^o are the ideal ones, $\overset{1}{\mathcal{G}}$ and $\overset{o}{\mathcal{G}}$ are the corresponding Green's functions. Equation (8.18) is not suitable for practical calculations, since the Green's function $\mathcal{G}_{ij}^{mn}$ has an infinite range and one would have to calculate infinite lattice sums. This difficulty can be overcome by making use of the relation

$$\overset{o}{\mathcal{G}}(\omega)\, \overset{o}{\mathcal{G}}(\omega) = \partial_{\omega^2} \overset{o}{\mathcal{G}}(\omega) \tag{8.19}$$

so that $\Delta z(\omega)$ is given by

$$\Delta z(\omega) = -\frac{1}{\pi} \operatorname{Im}\left\{\operatorname{Tr}\, t(\omega) \frac{\partial \overset{o}{\mathcal{G}}(\omega)}{\partial \omega}\right\} = -\frac{1}{\pi} \operatorname{Im}\left\{\sum_{\substack{mn\\ij}} t_{ij}^{mn}(\omega) \frac{\partial \overset{o}{\mathcal{G}}{}_{ji}^{nm}(\omega)}{\partial \omega}\right\}. \tag{8.20}$$

Here the summation is restricted to the perturbed region, where the t matrix has nonzero components. Instead of (8.20) two other equivalent expressions can be given for $\Delta z(\omega)$. By using the expression for the t matrix

$$t(\omega) = \Delta D \frac{1}{1 + \overset{o}{\mathcal{G}}(\omega)\, \Delta D} \tag{8.21}$$

where ΔD is the change of the dynamical matrix D near the defect, one obtains

$$\Delta z(\omega) = -\frac{1}{\pi} \operatorname{Im}\left\{\operatorname{Tr}\, \partial_\omega \ln\left(1 + \overset{o}{\mathcal{G}}(\omega)\, \Delta D(\omega)\right)\right\}. \tag{8.22}$$

Since Tr ln ... = ln Det ... and since Im ln z = arg z, this can also be written as

$$\Delta z(\omega) = -\frac{1}{\pi} \operatorname{Im}\{\partial_\omega \ln\{\operatorname{Det} \left|1 + \overset{o}{\mathcal{G}}(\omega)\, \Delta D(\omega)\right|\}\} \tag{8.23}$$

$$= -\frac{1}{\pi} \partial_\omega \arg\{\operatorname{Det} \left|1 + \overset{o}{\mathcal{G}}(\omega)\, \Delta D(\omega)\right|\}\ . \tag{8.24}$$

These expressions are important because one does not need to calculate the t matrix by inverting the matrix $1 + \overset{o}{\mathcal{G}}\Delta D$. The determinant of $1 + \overset{o}{\mathcal{G}}\Delta D$ is restricted to the defect subspace determined by the size of ΔD.

A very analogous and for practical purposes more useful result can be given in terms of the Green's function $\overset{o}{G}(\omega)$ which has been used for the perturbed phonons and which does not contain the $\sqrt{M}$ factors. Starting from (8.18), we have

$$\Delta z(\omega) = -\frac{1}{\pi}\partial_\omega \,\mathrm{Tr}\,\mathrm{Im}\left\{\ln\left(D - (\omega + i\eta)^2\right) - \ln\left(\overset{\circ}{D} - (\omega+i\eta)^2\right)\right\} \tag{8.25}$$

since $\mathcal{G}(\omega) = \left[D - (\omega + i\eta)^2\right]^{-1}$.

Now the first term can be rewritten as

$$\begin{aligned}\mathrm{Tr}\,\ln\left(D - (\omega+i\eta)^2\right) &= \ln\,\mathrm{Det}\left(D - (\omega+i\eta)^2\right)\\ &= \ln\left[\mathrm{Det}\,\frac{1}{M}\cdot \mathrm{Det}\left(\phi - M(\omega+i\eta)^2\right)\right].\end{aligned} \tag{8.26}$$

Due to the derivative ∂_ω, the mass determinant drops out, so that (8.25) is identical to

$$\begin{aligned}\Delta z(\omega) &= -\frac{1}{\pi}\partial_\omega\,\mathrm{Tr}\,\mathrm{Im}\left\{\ln\left(\phi - M(\omega+i\eta)^2\right) - \ln\left(\overset{\circ}{\phi} - \overset{\circ}{M}(\omega+i\eta)^2\right)\right\}\\ &= -\frac{1}{\pi}\partial_\omega\,\mathrm{Tr}\,\mathrm{Im}\left\{\ln\left[\overset{\circ}{G}(\omega)\left(\phi - M(\omega+i\eta)^2\right)\right]\right\}\end{aligned} \tag{8.27}$$

$$= -\frac{1}{\pi}\partial_\omega\,\mathrm{Tr}\,\mathrm{Im}\left\{\ln\left(1 + \overset{\circ}{G}(\omega)\,V(\omega)\right)\right\} \qquad \text{with } V(\omega) = \varphi - \Delta M\omega^2$$

$$= -\frac{1}{\pi}\,\mathrm{Tr}\,\mathrm{Im}\left\{\frac{1}{1+\overset{\circ}{G}(\omega)\,V(\omega)}\,\partial_\omega\left(\overset{\circ}{G}(\omega)\,V(\omega)\right)\right\}. \tag{8.28}$$

Whereas (8.26,27) in term of D and ϕ are quite analogous, (8.28) differs from the analogous equation (8.20), since in (8.28) also $V(\omega)$ has to be differentiated.

The foregoing results can be further simplified by using group theory, since, e.g., the determinant is a product of subdeterminates for the different irreducible subspaces ν. Thus

$$\Delta z(\omega) = -\frac{1}{\pi}\,\mathrm{Im}\left\{\partial_\omega \sum_\nu \ln\,\mathrm{Det}\left(1 + \overset{\circ}{G}_\nu(\omega)\,V_\nu(\omega)\right)\right\} \tag{8.29}$$

where $\overset{\circ}{G}_\nu$ and V_ν are the projections of $\overset{\circ}{G}$ and V in the νth irreducible subspace. Let us in particular discuss the nearest neighbour model for a substitutional defect in fcc. Using the cubic O_h symmetry the total representation is decomposed as

$$\Gamma_{fcc} = A_{1g} + A_{2g} + 2E_g + 2F_{1g} + 2F_{2g} + A_{2u} + E_u + 4F_{1u} + 2F_{2u}, \tag{8.30}$$

i.e., we have four one-dimensional (A_{1g}, A_{2g}, A_{2u}, E_u), four two-dimensional, and one four-dimensional submatrices. For the special case of a pure longitudinal nearest neighbour force constant as for the Mannheim model of Sect. 5.1, all subdeterminants become one-dimensional except that of F_{1u}, which is two-dimensional. The result has been given by AGRAWAL /8.2/, and TIWARI et al. /8.3/. Since it is quite lengthy, it will not be reproduced here.

The result simplifies considerably for an *isotopic defect*, for which

$$\Delta z(\omega) = \frac{3}{\pi}\,\mathrm{Im}\left\{\frac{1}{1-\overset{\circ}{G}{}^{(o)}_{xx}(\omega)\,\Delta M\omega^2}\,\partial_\omega\left(\overset{\circ}{G}{}^{(o)}_{xx}(\omega)\,\Delta M\omega^2\right)\right\}. \tag{8.31}$$

For instance, for $\omega \to 0$ we can use formula (2.68) for $\mathrm{Im}\{\overset{\circ}{G}{}^{(o)}_{xx}(\omega)\}$

$$\mathrm{Im}\{\overset{\circ}{G}{}^{(o)}_{xx}(\omega)\} \cong \frac{3\pi}{2M\omega_D^3}\,\omega \tag{8.32}$$

so that by neglecting the denominator in (8.31)

$$\Delta z(\omega) \cong 3\,\frac{3}{\omega_D^3}\,\omega^2\,\frac{3}{2}\,\frac{\Delta M}{\overset{\circ}{M}} = 3\overset{\circ}{z}(\omega)\,\frac{3}{2}\,\frac{\Delta M}{\overset{\circ}{M}} \quad . \tag{8.33}$$

The first factor of 3 arises from the different directions x, y and z. The second term is the phonon spectrum of the ideal lattice which is for $\omega \to 0$ changed by the factor $\frac{3}{2}\frac{\Delta M}{\overset{\circ}{M}}$. This is plausible since the Debye frequency should depend on the average mass,

$$\omega_D(c) \sim \left(\overset{\circ}{M} + c\Delta M\right)^{-1/2} \tag{8.34}$$

so that

$$\Delta\,\frac{1}{\left(\omega_D(c)\right)^3} = -\frac{3}{\omega_D^3}\,\frac{\Delta\omega_D}{\omega_D} = \frac{1}{\omega_D^3}\,c\,\frac{3}{2}\,\frac{\Delta M}{\overset{\circ}{M}} \; . \tag{8.35}$$

Low frequency behaviour of $\Delta z(\omega)$: In the general case the spectrum is for small frequencies $\omega \to 0$ also influenced by force-constant changes. In order to investigate this we start from (8.27) by noting, that $\mathrm{Im}\{G(\omega)\} \sim \omega$ for $\omega \to 0$ whereas $\mathrm{Re}\{\overset{\circ}{G}(\omega)\}$ approaches a constant value. By using the formula

$$\mathrm{Im}\,\ln z \cong \frac{\mathrm{Im}\,z}{\mathrm{Re}\,z} - \frac{1}{3}\left(\frac{\mathrm{Im}\,z}{\mathrm{Re}\,z}\right)^3 +- \dots \tag{8.36}$$

we see that the leading term for $\omega \to 0$ is given by

$$\Delta z(\omega) \cong -\frac{1}{\pi}\,\partial_\omega\,\mathrm{Tr}\,\frac{\mathrm{Im}\{\overset{\circ}{G}(\omega)\,V(\omega)\}}{1 + \mathrm{Re}\{\overset{\circ}{G}(\omega)\,V(\omega)\}} \; . \tag{8.37}$$

As we will see below, this term gives already a ω^2 behaviour, so that the higher order terms only give corrections $\sim \omega^5$. By using the low-frequency result for $\overset{\circ}{G}(\omega)$ (2.68)

$$\mathrm{Im}\{\overset{\circ}{G}{}^{(\underline{h})}_{ij}(\omega)\} \cong \alpha\omega(1 + \alpha'\omega^2)\delta_{ij} - \omega^3 \sum_{k\ell} R^{\underline{h}}_k\, T_{ijk\ell}\, R^{\underline{h}}_\ell \tag{8.38}$$

and by multiplying with $V(\omega) = \varphi - \Delta M\omega^2$, the first term gives no contribution when multiplied by φ, since

$$\sum_{m'i'} \mathrm{Im}\{\overset{\circ}{G}{}^{mm'}_{ii'}(\omega)\,\varphi^{m'n}_{i'j}\} \cong \alpha\omega(1 + \alpha'\omega^2) \sum_{m'} \varphi^{m'n}_{ij} = 0 \tag{8.39}$$

due to the translation invariance of φ. Thus the first term of $\mathrm{Im}\{\overset{\circ}{G}{}^{(\underline{h})}_{ij}(\omega)\}$ gives only a contribution with the mass matrix $\Delta M\omega^2$, whereas in the second term $\Delta M\omega^2$ can be neglected. Since both terms are then proportional to ω^3, they both give ω^2 contribution to the spectrum, the first being due to the mass change ΔM, the second one due to the force-constant change φ. For the first term we can neglect the denominator in (8.37), since the $\Delta M\omega^2$ in the denominator gives a higher order correction and since the force constant matrix φ drops out due to translation invariance. For the second term only the $\omega = 0$ value of the denominator is retained. Then one obtains the result

$$\Delta z(\omega) \cong 3\,\frac{3}{\omega_D^3}\,\omega^2\,\frac{3}{2}\,\frac{\Delta M}{\overset{o}{M}} + \frac{3}{\pi}\,\omega^2 \sum_{\substack{mn\\ ijk\ell}} R_k^m\, t_{ij}^{mn}(0)\, R_\ell^n\, T_{ijk\ell} \tag{8.40}$$

where t(0) is the static t matrix, which is connected with the change of the elasticity tensor (7.122)

$$\Delta z(\omega) \cong 3\,\frac{3}{\omega_D^3}\,\omega^2\,\frac{3}{2}\,\frac{\Delta M}{\overset{o}{M}} + \frac{3}{\pi}\,\frac{V_c}{c}\,\omega^2 \sum_{ijk\ell} \Delta C_{ijk\ell}\, T_{ijk\ell}\ . \tag{8.41}$$

Note that the mass- and force-constant contributions are additive, so that there is no interference term. By defining a change $\Delta\omega_D$ of the Debye frequency by

$$\Delta Z(\omega) = 3N\; 3\omega^2\, \Delta \frac{1}{\omega_D^3} = -\,\frac{3\Delta\omega_D}{\omega_D}\, 3N\, \frac{3\omega^2}{\omega_D^3} \tag{8.42}$$

we obtain for $\Delta\omega_D$ or the corresponding change of Debye temperature Θ_D due to a concentration c of point defects

$$\frac{\Delta\omega_D}{\omega_D} = \frac{\Delta\Theta_D}{\Theta_D} = -c\,\frac{1}{2}\,\frac{\Delta M}{\overset{o}{M}} - \frac{1}{3\pi}\,V_c \sum_{ijk\ell} \Delta C_{ijk\ell}\, T_{ijk\ell}\ . \tag{8.43}$$

When comparing with the result (2.70a) for the ideal crystal, one realizes that to linear order in c the Debye temperature $\Theta(c)$ can be calculated as for an ideal crystal, only the mass has to be identified with the average mass $M + c\Delta M$ and the elastic data with the true elastic constants $C_{ijk\ell} + \Delta C_{ijk\ell}$. Presumably this statement is also true for higher concentrations.

At first sight the change $\Delta Z(\omega)$ of the total spectrum for small frequencies seems to contradict the exact result of Sect. 4.4 (4.62) that the defect Green's function $\mathrm{Im}\{G_{ij}^{mn}(\omega)\}$ coincides with $\mathrm{Im}\{\overset{o}{G}{}_{ij}^{mn}(\omega)\}$ for small frequencies or that the local spectra are equal up to a trivial mass factor

$$z_i^m(\omega) \approx \frac{M^m}{\overset{o}{M}}\,\overset{o}{z}{}_i^m(\omega) \qquad \text{for } \omega \to 0\ . \tag{8.44}$$

Since $Z(\omega) = \sum_{mi} z_i^m(\omega)$, one would expect a similar result for $Z(\omega)$. However (4.62) and (8.44) are only valid if $|\underline{R}^m| \ll \overline{c}/\omega$ where $\overline{c}$ is an average velocity of sound. Whereas this is always valid, if $|\underline{R}^m|$ is finite and ω small enough, this result cannot be used for $Z(\omega)$, since we have there to sum over *all* $\underline{R}^m$ for finite ω.

Just the atoms far away, for which (8.44) does not hold, lead to a renormalization of the elastic constants.

Similar to the local spectra $z_i^m(\omega)$, also $\Delta z(\omega)$ obeys some simple *sum rules*. If the defect does not introduce additional degrees of freedom into the lattice, as it is the case for a substitutional defect, we have

$$\int d\omega\ \Delta z(\omega) = 0 \tag{8.45}$$

so that the positive and negative contributions are cancelling each other. Further for the second moment one has

$$\langle\omega^2\rangle = \int d\omega\ \omega^2\ \Delta z(\omega) = \sum_{mi} \Delta D_{ii}^{mm} = \sum_{mi} \left(\phi_{ii}^{mm}/M^m - \overset{\circ}{\phi}{}_{ii}^{mm}/\overset{\circ}{M}{}^m\right) . \tag{8.46}$$

Thus $\langle\omega^2\rangle$ is given by the difference of all Einstein frequencies $\omega_E^2 = \phi_{ii}^{mm}/M^m$ in the defect and ideal crystal.

For the case of a *localized vibration* we can use for the t matrix the representation

$$t \cong -\Delta D\ |\ell\rangle \frac{1}{\omega_\ell^2 - (\omega + i\eta)^2} \langle\ell|\ \Delta D \tag{8.47}$$

where $|\ell\rangle$ is the localized mode determined by $|\ell\rangle = \overset{\circ}{\mathcal{G}}(\omega_\ell)\Delta D\ |\ell\rangle$. By inserting this into (8.28) we obtain (since $\mathrm{Im}\{\overset{\circ}{\mathcal{G}}(\omega)\} = 0$ for $\omega > \omega_{max}$)

$$\begin{aligned} \Delta Z(\omega) &= N_d\ 2\omega\ \delta(\omega^2 - \omega_\ell^2)\ \mathrm{Tr}\ \Delta D\ |\ell\rangle \langle\ell|\ \Delta D\ \partial_{\omega^2}\overset{\circ}{\mathcal{G}}(\omega) \\ &= N_d\ \delta(\omega - \omega_\ell)\ \mathrm{Tr}\ \underbrace{\overset{\circ}{\mathcal{G}}\Delta D\ |\ell\rangle \langle\ell|\ \Delta D\overset{\circ}{\mathcal{G}}}_{|\ell\rangle\langle\ell|} = N_d\ \delta(\omega - \omega_\ell)\ . \end{aligned} \tag{8.48}$$

Thus each localized mode gives a contribution $N_d\delta(\omega - \omega_\ell)$ to the total spectrum. Due to the normalization of $\Delta Z(\omega)$, a corresponding amount has to be missing in the region of the ideal spectrum $0 < \omega < \omega_{max}$.

A similar result is obtained for the case of a *resonance mode* of low frequencies. With

$$t \cong -\Delta D\ |R\rangle \frac{1}{\omega_{res}^2 - \omega^2 - i\gamma\omega} \langle R|\ \Delta D \tag{8.49}$$

we obtain by similar arguments

$$\Delta Z(\omega) = N_d \frac{2\omega}{\pi} \frac{\gamma\omega}{(\omega^2 - \omega_{res}^2)^2 + \gamma^2\omega^2} \cong N_d\ \delta(\omega - \omega_{res}) \quad \text{if} \quad \gamma \ll \omega_{res}\ . \tag{8.50}$$

For the calculation of thermodynamic properties, the Lorentzian can be replaced by a δ function, if the damping is sufficiently small.

As a illustrative example Fig. 62 shows the changed density of state $\Delta z(\omega)$ for heavy isotopic defects with masses $M/\overset{\circ}{M} = 2, 4$ and 8. Here the central force nearest

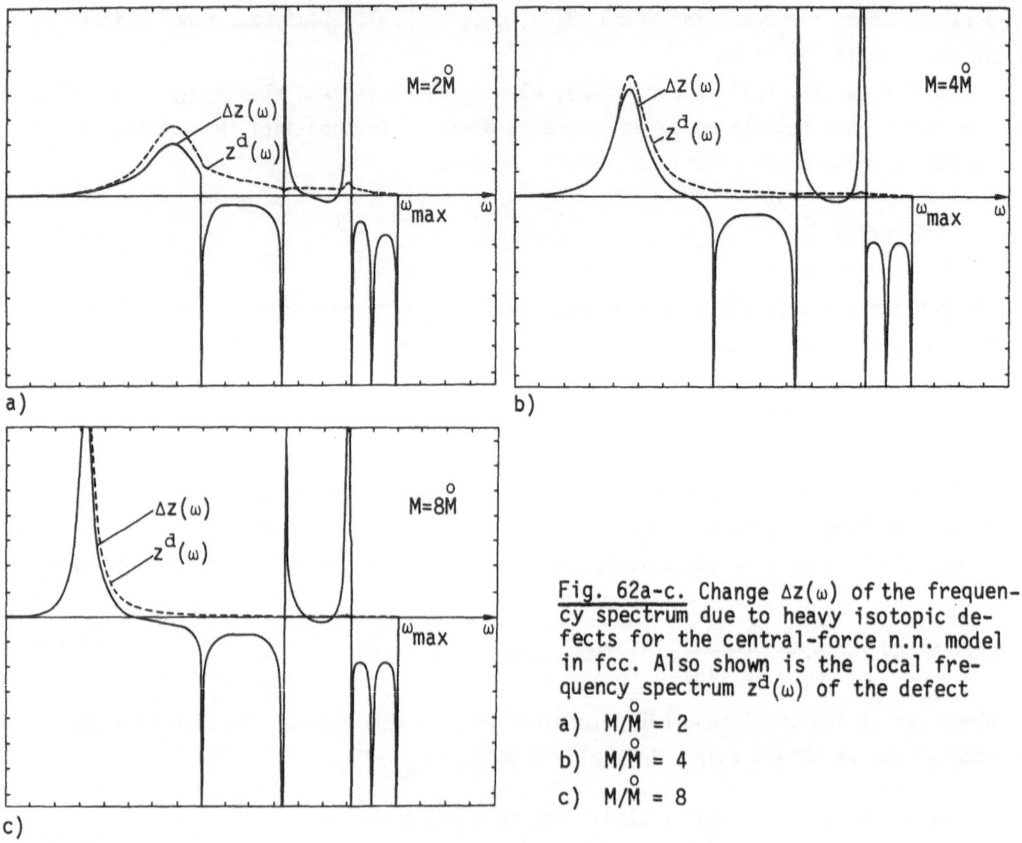

Fig. 62a-c. Change $\Delta z(\omega)$ of the frequency spectrum due to heavy isotopic defects for the central-force n.n. model in fcc. Also shown is the local frequency spectrum $z^d(\omega)$ of the defect
a) $M/\overset{\circ}{M} = 2$
b) $M/\overset{\circ}{M} = 4$
c) $M/\overset{\circ}{M} = 8$

neighbour model for fcc has been employed. Also shown are the corresponding local spectra of the isotopes which are the same as in Fig. 13. Both curves agree very well near the resonance frequency (for convenience $\Delta z(\omega)/3$ has been plotted, referring to one degree of freedom). However at higher frequencies, $\Delta z(\omega)$ shows extremely sharp peaks which are located at the critical frequencies of the host lattice (see below). Figure 63 shows the corresponding change $\Delta z(\omega)$ of the frequency spectrum for light isotopes with masses $M/\overset{\circ}{M} = 1/2$ and $1/4$. Also here the local spectra of the isotopes are shown for comparison. Very striking are again the singularities of $\Delta z(\omega)$ at the critical frequencies.

These singularities arise from the behaviour of $\overset{\circ}{G}(\omega)$ at critical points. Since according to (2.88,90) $\overset{\circ}{G}(\omega)$ behaves as $\sqrt{\omega^2 - \omega_c^2}$ at a critical frequency ω_c, the derivative $\partial G/\partial\omega^2$ entering in $\Delta z(\omega)$ (8.20,28) diverges at ω_c

$$\frac{\partial\overset{\circ}{G}}{\partial\omega^2} \sim \frac{1}{2}\,\frac{1}{\sqrt{\omega^2 - \omega_c^2}} \tag{8.51}$$

which leads to the spikes in Figs. 62,63 at the critical frequencies.

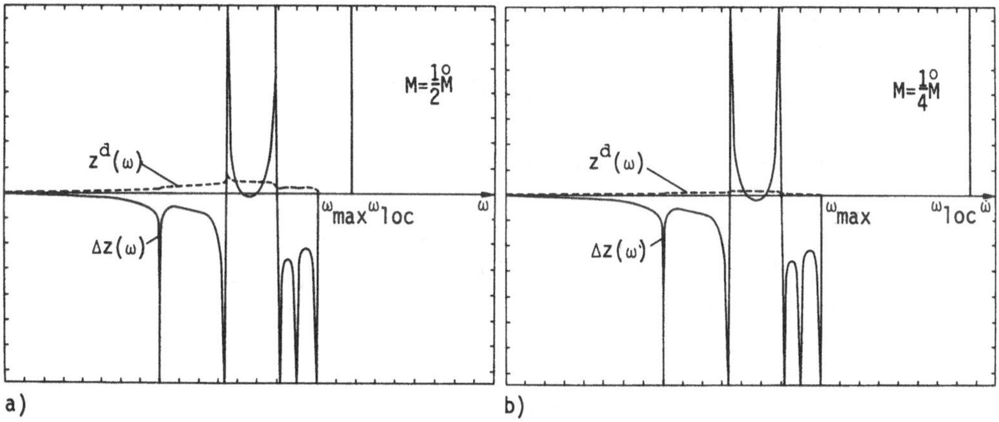

Fig. 63a,b. Change $\Delta z(\omega)$ of the frequency spectrum due to light isotopic defects (central-force n.n. model in fcc) Also shown is the local spectrum $z^d(\omega)$ of the defect. a) $M/\overset{o}{M} = 1/2$, b) $M/\overset{o}{M} = 1/4$

These divergencies signalize a breakdown of the low concentration expansion of the spectrum $Z(\omega)$. In reality the phonon dispersion curves are broadened due to the scattering at impurities (see Sect. 7.3) or due to anharmonic effects. Then the critical singularities are smeared out leading to more or less pronounced peaks in $\Delta z(\omega)$. In a simple model /7.5/ one can assume a constant damping $\Sigma''_\sigma(\underline{k},\omega) = \Gamma_o$ at the critical point. For a minimum of $\omega_\sigma(\underline{k})$ this leads to the integral

$$\begin{aligned}\delta z(\omega) &= \frac{1}{3V_B}\int d(\delta\underline{k})\,\frac{1}{\pi}\,\frac{\Gamma_o}{(\omega^2-\omega_c^2-\sum\limits_{\alpha=1}^{3}\lambda_\alpha^2\delta k_\alpha^2)^2+\Gamma_o^2}\\ &= \frac{2\pi}{3V_B}\,\frac{1}{\sqrt{\lambda_1\lambda_2\lambda_3}}\int\limits_0^\infty \sqrt{x}\,dx\,\frac{1}{\pi}\,\frac{\Gamma_o}{(\omega^2-\omega_c^2-x^2)^2+\Gamma_o^2}\end{aligned} \tag{8.52}$$

instead of (2.83). The integral can be calculated by residuum calculus, with the result

$$\delta\overset{o}{z}(\omega) = \frac{2\pi}{3V_B}\,\frac{1}{\sqrt{\lambda_1\lambda_2\lambda_3}}\,\frac{1}{\sqrt{2}}\left[\left((\omega^2-\omega_c^2)^2+\Gamma_o^2\right)^{1/2}+\omega^2-\omega_c^2\right]^{1/2} \tag{8.53}$$

$$\approx \frac{2\pi}{3V_B}\,\frac{1}{\sqrt{\lambda_1\lambda_2\lambda_3}}\begin{cases}\sqrt{\omega^2-\omega_c^2} & \text{for } \omega^2-\omega_c^2 \gg \Gamma_o\\ \sqrt{\dfrac{\Gamma_o}{2}} & \text{for } \omega^2=\omega_c^2\\ \dfrac{\Gamma_o}{2}\,\dfrac{1}{\sqrt{\omega_c^2-\omega^2}} & \text{for } \omega_c^2-\omega^2 \gg \Gamma_o\,.\end{cases} \tag{8.54}$$

Thus the singularities of $\delta\overset{\circ}{z}(\omega)$ and its derivative are smeared out over the frequency range Γ_o. Therefore the maxima for $\partial\overset{\circ}{G}/\partial\omega^2$ and $\Delta z(\omega)$ are proportional to $1/\sqrt{\Gamma_o}$. Since for small concentrations the damping Γ_o should be proportional to c, we obtain for the above model that at the critical points $\Delta z(\omega_c) \sim 1/\sqrt{c}$ or that the total change $\Delta Z(\omega_c) \sim \sqrt{c}$ instead of $\sim c$. Thus at critical points the low c expansion for $\Delta Z(\omega)$ is not valid. Nevertheless the expansion is still valid for the free energy, since due to the integration over ω the singularities are no longer important.

Let us briefly treat the case of *vacancies and interstitials*, where the defect crystal contains fewer or more degrees of freedom than the ideal crystal. In this case the sum rule (8.45) no longer holds. Instead we have, e.g., for an interstitial

$$\int d\omega\ \Delta z(\omega) = \int d\omega\left(\sum_{\alpha}\delta(\omega-\omega_\alpha) - \sum_{\alpha'}\delta(\omega-\overset{\circ}{\omega}_{\alpha'})\right) = 3 \tag{8.55}$$

since the interstitial introduces three new degrees of freedom into the lattice. For a vacancy we have -3 instead. Then the matrices ϕ, $\overset{\circ}{\phi}$ and G, $\overset{\circ}{G}$ are of different dimensions. Starting again from (8.27) we have

$$\begin{aligned}\Delta z(\omega) &= \frac{1}{\pi}\,\mathrm{Im}\{G(\omega) - \overset{\circ}{G}(\omega)\} = \frac{1}{\pi}\,\partial_\omega\,\mathrm{Im}\left\{\mathrm{Tr}\,\ln\left(\overset{\circ}{\phi} - \overset{\circ}{M}(\omega+i\eta)^2\right) - \mathrm{Tr}\,\ln\left(\phi - M(\omega+i\eta)^2\right)\right\} \\ &= \frac{1}{\pi}\,\partial_\omega\,\mathrm{Im}\left\{\ln\frac{\mathrm{Det}\left|\overset{\circ}{\phi} - \overset{\circ}{M}(\omega+i\eta)^2\right|}{\mathrm{Det}\left|\phi - M(\omega+i\eta)^2\right|}\right\}.\end{aligned} \tag{8.56}$$

For an interstitial, ϕ has the structure

$$\phi - M\omega^2 = \begin{Bmatrix}\phi_{ii} - M\omega^2 & \phi_{ih} \\ \phi_{hi} & \phi_{hh} - \overset{\circ}{M}\omega^2\end{Bmatrix} \tag{8.57}$$

where the index i refers to the interstitial subspace and h to the subspace of the host atoms. The submatrix $\overset{\circ}{\phi} - \overset{\circ}{M}\omega^2$ can be enlarged by adding a unit matrix in the interstitial subspace, so that the Det$\left|\overset{\circ}{\phi} - \overset{\circ}{M}\omega^2\right|$ is unchanged. Thus

$$\overset{\circ}{\phi} - \overset{\circ}{M}\omega^2 = \begin{Bmatrix}1 & 0 \\ 0 & \overset{\circ}{\phi}_{hh} - \overset{\circ}{M}\omega^2\end{Bmatrix}, \qquad \overset{\circ}{G} = \frac{1}{\overset{\circ}{\phi} - \overset{\circ}{M}\omega^2} = \begin{Bmatrix}1 & 0 \\ 0 & \dfrac{1}{\overset{\circ}{\phi}_{hh} - \overset{\circ}{M}\omega^2}\end{Bmatrix}. \tag{8.58}$$

Inserting this into (8.56) for $\Delta z(\omega)$ we obtain

$$\Delta z(\omega) = -\frac{1}{\pi}\,\partial_\omega\,\mathrm{Im}\left\{\ln\,\mathrm{Det}\left|\overset{\circ}{G}(\omega)\left(\phi - M\omega^2\right)\right|\right\} \tag{8.59}$$

where in generalization to (8.28) $\overset{\circ}{G}(\phi - M\omega\ )$ has the form

$$\overset{\circ}{G}(\phi - M\omega^2) = \begin{Bmatrix} \phi_{ii} - M\omega^2 & \phi_{ih} \\ \dfrac{1}{\overset{\circ}{\phi}_{hh} - \overset{\circ}{M}\omega^2}\,\phi_{hi} & 1 + \dfrac{1}{\overset{\circ}{\phi}_{hh} - \overset{\circ}{M}\omega^2}\,V_{hh} \end{Bmatrix} \quad \text{with } V_{hh} = \phi_{hh} - \overset{\circ}{\phi}_{hh} \tag{8.60}$$

and can be calculated by the Green's function technique just as (8.28). A somewhat different description using the effective force constant introduced in Chap. 4 has been given by MAHANTY and SACHDEV /8.4/.

8.3 Change of the Specific Heat

According to (8.14) the change of the lattice specific heat is given by

$$\Delta c_L(T) = N_d\, k \int d\omega\, \Delta z(\omega)\, \frac{x^2 e^x}{(e^x - 1)^2}\,, \qquad x = \hbar\omega/kT\,. \tag{8.61}$$

(In addition to the lattice specific heat, there is also a change of the electronic specific heat $c_e(T) = \gamma T$. Here the change $\Delta\gamma$ is proportional to the change of the electronic density of states at the Fermi energy.)

For low T, the small frequency expansion (8.41) of $\Delta z(\omega)$ can be used. Then $\Delta c_L(T)$ is like the ideal lattice heat proportional to T^3 where the proportionality factor is determined by the changed Debye temperature (8.43)

$$\Delta c_L(T) = 3Nk\,\frac{4\pi^4}{5}\,T^3\,\Delta\left(\frac{1}{\Theta_D^3}\right) = -3\overset{\circ}{c}_L(T)\,\frac{\Delta\Theta_D}{\Theta_D} \quad \text{with} \quad \Theta_D = \frac{\hbar\omega_D}{k}\,. \tag{8.62}$$

For high T, $\Delta c_L(T)$ vanishes due to the normalization of $\Delta z(\omega)$ (8.45). The first non-vanishing term goes as $1/T^2$ and is determined by the changed Einstein frequencies due to the sum rule (8.46)

$$\Delta c_L(T) = -N_d\, k\,\frac{1}{12}\left(\frac{\hbar}{kT}\right)^2 \int d\omega\, \Delta z(\omega)\, \omega^2 = -N_d k\,\frac{1}{12}\left(\frac{\hbar}{kT}\right)^2 \sum_{mi} \left(\phi_{ii}^{mm}/M^m - \overset{\circ}{\phi}{}_{ii}^{mm}/\overset{\circ}{M}\right). \tag{8.63}$$

Especially interesting is the case of a low-lying resonance for which $\Delta z(\omega)$ can be represented by (8.50). If the resonance is strongly peaked, we may replace $\Delta z(\omega)$ by a δ function, so that

$$\Delta c_L(T) \approx N_d\, k\,\frac{x_{res}^2 \cdot \exp(x_{res})}{\left(\exp(x_{res} - 1)\right)^2} \quad \text{with} \quad x_{res} = \hbar\omega_{res}/kT\,. \tag{8.64}$$

Thus for relatively low temperatures $T \gtrsim h\omega_{res}/k$ we obtain already the classical value $N_d k$, whereas the ideal crystal still observes the T^3 law. The relative change

$$\frac{\Delta c_L(T)}{\overset{\circ}{c}_L(T)} = \frac{N_d}{3N}\,\frac{5}{4\pi^4}\left(\frac{\omega_D}{\omega_{res}}\right)^3 \frac{x_{res}^5 \cdot \exp(x_{res})}{\left(\exp(x_{res} - 1)\right)^2} \tag{8.65}$$

has for $x_{res} = 4.93$ or $T \approx \hbar\omega_{res}/5k$ a maximum with the value

$$\frac{\Delta c_L(T)}{\overset{o}{c}_L(T)} = c \cdot 0.091 \left(\frac{\omega_D}{\omega_{res}}\right)^2 . \tag{8.66}$$

Thus the relative change can be appreciably larger than the defect concentration.

In addition to the contributions from the force-constant changes localized to the defects one also obtains a nonlocalized contribution due to the image expansion. This can be calculated from the volume or pressure derivative of the ideal spectrum or the ideal phonons, respectively, leading to a change Δz^{Im} of the spectrum

$$\Delta z^{Im}(\omega) = \frac{\partial \overset{o}{z}(\omega)}{\partial V} \Delta V^{Im} = - \frac{\partial \overset{o}{z}(\omega)}{\partial p} \frac{K}{V} \Delta V^{Im} \tag{8.67}$$

and an analogous contribution to $\Delta c_L(T)$.

Calculations of $\Delta c_L(T)$ for the central force n.n. model in fcc (Mannheims's model) have recently been performed by PERESADA and TOLSTOLUZHSKII /8.5/.

The effect of heavy impurities on the specific heat has been predicted by KAGAN and IOSILEVSKII /8.6/. The first measurements were performed by PANOVA and SAMOILOV /8.7/ for the alloy Mg-2.8 at% Pb being characterized by an extremely high mass ratio of $M_{Pb}/M_{Mg} = 8$. Their results are shown in Fig. 64 (note the two different scales for the lower and higher temperature range). At about 10 K, the specific heat of Mg $Pb_{0.028}$

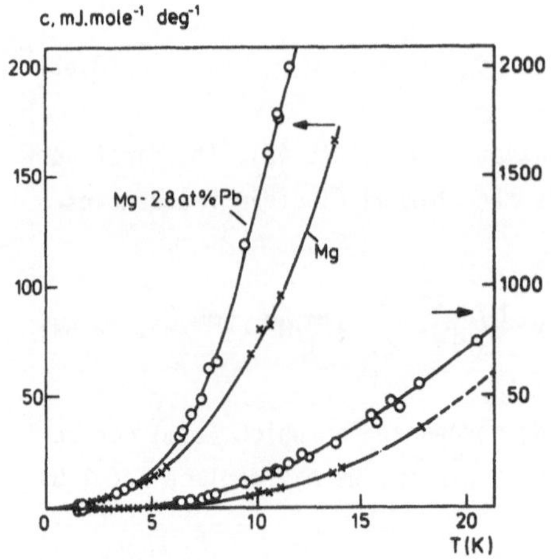

Fig. 64. Temperature dependence of the specific heat of Mg and the alloy Mg-2.8 at% Pb. Note the different scales on the left for the curves I and II (according to /8.7/)

is about twice the specific heat of pure Mg. The results are in close agreement with the mass defect theory, i.e., force constant changes are apparently not important.

CAPE et al. /8.8/ have made specific heat measurements for dilute Mg-Pb alloys and Mg-Cd alloys (mass ratio $M_{Cd}/M_{Mg} = 4.6$). For Mg-Pb their maximum change is about

30% smaller than the one of PANOVA and SAMOILOV /8.7/.

Al-Ag alloys (mass ratio 4.0) have been studied by CULBERT and HOBENER /8.9/, HARTMANN et al. /8.10/, and KRYLOVSKII et al. /8.11/. Fig. 65 shows the results of measurements by HARTMANN et al. for Al-0.95 at% Ag and Al-0.5 at% Ag. The results

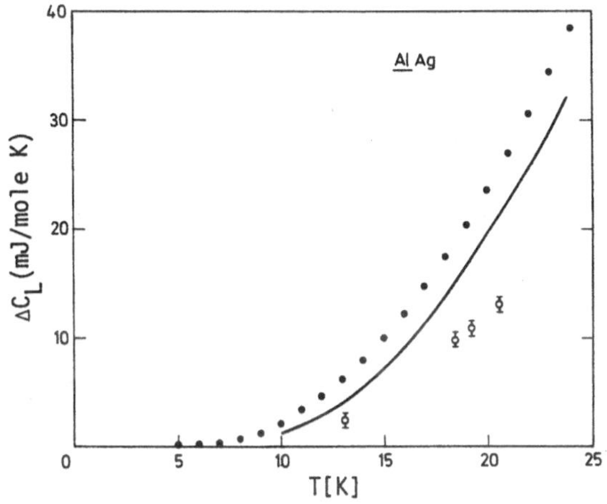

Fig. 65. Enhanced specific heat of Al-0.95 at% Ag (•) and Al-0.5 at% Ag (ǭ). The solid line is the result for isotopic defects for c = 0.95 at% (according to /8.10/)

for 0.95% Ag are compared with calculated values for isotopic defects (full line), giving considerably lower values. This indicates a local force-constant softening. A theoretical analysis of these measurements using a nearest neighbour central force model has been performed by TIWARI et al. /8.3/ [5]. A reasonable fit to the experimental data is obtained by a force-constant softening by 6% for the nearest neighbour. Neutron measurements of the phonon dispersion curves /7.30/ give a somewhat larger softening by 20% (see Sect. 7.5.1).

Measurements at very low temperatures (below 5 K) were performed by DELINGER et al. /8.12/ for Cu-Au alloys and GREEN and VALLADARES /8.13/ for Ag-Au alloys. The latter case has been analyzed by TIWARI and AGRAWAL /8.14/. In this low-temperature region both $\overset{o}{c}_L(T)$ and $\Delta c_L(T)$ obey a T^3 law and can be characterized by a concentration-dependent Debye temperature. For illustration, Fig. 66 shows the resulting Debye temperature of Cu-Au as a function of decomposition.

For further references see: VERBECK et al. (Au-V and Au-Ni) /8.16/, TIWARI and AGRAWAL (analysis of Au-V) /8.15/, VEAL and RAYNE (Cu-Zn) /8.17/, WILL and GREEN

[5] The analysis of the specific heat data by TIWARI et al. should be considered with caution. Apparently these authors do not get a T^3 law at low temperatures, e.g., in /8.15/ they claim that a nearest neighbour model with longitudinal force constants gives a T^2 law! Further, the image terms are not treated correctly. It would be good if these calculations were repeated.

DEBYE TEMPERATURE Θ (K)

COMPOSITION at.% Au

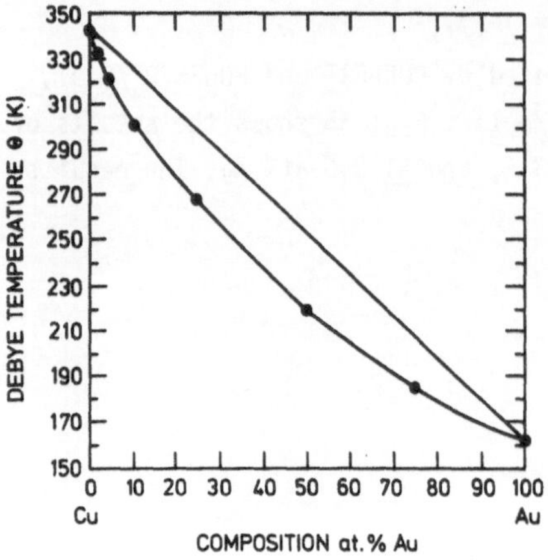

<u>Fig. 66.</u> Debye temperature of the Cu-Au alloy as a function of the decomposition (according to /8.12/)

(A̲u-Sn) /8.18/, MASSALSKI and ISSACS (A̲g-Sn) /8.19/, K.H. RIEDER (localized modes and specific heat) /8.20/, CHERNOPLEKOV et al. (T̲i-U) /7.39/, CHERNOPLEKOV et al. (V̲-Ta) /7.42/, SHIKOV et al. (V̲-Be, V̲-U) /8.21/.

8.4 Formation Entropy of Point Defects

As derived in Sect. 8.1 (8.17), the formation entropy of a substitutional defect is in the classical limit identical with the vibrational part of the free energy, besides a factor $-1/T$, and given by

$$s = - \frac{1}{T} f_{vibr} = - k \int d\omega \ln \Delta z(\omega) \ . \tag{8.68}$$

Several other representations can be given using equivalent formulas for $\Delta z(\omega)$

$$\Delta z(\omega) = \sum_\alpha \left(\delta(\omega - \omega_\alpha) - \delta(\omega - \overset{o}{\omega}_\alpha) \right) = 2\omega \ Tr \left(\delta(\omega^2 - D) - \delta(\omega^2 - \overset{o}{D}) \right) \tag{8.69}$$

$$\text{with} \quad D = \frac{1}{\sqrt{M}} \phi \frac{1}{\sqrt{M}} \ .$$

Thus we obtain for s

$$s = \frac{k}{2} \sum_\alpha \ln \frac{\overset{o}{\omega}_\alpha^2}{\omega_\alpha^2} = \frac{k}{2} \ Tr \left(\ln \overset{o}{D} - \ln D \right) = \frac{k}{2} \ln \frac{Det \ \overset{o}{D}}{Det \ D} \tag{8.70}$$

where for the last step we have used

$$\sum_\alpha \ln \omega_\alpha^2 = \ln \prod_\alpha \omega_\alpha^2 = \ln Det \ D \ .$$

Note that the formation entropy is positive if on the average $\omega_\alpha^2 < \overset{o}{\omega}_\alpha^2$, whereas s is negative, if $\omega_\alpha^2 > \overset{o}{\omega}_\alpha^2$.

The determinant formula is especially interesting since it allows a separation of the mass dependence and the force-constant dependence of s. Since

$$\mathrm{Det}\left(\frac{1}{\sqrt{M^m}}\phi^{mn}_{ij}\frac{1}{\sqrt{M^n}}\right) = \frac{1}{(M^1 M^2 \dots M^N)^3}\ \mathrm{Det}\ \phi^{mn}_{ij} \tag{8.71}$$

the mass dependence can be split off with the result

$$s = \frac{k}{2}\left[3\ \ln\frac{M}{\overset{\circ}{M}} + \ln\frac{\mathrm{Det}\ \overset{\circ}{\phi}}{\mathrm{Det}\ \phi}\right] . \tag{8.72}$$

Thus for a defect, where force-constant changes are not important like, e.g., for Ag in Al, s can be approximated by the first mass term only yielding, e.g., for Ag in Al: $s \approx 2.1k$. However force-constant changes cannot as easily be taken into account. Several methods can be used for this problem:

a) *Green's function method:* In (8.70) the determinants can be written as

$$\frac{\mathrm{Det}\ \phi}{\mathrm{Det}\ \overset{\circ}{\phi}} = \mathrm{Det}\left(\frac{1}{\overset{\circ}{\phi}}\ \phi\right) = \mathrm{Det}\left(1 + \frac{1}{\overset{\circ}{\phi}}\ (\phi - \overset{\circ}{\phi})\right) = \mathrm{Det}\left(1 + \overset{\circ}{G}(0)\ \varphi\right) \tag{8.73}$$

where G(0) is the static Green's function for frequency $\omega = 0$ and φ the force-constant change, thus

$$s = \frac{k}{2}\ \ 3\left[\ln\frac{M}{\overset{\circ}{M}} - \ln\ \mathrm{Det}\left(1 + G(0)\ \varphi\right)\right]. \tag{8.74}$$

Since the range of the determinant is determined by the range of φ, one has only to calculate a relatively low-dimensional determinant. Using group theory the determinant can be further reduced by splitting up into subdeterminants for each irreducible representation ν, so that

$$s = \frac{k}{2}\left[3\ \ln\frac{M}{\overset{\circ}{M}} - \sum_{\nu}\ln\ \mathrm{Det}\left(1 + \overset{\circ}{G}_{\nu}(0)\ \varphi_{\nu}\right)\right] . \tag{8.75}$$

This method has been used recently by GOVINDARAJAN et al. /8.22/ for a calculation of s for substitutional defects in alkali halides and by HATCHER et al. /8.23/ for a calculation of s for vacancies in fcc metals (see below).

b) *Einstein approximation:* A quick estimate is obtained, if all frequencies $\overset{\circ}{\omega}{}^2_{\alpha}$ and ω^2_{α} are calculated in the Einstein approximation. For example, let us consider a substitutional defect in an fcc lattice with central force nearest neighbour interaction (Mannheim's model). Then the Einstein frequency for the defect is $\left(\omega^{d}_{E}\right)^2 = 4f/M$ compared with $4\overset{\circ}{f}/\overset{\circ}{M}$ in the ideal lattice. This frequency is threefold degenerate. For each nearest neighbour we obtain one frequency $\left(\omega^{n.n.}_{E}\right)^2 = (3\overset{\circ}{f} + f)/\overset{\circ}{M}$ for vibration into the direction of the defect and two perturbed Einstein frequencies for vibrations perpendicular to the nearest neighbour direction. Thus

$$s_{\mathrm{Einstein}} \approx \frac{k}{2}\left[3\,\ln\frac{4\overset{\circ}{f}/\overset{\circ}{M}}{4f/M} + 12\,\ln\frac{4\overset{\circ}{f}/M}{(3\overset{\circ}{f}+f)/\overset{\circ}{M}}\right]$$
$$= k\,\frac{3}{2}\,\ln\left[\frac{M}{\overset{\circ}{M}}\frac{\overset{\circ}{f}}{f}\left(1+\frac{1}{4}\,\frac{f-\overset{\circ}{f}}{\overset{\circ}{f}}\right)^{-4}\right]. \tag{8.76}$$

It is seen that force-constant changes are somewhat more important than comparable mass changes, since also the frequencies of the nearest neighbours are perturbed.

c) *Cluster method:* By generalizing the Einstein approximation, one can consider the dynamics of a cluster of atoms by fixing all other atoms. For instance, the cluster may consist of the defect and the atoms in the first shell or in the first few shells around the defect. In this case one would have to calculate all ideal and perturbed frequencies $\overset{\circ}{\omega}_\alpha$ and ω_α for such a cluster as for instance has been done by BURTON /8.24/ for the case of vacancies in fcc and bcc metals. Instead one may also directly calculate the determinants of $\overset{\circ}{\phi}$ and ϕ for these clusters (8.70) which is equivalent but more economic from the numerical point of view. The latter method has been used by HATCHER et al /8.23/ for vacancies. Here clusters containing up to 400 atoms have been considered.

d) *Moment method:* Another generalization of the Einstein approximation can be obtained by splitting up the coupling matrices ϕ_{ij}^{mn} and $\overset{\circ}{\phi}{}_{ij}^{mn}$ into diagonal parts $\phi_{E\,ij}^{\;mn}$ and $\overset{\circ}{\phi}{}_{E\,ij}^{\;mn}$, the eigenvalues of which are the Einstein frequencies, and into nondiagonal parts $\Delta\phi_{ij}^{mn}$ and $\Delta\overset{\circ}{\phi}{}_{ij}^{mn}$. Then one expands the terms $\ln\phi$ and $\ln\overset{\circ}{\phi}$ into powers of $\Delta\phi$ and $\Delta\overset{\circ}{\phi}$

$$\ln\phi = \ln\,(\phi_E+\Delta\phi) = \ln\,\phi_E + \frac{1}{\sqrt{\phi_E}}\,\Delta\phi\,\frac{1}{\sqrt{\phi_E}} - \frac{1}{2}\,\frac{1}{\sqrt{\phi_E}}\,\Delta\phi\,\frac{1}{\phi_E}\,\Delta\phi\,\frac{1}{\sqrt{\phi_E}} + - \ldots \tag{8.77}$$

Then one has for s

$$s = \frac{k}{2}\left[\mathrm{Tr}\,(\ln\overset{\circ}{\phi}_E - \ln\phi_E) - \frac{1}{2}\,\mathrm{Tr}\left(\frac{1}{\overset{\circ}{\phi}_E}\,\Delta\overset{\circ}{\phi}\,\frac{1}{\overset{\circ}{\phi}_E}\,\Delta\overset{\circ}{\phi} - \frac{1}{\phi_E}\,\Delta\phi\,\frac{1}{\phi_E}\,\Delta\phi\right) + - \ldots\right]. \tag{8.78}$$

The first term is again the Einstein approximation. The term linear in $\Delta\phi$ vanishes since $\Delta\phi$ and $\Delta\overset{\circ}{\phi}$ are nondiagonal. Most essential for this method is that the matrices $\Delta\phi$ and $\Delta\overset{\circ}{\phi}$ are of short range. The method has been proposed by HUCKABY /8.25/.

Let us briefly discuss a modification which has to be applied for the case of *interstitials* and *vacancies*. For an interstitial, the sum rule (8.45) is no longer valid, instead

$$\int d\omega\;\Delta z(\omega) = 3 \tag{8.79}$$

since the interstitial introduces three new degrees of freedom into the lattice. Therefore s remains temperature dependent also at high temperature. Instead of (8.68) we have to go back to the original expression (8.17)

$$s(T) = k \int d\omega\ (1 + \ln \frac{kT}{\hbar\omega})\ \Delta z(\omega) = 3k\ (1 + \ln \frac{kT}{\hbar}) - k \int d\omega\ \ln\omega\ \Delta z(\omega)\ . \quad (8.80)$$

In addition to the local force-constant changes around the defect, we also have nonlocalized changes due to the *image field*. Since the image fields of many defects superimpose and lead to a homogeneous expansion of the lattice, the corresponding change of the entropy can be calculated from the volume dependence of the entropy of the ideal crystal. Thus

$$s^{\mathrm{Im}} = \left.\frac{\partial S}{\partial V}\right|_T \Delta V^{\mathrm{Im}} = \left.\frac{\partial S}{\partial V}\right|_T \frac{2}{3}\frac{1-2\nu}{1-\nu}\ \Delta V^{\mathrm{total}} \quad (8.81)$$

where we have used the relation (7.113) between the volume change due to image forces and the total volume change $\Delta V^{\mathrm{total}}$. Further $\partial S/\partial V$ can be expressed by the compression modulus K and the volume expansion coefficient α by using the thermodynamic relation

$$\left.\frac{\partial S}{\partial V}\right|_T = \left.\frac{\partial P}{\partial T}\right|_V = -V \left.\frac{\partial P}{\partial V}\right|_T \left.\frac{1}{V}\frac{\partial V}{\partial T}\right|_P = K\alpha\ . \quad (8.82)$$

There are rather few calculations of the formation entropy of point defects. Despite the fact that there seems to be a wealth of experimental information about thermodynamic proporties of alloys /8.23/, apparently these data have not been analyzed. We only know of two efforts to calculate entropies: the calculation of the *formation entropy of vacancies* in fcc metals by HATCHER et al. /8.24/ and others (see references in /8.24/), and the calculation of the *solution entropy of hydrogen* in metals by MAGERL et al. /8.27/.

MAGERL et al. have split up the vibrational contribution to the entropy into two parts. First, the contribution from the three localized modes is given by

$$s_{\ell oc} = k \sum_{\alpha=1}^{3} \left(\frac{\hbar\omega_\alpha/kT}{\exp(\hbar\omega_\alpha/kT) - 1} - \ln\left(1 - \exp(-\hbar\omega_\alpha/kT)\right)\right)\ . \quad (8.83)$$

The second contribution comes from the frequencies in the continuous spectrum. In the approximation where phonon broadening is neglected, it can be expressed by the concentration dependence of the phonons $\omega_\sigma(\underline{q},c)$ of the metal - hydrogen system. For temperatures above the Debye temperature it is given by

$$s_{\mathrm{band}} = -\frac{k}{N_d} \sum_{q\sigma} \ln \frac{\omega_\sigma(\underline{q},c)}{\omega_\sigma(\underline{q},0)} = -3k\,\frac{1}{N} \sum_{q\sigma} \frac{\partial \ln \omega_\sigma(\underline{q},c)}{\partial c}\ . \quad (8.84)$$

In this way $s_{\ell oc}$ and s_{band} have been calculated from frequencies of localized modes and the phonon dispersion curves as determined by neutron scattering. In addition an electronic contribution s_{elect} has been added being calculated from the electronic specific heat coefficient $s_{\mathrm{elect}} = \left(\partial\gamma(c)/\partial c\right)\cdot T$. The so calculated excess entropy s^{exc}

$$s^{exc} = s_{\ell oc} + s_{band} + s_{elect} + k \ln\beta \tag{8.85}$$

is then compared with experimental values. The term $k\ln\beta$ is a configurational entropy with β the ratio of the number of the interstitial sites per host atom. For bcc one has $\beta = 3$ for octahedral sites and $\beta = 6$ for tetrahedral sites. For fcc $\beta = 1$ for octahedral and $\beta = 2$ for tetrahedral sites.

As an example, MAGERL et al. obtained for the excess entropy of Ta at 400 K: $s^{exc} = 1.04\,k$ for tetrahedral sites and $s^{exc} = 0.45\,k$ for octahedral sites, where the difference of about 0.60 arises practically only due to the different $\ln\beta$-term. Since the experimental value is 1.06 K, this clearly suggests that hydrogen occupies the tetrahedral position in Ta. A similar conclusion is reached for Nb. For the fcc Pd the calculations indicate both for H and D the octahedral site.

The formation entropies of vacancies have been calculated by HUNTINGTON et al. /8.28/, and SCHOTTKY et al. /8.29/. A careful analysis has recently been given by HATCHER et al. /8.24/ where also a complete list of references can be found. The authors assume a pair potential interaction. Three potentials, a Born-Mayer potential, a short-range Morse potential and a nearest neighbour Morse potential have been employed. All three were fitted to Cu. The authors have used the cluster method based on the determinant formula (8.72) and the Green's function method to calculate s. However, the easiest method is the Einstein approximation which we briefly discuss.

By the formation of the vacancy the atom at the vacancy site is transferred to the surface where it is incorporated preferentially in a step or kink, so that the total number of surface and bulk atoms is the same as before. Thus by making the transition to the infinite crystal we must always assure that the ideal crystal has the same number of atoms as the defect one. This can be achieved by representing $\overset{o}{Z}(\omega)$ as sum of local spectra $\overset{o}{z}{}^m_i(\omega)$ and leaving out the $m = 0$ term. Thus

$$s = k \int d\omega \ln\omega \sum_{\substack{mi \\ (m\neq 0)}} \left(\overset{o}{z}{}^m_i(\omega) - z^m_i(\omega) \right) . \tag{8.86}$$

The most drastic differences between the spectra are expected at the nearest neighbour sites. The corresponding spectra are shown in Fig. 28. The main effect is a shift to lower frequency for vibrations towards the vacancy.

In a central-force nearest neighbour model only the force constant between a nearest neighbour and the vacancy is missing. Thus of all Einstein frequencies only the ones of the nearest neighbours and here only the one for vibrations towards the vacancy is changed and in fcc crystals given by $\omega_E^2 = 3\overset{o}{f}/\overset{o}{M}$ compared to $4\overset{o}{f}/\overset{o}{M}$ for all other (unchanged) Einstein frequencies (see Sect. 5.3 and Fig. 27). Since we have twelve neighbours the Einstein model predicts

$$s_{1v} = \frac{k}{2} 12 \ln \frac{4\overset{\circ}{f}/\overset{\circ}{M}}{3\overset{\circ}{f}/\overset{\circ}{M}} \approx 1.73\,k \quad . \tag{8.87}$$

This compares favourably with the exact result s = 2.05 k as obtained in /8.24/ for the same model by the Green's function method.

The effects of longer ranging force constants are much more difficult to take into account. In /8.24/ clusters containing up to 150 - 400 atoms have been studied before a reasonable convergence could be obtained. However this problem is less severe if the corresponding static displacements are relatively small, as it is, e.g., the case for the chosen short-ranged Morse potential. Here a value of 2.3 k has been obtained for Cu. This agrees very well with recent experimental results as obtained by quenching studies /8.30/ and positron annihilation measurements /8.31/ (see the discussion in /8.24/).

References

1.1 I.M. Lifshits: J. Phys. USSR 7, 215, 249 (1943); J. Phys. USSR 8, 89 (1948)

1.2 W. Ludwig: *Ergebnisse der Exakten Naturwiss.* Vol. 35 (Springer, Berlin, Göttingen, Heidelberg 1964)

1.3 W. Ludwig: "Recent Developments in Lattice Theory", in Springer Tracts in Modern Physics, Vol. 43 (Springer, Berlin, Heidelberg, New York 1967)

1.4 A.A. Maradudin: "Theoretical and Experimental Aspects of the Effects of Point Defects and Disorder on the Vibrations of Crystals", in *Solid State Physics* Vol. 18, 273, Vol. 19, 1 (Academic Press, New York 1966)

1.5 A.A. Maradudin, E.W. Montroll, G.H. Weiss, I.P. Ipatova: "Theory of Lattice Dynamics in the Harmonic Approximation", in *Solid State Physics*, Suppl. 3, 2nd ed. (Academic Press, New York 1971)

1.6 R.J. Elliott: "Vibrations of Defects in Lattices", in Proc. 6th Scottish Universities Summer School 1965 (Oliver and Boyd , Edinburgh 1966) p. 377

1.7 M.V. Klein: "Localized Modes and Resonance States in Alkali Halides", in *Physics of Colour Centers*, ed. by W.B. Fowler (Academic Press, New York 1968) p. 430

1.8 I.M. Lifshits, A.M. Kosevich: Rep. Prog. Phys. 29, 217 (1966)

1.9 R.E. Wallis (ed.): *Localized Excitations in Solids* (Plenum Press, New York 1968)

1.10 R.J. Elliott, J.A. Krumhansl, P.L. Leath: Rev. Mod. Phys. 46, 465 (1974)

1.11 D.W. Taylor: "Dynamics of Impurities in Crystals", in *Dynamical Properties of Solids*, Vol. 2, ed. by G.K. Horton, A.A. Maradudin (North Holland, Amsterdam 1975) p. 285

1.12 A.S. Barker, Jr., A.J. Sievers: Rev. Mod. Phys. 47, Suppl. No. 2, S1 (1975)

1.13 R.F. Wood: Methods Comput. Phys. 15, 119 (1976)

1.14 R.M. Nicklow: "Phonons and Defects", to be published in: *Neutron Scattering in Materials Science*, ed. by G. Kostortz (in series: Treatise on Material Science and Technology, ed. by H. Hermans)

2.1 M. Born, K. Huang: *Dynamical Theory of Crystal Lattices* (University Press, Oxford 1954)

2.2 G. Leibfried: "Gittertheorie der mechanischen and theoretischen Eigenschaften der Kristalle", in *Encyclopedia of Physics*, Vol. 7, Part 1 (Springer, Berlin, Göttingen, Heidelberg 1955)

2.3 A.A. Maradudin, E.W. Montroll, G.H. Weiss, I.P. Ipatova: "Theory of Lattice Dynamics in the Harmonic Approximation", in *Solid State Physics*, Suppl. 3., 2nd ed. (Academic Press, New York 1971)

2.4 G. Leibfried, N. Breuer: "Point Defects in Metals I", in Springer Tracts in Modern Physics, Vol. 81 (Springer, Berlin, Heidelberg, New York 1978)
2.5 L. van Hove: Phys. Rev. 89, 1189 (1953)
2.6 J.C. Phillips: Phys. Rev. 104, 1263 (1956)
2.7 H.B. Rosenstock: Phys. Rev. 97, 290 (1955)
2.8 G.S. Joyce: J. Math. Phys. 12, 1390 (1971)
2.9 K. Schroeder: *Diffusion Reactions of Point Defects*. Ber. Kernforschungsanlage Jülich, JOL-1083-FF (1974)
2.10 E.W. Montroll, G.H. Weiss: J. Math. Phys. 6, 167 (1965)
2.11 S. Katsura, T. Horiguchi: J. Math. Phys. 12, 230 (1971)
2.12 G.S. Joyce: J. Phys. C 4, L 53 (1971)
2.13 M. Inoue: J. Math. Phys. 15, 704 (1974)
2.14 G.S. Joyce: J. Phys. A 5, L 65 (1972)
2.15 S. Katsura, S. Inawashivo, Y. Abe: J. Math. Phys. 12, 895 (1971)
2.16 G. Gilat, L.J. Raubenheimer: Phys. Rev. 144, 390 (1966)
2.17 H.R. Schober, M. Mostoller, P.H. Dederichs: Phys. Status Solidi (b) 64, 173 (1974)
2.18 O. Jepson, O.K. Andersen: Solid State Commun. 9, 1763 (1971)
2.19 G. Lehmann, M. Taut: Phys. Status Solidi B 54, 469 (1972)
2.20 Y. Endoh, G. Shirane, J. Skalyo: Phys. Rev. 11, 1687 (1975)
2.21 G. Gilat, R.M. Nicklow: Phys. Rev. 143, 487 (1966)
2.22 R.M. Nicklow, G. Gilat, H.G. Smith, L.J. Raubenheimer and M.K. Wilkinson: Phys. Rev. 164, 922 (1967)
2.23 J.W. Lynn, H.G. Smith, R.M. Nicklow: Phys. Rev. (preprint: April 1973)
2.24 A.J. Millington, G.L. Squires: J. Phys. F 1, 244 (1971)
2.25 V.J. Minkiewicz, G. Shirane, R. Nathans: Phys. Rev. 162, 528 (1967)
2.26 Y. Nakagawa, A.D.B. Woods: Phys. Rev. Lett. 11, 271 (1963)
3.1 R. Brout, W.M. Visscher: Phys. Rev. Lett. 9, 54 (1962)
3.2 Yu. M. Kagan, Ya.A. Isoleoskii: Zh. Eksp. Teor. Fiz. 42, 259 (1962) [Sov. Phys.-JETP 15, 182 (1962)]
3.3 S. Takano: Prog. Theor. Phys. Jpn. 29, 191 (1963)
3.4 B. Lengeler, W. Ludwig: Z. Phys. 171, 273 (1963)
4.1 O. Litzmann, P. Rosza: Proc. Phys. Soc. Lond. 85, 285 (1965)
4.2 J.A. Krumhansl, J.A.D. Matthews: Phys. Rev. 166, 856 (1968)
4.3 J.A. Krumhansl: "Asymptotic Description of Defect Excitations", in *Localized Excitation in Solids*, ed. by R.F. Wallis (Plenum Press, New York 1968) p. 17
4.4 K. Kunc: Czech. J. Phys. B 15, 883 (1965)
4.5 J. Mahanty: Phys. Lett. A 29, 583 (1969)
4.6 M. Sachdew, J. Mahanty: J. Phys. C 3, 1225 (1970)
4.7 J. Mahanty: Phys. Lett. 29 A, 583 (1969)
4.8 K. Dettmann, W. Ludwig: Phys. Condens. Matter 2, 241 (1964)
4.9 P. Dean: J. Phys. C 1, 22 (1968)
4.10 T. Fujita: Prog. Theor. Phys. Suppl. 45, 1 (1970)
4.11 J.B. Page: Phys. Rev. B 10, 719 (1974)
4.12 R. Zeller: *Schwingungsverhalten von Zwischengitteratomen*. Ber. Kernforschungsanlage Jülich, JOL-1259 (1975)
4.13 P.H. Dederichs, R. Zeller: Phys. Rev. B 14, 2314 (1976)
4.14 M. Wagner: Phys. Rev. 131, 2520 (1963); 133 A, 750 (1964)
4.15 B.K. Agrawal, P.N. Ram: Phys. Rev. B 5, 3308 (1972)
4.16 P.N. Ram, B.K. Agrawal: Solid State Commun. 13, 1671 (1973)
4.17 A.M. Stoneham: *Theory of Defects in Solids*, (Clarendon Press, Oxford 1975)
5.1 P.D. Mannheim: Phys. Rev. 165, 1011 (1968)
5.2 P.D. Mannheim, S.S. Cohen: Phys. Rev. B 4, 3748 (1971)
5.3 P.D. Mannheim: Phys. Rev. B 5, 745 (1972)
5.4 K. Lakatos, J.A. Krumhansl: Phys. Rev. 175, 841 (1968); 180, 729 (1969)
5.5 B.K. Agrawal: Phys. Rev. 186, 712 (1969)
5.6 L. van Hove: Phys. Rev. 95, 249 (1954)
5.7 M. Marshall, S.W. Lovesey: *Theory of Thermal Neutron Scattering* (Clarendon Press, Oxford 1971)
5.8 K.S. Singwi, A. Sjölander: Phys. Rev. 120, 1093 (1960)
5.9 M.A. Krivoglaz: Sov. Phys.-Solid State 6, 1340 (1964)
5.10 R. Kubo: J. Phys. Soc. Jpn. 17, 1100 (1962)

5.11 A.A. Maradudin, P.A. Flynn: Phys. Rev. 129, 2529 (1963)
5.12 G. Kaindl, D. Salomon, G. Wortmann: "Mössbauer Isomer Shifts of the 6.2 KeV Gamma Rays of ^{181}Tantalum", in Proc. 8th Symp. on Mössbauer Effect Methology (Plenum Press, New York 1973) p. 211
5.13 M. Sachdew and V.K. Tewary: J. Phys. F 3, 1256 (1973)
5.14 J.F. Prince, L.D. Roberts, D.J. Erickson: Phys. Rev. B 13, 24 (1976)
5.15 G.M. Rothberg, S. Guimard, N. Benezer-Koller: Phys. Rev. B 1, 136 (1970)
5.16 D.A. O'Connor, R.W. Reeks, G. Skyrme: J. Phys. F 2, 1179 (1972)
5.17 D.G. Howard, R.H. Nussbaum: Phys. Rev. B 9, 794 (1974)
5.18 J.M. Grow, D.G. Howard, R.H. Nussbaum, M. Takeo; Phys. Rev. B 17, 15 (1978)
5.19 C.F. Steen, D.G. Howard, R.H. Nussbaum: Solid State Commun. 10 (E) 584 (a) (1972)
5.20 B.F. Brace, D.G. Howard, R.H. Nussbaum: Phys. Lett. 43A, 336 (1973)
5.21 R.H. Nussbaum, D.G. Howard, W.L. Nees, C.F. Steen: Phys. Rev. 173, 653 (1968)
5.22 R.D. Taylor, D.J. Erickson, T.A. Kitchens: J. Phys. (Paris) Suppl. 37, C6-35 (1976)
5.23 D.G. Howard, J.G. Dash: J. Appl. Phys. 38, 991 (1967)
5.24 O.P. Balkanski, V.V. Chekin: Sov. Phys.-Solid State 12, 2919 (1971)
5.25 V.A. Brynkhanov, N.N. Delyagin, V.S. Shpinel: Sov. Phys.-JETP 20, 55 (1965)
5.26 G. van Landuyt, C.W. Kimball, F.Y. Fradin: Phys. Rev. B 15, 5119 (1977)
5.27 R.D. Hatcher, R. Zeller, P.H. Dederichs: Phys. Rev. B 19, 5083 (1979)
6.1 W. Schilling: J. Nucl. Mater. 69/70, 465 (1978); Proc. of the Conf. "Properties of Atomic Defects in Metals", Argonne (Oct. 1976)
6.2 G. Verdan, R. Rubin, W. Kley: "The Dynamics of Hydrogen Impurities in Niobium and Vanadium", in *Neutron Inelastic Scattering*, Proc. IAEA Symp. Copenhagen, Vol. I (IAEA, Vienna, 1968) p. 223
6.3 J.H.L. Birchall, D.K. Ross: "A Neutron Scattering Study of Hydrogen Motions in Nb", *International Meeting on Hydrogen in Metals*, Jülich, March 1972. Ber. Kernforschungsanlage Jülich, JÜL-Conf. 6, Vol. 1, p. 313
6.4 J.J. Rush, R.C. Lionigston, L.A. De Graaf, H.E. Flotow, J.M. Rowe: J. Chem. Phys. 59, 6570 (1973)
6.5 K. Sköld, G. Nelin: J. Chem. Phys. 28, 2369 (1967)
6.6 R. Rubin, Y. Claessen: Solid State Commun. 8, 1321 (1970)
6.7 H.D. Carstanjen, R. Sizmann: Intern. Meeting on Hydrogen in Metals, Jülich, March 1972. Ber. Kernforschungsanlage Jülich, JÜL-Conf. 6, 1, 118 (1972)
6.8 G.S. Bauer, W. Schmatz, W. Just: "The Strain Field around Deuterium in Niobium", in 2nd Information Conf. on Hydrogen in Metals, Paris 1977, Paper 2 C 15 (Pergamon Press, New York 1977)
6.9 K.W. Kehr: *Diffusion von Wasserstoff in Metallen*. Ber. Kernforschungsanlage Jülich, JÜL-1211 (1975)
6.10 Intern. Meeting on Hydrogen in Metals, Jülich, March 1972. Ber. Kernforschungsanlage Jülich, JÜL-Conf. 6 1 and 2
6.11 2nd Intern. Conf. on Hydrogen in Metals, Paris 1977 (Pergamon Press, New York 1977)
6.12 V. Lottner, H.R. Schober, W.J. Fitzgerald: Phys. Rev. Lett. 42, 1162 (1979) H.R. Schober, V. Lottner: Z. Phys. Chem. (NF) in press
6.13 A. Scholz, C. Lehmann: Phys. Rev. B 6, 813 (1973)
6.14 P.H. Dederichs, C. Lehmann, A. Scholz: Phys. Rev. Lett. 31, 1130 (1973)
6.15 R.M.J. Cotterill, M. Doyama: Energies and Atomic Configurations of Point Defects in FCC Metals", in *Lattice Defects and their Interactions*, ed. by R.R. Hasigutti (Gordon and Breach, New York 1966)
6.16 P. Ehrhart, H.G. Haubold and W. Schilling: Adv. Solid State Physics XIV, 87 (1974)
6.17 H.G. Haubold: "Study of Irradiation Induced Point Defects by Diffuse Scattering", in *Proc. Conf. on Fundamental Aspects of Radiation Damage in Metals*, Gatlinburg, Vol. 1 (Oct. 1975) p. 268
6.18 P. Ehrhart: J. Nucl. Mater. 69/70, 200 (1978); Proc. of the Conf. "Properties of Atomic Defects in Metals", Argonne (Oct. 1976)
6.19 V. Spiric, L.E. Rehn, K.-H. Robrock, W. Schilling: Phys. Rev. B 15, 672 (1977)
6.20 K. Forsch, J. Hemmerich, H. Knöll, G. Lucki: Phys. Status Solidi (a) 23, 223 (1974)

6.21 R.A. Johnson: J. Phys. F 3, 295 (1973)
6.22 P.H. Dederichs, C. Lehmann, H.R. Schober, A. Scholz, R. Zeller: J. Nucl. Mater. 69/70, 176 (1978); Proc. of the Conf. "Properties of Atomic Defects in Metals", Argonne (Oct. 1976)
6.23 H.R. Schober, R. Zeller: J. Nucl. Mater. 69/70, 341 (1978)
6.24 G. Vogl, M. Mansel, W. Vogl: J. Phys. F 4, 2321 (1974)
6.25 G. Vogl, M. Mansel: "Mössbauer Studies of Interstitials in FCC Metals", in *Proc. Conf. on Fundamental Aspects of Radiation Damage in Metals*, Gatlinburg, Vol. 1 (Oct. 1975) p. 349
6.26 G. Vogl, M. Mansel, P.H. Dederichs: Phys. Rev. Lett. 36, 1497 (1976)
7.1 R.J. Elliott, J.A. Krumhansl, P.L. Leath: Rev. Mod. Phys. 46, 465 (1974)
7.2 R.J. Elliott, D.W. Taylor: Proc. Roy. Soc. London A 296, 161 (1967)
7.3 W. Marshall, S.W. Lovesey: *Theory of Thermal Neutron Scattering* (Clarendon Press, Oxford 1971)
7.4 P.A. Egelstaff (ed.): Thermal Neutron Scattering (Academic Press, New York 1965)
7.5 M.A. Krivoglaz: *Theory of X-Ray and Thermal-Neutron Scattering by Real Crystals* (Plenum Press, New York 1969)
7.6 P.H. Dederichs: J. Phys. F 3, 47 (1973)
7.7 K. Lakatos, J.A. Krumhansl: Phys. Rev. 175, 841 (1968); Phys. Rev. 180, 729 (1969)
7.8 P.D. Mannheim: Phys. Rev. 165, 1011 (1968)
7.9 R.G. Augst, A.P. Zhernov: Sov. Phys. - Solid State 11, 634 (1969))
7.10 S.S. Cohen: Phys. Rev. B 10, 384 (1974)
7.11 H.R. Schober, V.K. Tewary, P.H. Dederichs: Z. Phys. B 21, 255 (1975)
7.12 R.F. Wood, M. Mostoller: Phys. Rev. Lett. 35, 45 (1975)
7.13 A.P. Zhernov, T.N. Kulagina: Sov. Phys. - Solid State 17, 941 (1975)
7.14 Yu. Kagan, A.P. Zhernov: Sov. Phys. - Solid State 9, 1724 (1968)
7.15 H.B. Møller, A.R. Mackintosh: "Investigation of Localized Excitations by Inelastic Neutron Scattering", in *Localized Excitations in Solids*, ed. by R.F. Wallis (Plenum Press, New York 1968)
7.16 R.M. Cunningham, L.D. Muhlestein, W.M. Schaw, C.W. Thompson: Phys. Rev. B 2, 4864 (1970)
7.17 K.M. Kesharwani, B.K. Agrawal: Phys. Rev. B 6, 2178 (1972)
7.18 E.C. Svenson, B.N. Brockhouse: Phys. Rev. Lett. 18, 858 (1967)
7.19 E.C. Svenson, W.A. Kamitakahara: Can. J. Phys. 49, 2291 (1971)
7.20 R. Bruno, D.W. Taylor: Can. J. Phys. 49, 2496 (1971)
7.21 M. Mostoller, T. Kaplan, N. Wakabayashi, R.M. Nicklow: Phys. Rev. B 10, 3144 (1974)
7.22 D. Walton, H.A. Mook, R.M. Nicklow: Phys. Rev. Lett. 33, 412 (1974)
7.23 V. Narayanamurhi, R.O. Pohl: Rev. Mod. Phys. 42, 201 (1970)
7.24 N.E. Byer, H.S. Sack: Phys. Status Solidi 30, 569 (1968)
7.25 M.V. Klein: Phys. Rev. 186, 839 (1969)
7.26 A. Zinken, U. Buchenau, H.J. Frenzl, H.R. Schober: Solid State Commun. 22, 693 (1977)
7.27 H.G. Smith, N. Wakabayashi, M. Mostoller: In *Superconductivity in d- and f-Band Metals*, ed. by D.H. Douglass (Plenum Press, New York 1976)
7.28 N. Kunitomi, Y. Tsunoda, N. Wakabayashi, R.M. Nicklow, H.G. Smith: Solid State Commun. 25, 921 (1978)
7.29 R.F. Wood, M. Mostoller: Phys. Rev. Lett. 35, 45 (1975)
7.30 H.R. Schober, P.H. Dederichs, V.K. Tewary: Z. Phys. B 21, 255 (1975)
7.31 a) R.M. Nicklow, R.R. Coltman, F.W. Young, R.F. Wood: Phys. Rev. Lett. 35, 444 (1975)
b) R.M. Nicklow, W.P. Crummett, J.M. Williams: To be published
7.32 K. Böning, G.S. Bauer, H. Fenzl, R. Scherm, W. Kaiser: Phys. Rev. Lett. 38, 852 (1977)
7.33 R.M. Nicklow, P.R. Vijayaraghavan, H.G. Smith, M.K. Wilkinson: Phys. Rev. Lett. 20, 1245 (1968)
R.M. Nicklow, P.R. Vijayaraghavan, H.G. Smith, D. Dolling, M.K. Wilkinson: "Coherent Inelastic Scattering Studies of Local and In-Band Modes in $Cu_{1-x}Al_x$

Crystals", in: *Neutron Inelastic Scattering*, Vol. 1, Symp. Copenhagen 1968 (IAEA Vienna 1968)
7.34 N. Wakabayashi, R.M. Nicklow, H.G. Smith: Phys. Rev. B 4, 2558 (1971)
7.35 N. Wakabayashi: Phys. Rev. B 8, 6015 (1973)
7.36 J. Als-Nielson: "Localized and Band Phonons in Ta_{88}-Nb_{12}", in *Neutron Inelastic Scattering*, Vol. 1, Symp. Copenhagen 1968, (IAEA, Vienna 1968)
7.37 H.G. Smith, N. Wakabayashi, R.M. Nicklow: "Localized Torsional and Translational Modes in KCl (NH_4)", in *Neutron Inelastic Scattering* (IAEA, Vienna 1972)
7.38 N.A. Chernoplekov, M.G. Zemlyanov: Sov. Phys. JETP 22, 315 (1966)
7.39 N.A. Chernoplekov, G.K. Panova, M.G. Zemlyanov, B.N. Samoilov, V.I. Kutaitsev: Phys. Status Solidi 20, 767 (1967)
7.40 B. Mozer, K. Otnes, C. Thaper: Phys. Rev. 152, 535 (1966)
7.41 G.F. Syrykh, A.P. Zhernov, M.G. Zemlyanov, S.P. Mironov, N.A. Chernoplekov, Yu.L. Shitikov: Sov. Phys. JETP 49, 183 (1976)
7.42 N.Y. Chernoplekov, G.K. Panova, B.N. Samoilov, A.A. Shikov: Sov. Phys. JETP 36, 731 (1973)
7.43 G.F. Syrykh, A.P. Zhernov, M.G. Zemlyanov, G.I. Solovev: Phys. Status Solidi (b) 79, 105 (1977)
7.44 H.G. Purwins, R. Labusch, P. Haasen: Z. Metallkunde 57, 867 (1966)
7.45 W. Köster: Z. Metallkunde 62, 123 (1971)
7.46 J.T. Lenkken, E. Lähteenkorva: J. Phys. F 8, 1643 (1978)
7.47 W.C. Hubbell, F.R. Brotzen: J. Appl. Phys. 43, 3306 (1972)
7.48 K. Fuchs: Proc. Soc. London A 151, 585 (1935); A 153, 622 (1936)
7.49 L.E. Rehn, J. Holder, A.V. Granato, R.R. Coltman, F.W. Young, Jr.: Phys. Rev. B 10, 349 (1974)
J. Holder, A.V. Granato, L.E. Rehn: Phys. Rev. B 10, 363 (1974)
J. Holder, A.V. Granato, L.E. Rehn: Phys. Rev. Lett. 32, 1053 (1974)
7.50 K.H. Robrock, W. Schilling: J. Phys. F 6, 303 (1976)
7.51 P.H. Dederichs, C. Lehmann, A. Scholz: Z. Phys. B 20, 155 (1975)
7.52 W. Ludwig: "Influence of Defects on Elastic Constants", in Calculations of Properties of Vacancies and Interstitials. Nat. Bur. Stand., Misc Publ. 287, 151 (1966)
7.53 R.J. Elliott, J.A. Krumhansl, T.H. Merrett: "Elastic Strains in Imperfect Crystals", in *Localized Excitations in Solids*, ed. by R.F. Wallis (Plenum Press, New York 1968) p. 709
7.54 M.M. Bredov, B.A. Kotov, M.N. Okuneva, V.S. Oskotskii, A.L. Shakh-Budagov: Sov. Phys.-Solid State 9, 214 (1967)
N. Breuer: *On the Determination of the Phonon Spectrum from Coherent Inelastic Neutron Scattering by Polycrystals*. Ber. Kernforschungsanlage Jülich, JÜL-1083-FF (1974)
7.55 M.A. Ivanov: Sov. Phys.-Solid State 12, 1508 (1971)
7.56 I. Natkaniec, K. Parlinski, A. Bajorek, M. Sudnik-Hrynkiewicz: Physics Lett. 24 A, 517 (1967)
8.1 I.M. Lifshits: Zh. Eksp. Teor. Fiz. 17, 1071, 1076 (1947); Nuovo Cimento, Suppl. A 13, 591 (1956)
8.2 B.K. Agrawal: Phys. Rev. 186, 712 (1969)
8.3 M.D. Tiwari, K.M. Kesharwani, B.K. Agrawal: Phys. Rev. B 7, 2378 (1973)
8.4 J. Mahanty, M. Sachdev: J. Phys. C 3, 773 (1970)
8.5 V.I. Peresada, V.P. Tolstoluzhskii: Sov. J. Low Temp. Phys. 3 (6), 383 (1977)
8.6 Yu. Kagan, Ya. Isoilevskii: Sov. Phys. JETP 18, 562 (1964)
8.7 G.K. Panova, B.N. Samoilov: Sov. Phys. JETP 22, 320 (1966)
8.8 J.A. Cape, G.W. Lehmann, W.V. Johnson, R.W. DeWames: Phys. Rev. Lett. 16, 892 (1966)
8.9 H. Culbert, R.P. Hübener: Physics Lett. 24 A, 530 (1967)
8.10 W.M. Hartmann, H.V. Culbert, R.P. Hübener: Phys. Rev. B 1, 1486 (1969)
8.11 V.S. Kryloskiy, V.I. Ovcharenko, V.A. Pervakov, V.I. Khotkevich: Fiz. Met. Metalloved. 35, 1325 (1973)
8.12 W.G. Dehlinger, W.R. Savage, J.W. Schweitzer: Phys. Rev. B 6, 338 (1972)
8.13 B.A. Green, Jr., A.A. Valladares: Phys. Rev. 142, 379 (1966)
8.14 M.D. Tiwari, B.K. Agrawal: J. Phys. F 3, 2051 (1973)

8.15 M.D. Tiwari, B.K. Agrawal: Phys. Status Solidi (b) 58, 209 (1973)
8.16 B.H. Verbeck, M.F. Pikart, G.J. van den Berg: Physica 75, 305 (1974)
8.17 B.M. Veal, J.A. Rayne: Phys. Rev. 130, 2156 (1963)
8.18 A.T. Will, B.A. Green, Jr.: Phys. Rev. 150, 519 (1966)
8.19 T.B. Massalski, L. Isaacs: Phys. Rev. 138, A 139 (1965)
8.20 K.H. Rieder: Acta Phys. Austriaca, 32, 290 (1970)
8.21 A.A. Shikov, N.A. Chernoplekov, G.K. Panova, B.N. Samoilov, A.P. Zhernov: Sov. Phys.-JETP 43, 354 (1976)
8.22 J. Govindarajan, P.W.M. Jacobs, M.A. Nerenberg: J. Phys. C 9, 3911 (1976)
8.23 A. Hultgreen, P.D. Desai, D.T. Hawkins, M. Gleiser, K.K. Kelley: *Selected Values of the Thermodynamic Properties of Binary Alloys* (American Soc. of Metals 1973)
8.24 R.D. Hatcher, R. Zeller, P.H. Dederichs: Phys. Rev. B 19, 5083 (1979)
8.25 J.J. Burton: Phys. Rev. B 5, 2948 (1972)
8.26 D.A. Huckaby: J. Chem. Phys. 65, 607 (1976)
8.27 A. Magerl, N. Stump, H. Wipf, G. Alefeld: J. Phys. Chem. Sol. 38, 683 (1977)
8.28 H.B. Huntington, G.A. Shirn, E.S. Wajda: Phys. Rev. 99, 1085 (1955)
8.29 G. Schottky, A. Seeger, G. Schmid: Phys. Status Solidi 4, 439 (1964)
8.30 R.R. Bourassa, B. Lengeler: J. Phys. F 6, 1405 (1976)
8.31 S. Mantl, W. Triftháuser: Phys. Rev. B 17, 1645 (1978)

Theory of Diffusion Controlled Reactions of Point Defects in Metals

K. Schröder

1. Introduction

The importance of point defects like vacancies, interstitials and impurity atoms for the understanding of many features of real crystals has motivated extensive research /1.1/. The structure and vibrational behaviour of point defects, the influence of mechanical properties and their role in diffusion and solid state reactions have been studied. Many aspects can be found in Flynn's book on "Point Defects and Diffusion" /1.2/. The static structure and vibrational behaviour of point defects have recently been reviewed by DEDERICHS et al. /1.3/, and a microscopic picture for the elementary diffusional jumps of point defects has been discussed by YOUNG /1.4/ and FRANKLIN /1.5/. Diffusion in ideal crystals and the information about point defects obtained from tracer diffusion experiments and mechanical relaxation methods have been discussed by MANNING /1.6/, PETERSON /1.7/, and NOWICK and BERRY /1.8/. Microscopic methods like neutron scattering /1.9/, nuclear magnetic resonance /1.10/ and Mössbauer technique /1.11/ have proven particularly fruitful to study the behaviour of point defects.

We shall briefly review the work on diffusion in ideal lattices in Sect.2 and particularly emphasize the influence of the discrete lattice on point defect diffusion. We shall discuss the Green's function for diffusion on a lattice in some detail because its knowledge is necessary for the interpretation of experiments. The analytical properties, the asymptotic behaviour and the transition to continuum diffusion are discussed. We give a brief survey of experimental methods to determine the geometry and kinetics of diffusion.

Point defects can interact with lattice inhomogeneities like other point defects, dislocations, precipitations, etc. (Sect. 3). In metals the elastic interaction due to the strain field connected with point defects in solids is most important; in semiconductors and ionic crystals the Coulomb interaction has to be considered for charged defects. The elastic interaction always has attractive direction indicating a general trend towards agglomeration of defects.

In Sect. 4 we describe the diffusion of point defects in external fields. It turns out that one has to distinguish two interactions, one for the stable configuration and one for the saddle point configuration of the mobile defect /1.12/. Whereas the stability of a complex (binding energy) and the stationary distribution of point defects around sinks are determined by the interaction in the stable configuration, the change of the diffusion rate is determined by the interaction in the saddle point configuration alone; the diffusion tensor can become anisotropic even in cubic materials for defects with cubic symmetry in the equilibrium configuration.

Due to the interaction defects can react with each other (and with sinks) as soon as they become mobile. Such reactions are usually described by kinetic equations using phenomenological rate constants /1.13,14/. In Sect. 5 we review the description of diffusion limited reactions, limiting our presentation to stationary situations. First we discuss the phenomenological theory which describes the interaction by a reaction radius R_a and the motion of the defects by continuum diffusion with diffusion constant D_o. In the dilute limit (independent sink approximation) the reaction rate K can be expressed in terms of these parameters /1.15/ by $K = 4\pi D_o R_a$. The influence of larger sink densities is discussed briefly.

In Sect. 6 we show that the concept of a reaction radius remains valid if one takes into account the discrete nature of diffusion on a lattice. One obtains an explicit expression for the reaction radius for each set of reaction sites. If the reaction sites form a compact set, e.g., one site and its nearest neighbours, the results agree quite well with continuum theory with R_a being approximately equal to the separation of the outermost reaction site from the center of the set /1.16/.

Finally in Sect. 7 we discuss the influence of long range potentials between sinks and mobile defects on the reaction rate. Vibrational principles are set up which yield upper and lower limits for the reaction radii. Due to the interaction the effective reaction radius becomes temperature dependent. The explicit form is determined by the radial dependence of the interaction whereas the absolute value depends on the strength and the anisotropy of the interaction /1.17/. For potentials with attractive directions the reaction radius is approximately given by the distance at which the potential energy gained by the defect in the saddle point configuration equals the thermal energy. A critical discussion of the applicability of the available theories concludes the paper.

2. Diffusion in Ideal Crystals

Microscopically, defects move by jumping from one equilibrium site to neighbouring ones. Successive jumps then lead to long-range diffusion. The jump process can be very complicated since it involves the coherent motion of several atoms as is illustrated in the following by three examples. We will then discuss the lattice of the stable positions ("diffusion lattice") and give a short derivation of the jump frequency in order to establish which parameters determine the diffusion rate. Then we will derive the solution of the lattice diffusion equation and discuss experimental methods which yield information about the macroscopic continuum aspects and microscopic lattice aspects of diffusion.

Conceptually the simplest case is the diffusion of a small impurity interstitial. The elementary jump is schematically illustrated in Fig. 2.1 for an octahedral interstitial in an fcc crystal (e.g., H in Pd /2.1/). The impurity atom changes its

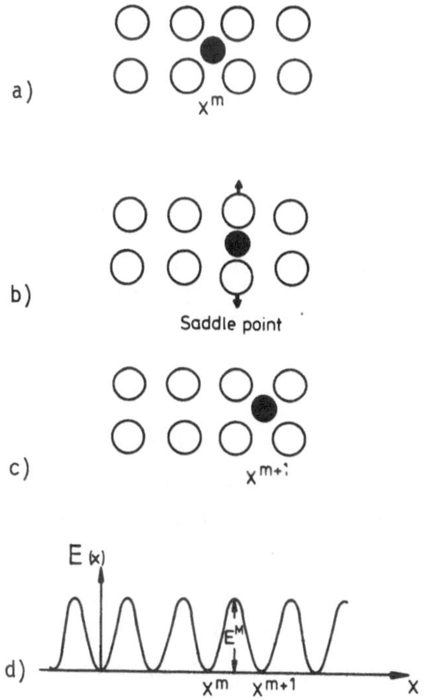

Fig. 2.1. (a-c) show schematically the configurations for a jump in (110)-direction of an octahedral interstitial in an fcc crystal from site x^m to a neighbouring site x^{m+1}. In the saddle point configuration (b) the displacements of the neighbouring atoms are indicated by the arrows. (d) One-dimensional [(110)-direction] periodic model potential for a jumping interstitial. x^m, x^{m+1} are neighbouring equilibrium positions, the energy of the saddle point is E^M

position to a nearest-neighbour site whereas the host atoms are more or less at the same positions before and after the jump (apart from the small static displacements). During the jump, however, some host atoms move considerably in order to open an easy jump path for the impurity atom. Thus the microscopic structure of the defect changes during the jump and consequently also the interaction with an external force field. We shall come back to this in Sect. 4.

Classically, one can in this case construct a potential energy surface as function of the impurity position by taking the minimum energy of the total crystal with the position of the impurity atom fixed and the host atoms relaxed accordingly. If one knows the impurity-host interaction and the host-host interaction one can obtain the potential energy surface, e.g., by computer simulation /2.2/. This potential energy surface is periodic with lattice symmetry. It exhibits minima at the stable and metastable positions of the impurity atom which are separated by saddle points. In order to jump from one equilibrium site to a neighbouring one the impurity atom has to cross the saddle point by thermal excitation. The energy is provided by the phonon system which is considered as a heat bath. The barrier height is called activation energy E^M for migration (see Fig. 2.1d). We shall discuss the classical approach to the calculation of the jump frequency briefly in Sect. 2.2.

A more complicated picture applies, e.g., to the motion of a <100>-split interstitial (see Fig. 2.2) which is the stable configuration of the self-interstitial in Al and Cu /2.3/ and probably also in other fcc metals. Using a phenomenological

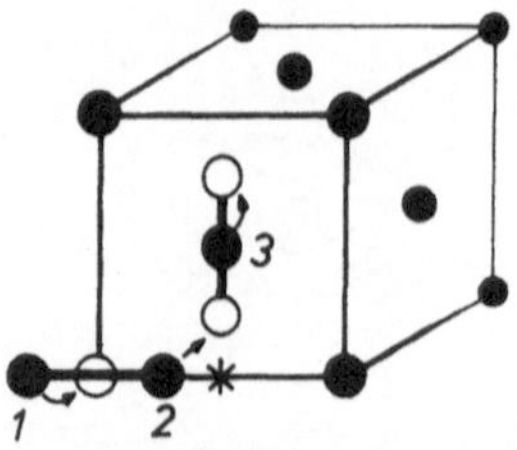

Fig. 2.2. Jump of a <100>-split interstitial. Indicated is the motion of atoms No. 1 - 3 which results in a displacement of the center of mass by a nearest-neighbour distance and a rotation of the axis

two-body interaction potential the potential energy surface for a self-interstitial in Cu has been calculated /2.4/. The dynamics /2.5/ of the <100>-split interstitial and the consequences for the diffusion /2.6/ have been discussed in two recent reviews.

The main difference to the case discussed above is that not a single atom jump can account for the motion of the <100>-split interstitial as is illustrated in Fig. 2.2. A <100>-split interstitial at the front lower left corner of the elementary cube consisting of atoms No.1 and 2 changes its position to the front face center, then consisting of atoms No.2 and 3. To complete this jump none of the atoms has to move a full nearest-neighbour distance but many atoms change their positions by a smaller amount so that in total one atomic mass is displaced by a nearest-neighbour distance. By successive jumps the structural defect (<100>-split interstitial) can move through the entire crystal whereas one particular atom is essentially confined to the sites neighbouring an octahedral interstitial site (for atom

2 indicated by the star in Fig. 2.2). During the jump the <100>-split interstitial changes its structure considerably which again results in a change of the interaction with external force fields.

A similar picture applies to the motion of a vacancy as is illustrated in Fig. 2.3 for an fcc crystal. The equilibrium configuration of a vacancy has cubic symmetry and the displacements of the neighbouring atoms are small. The vacancy moves

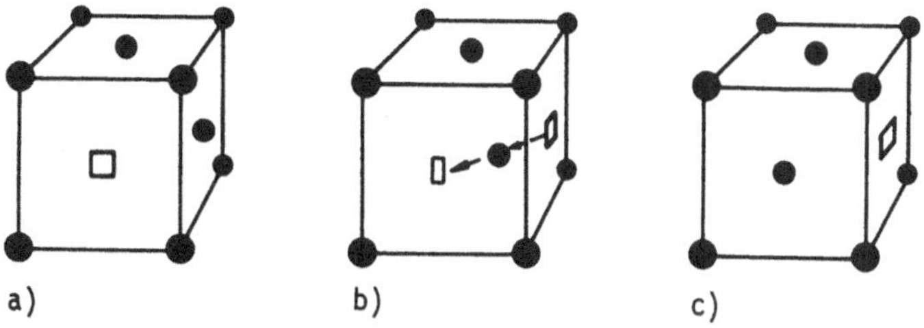

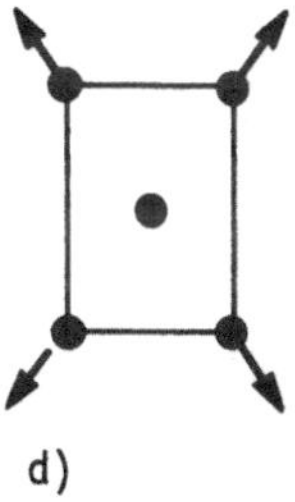

Fig. 2.3. a-c) Show schematically the configurations for a jump of a vacancy in an fcc crystal. d) Displacements of the four neighbouring atoms in the (110) plane for the saddle point configuration b)

if one of the neighbouring atoms jumps to the empty site leaving its original site vacant. During this jump the atom has to move through interstitial region. The saddle point configuration consists of two "half" vacancies and an interstitial atom (Fig. 2.3b). The neighbouring atoms are displaced considerably from their ideal lattice positions in order to minimize the repulsive interaction with the interstitial atom. The configuration is noncubic (orthorhombic in fcc crystals) and the interaction with external fields differs markedly from that of a vacancy in the equilibrium configuration.

In most cases the barrier height is rather large and jumps are very infrequent compared to typical vibrational frequencies. On the other hand the time to complete a jump is of the order of the reciprocal Debye frequency. Then the usual assumption is justified that before attempting a new jump a defect loses the information about the energy and momentum it carried during the previous jump. Thus successive jumps are uncorrelated. A possible exception is the motion of hydrogen in bcc metals and of self-interstitials at high temperatures. Here the jumps are so frequent (of the order of $10^{12}s^{-1}$) that a persistence of the jump direction due to the momentum carried during a jump seems possible /2.7/. This effect has also been seen in computer simulation of vacancy diffusion /2.8/. Its consequence for the interpretation of diffusion experiments has not yet been discussed thoroughly. We are not taking this effect into account and will always assume that the defects perform a random walk with kinetically uncorrelated jumps. In certain cases correlations appear for geometrical reasons as in the case of diffusion of substitutional impurities via vacancies /2.9/ (see also Sect. 2.1).

2.1 Lattice of Equilibrium Sites ("Diffusion Lattice")

Due to the periodic structure of the ideal host lattice the equilibrium sites of a mobile defect also form a lattice.

In the case of a vacancy this "diffusion lattice" coincides with the host lattice. This is also the case for substitutional impurities. However, these defects can only jump via a second defect, namely a vacancy. The presence of this additional defect lowers the symmetry of the neighbourhood of the impurity atom, and this gives rise to *geometrical* correlation between successive jumps as indicated in Fig. 2.4. After

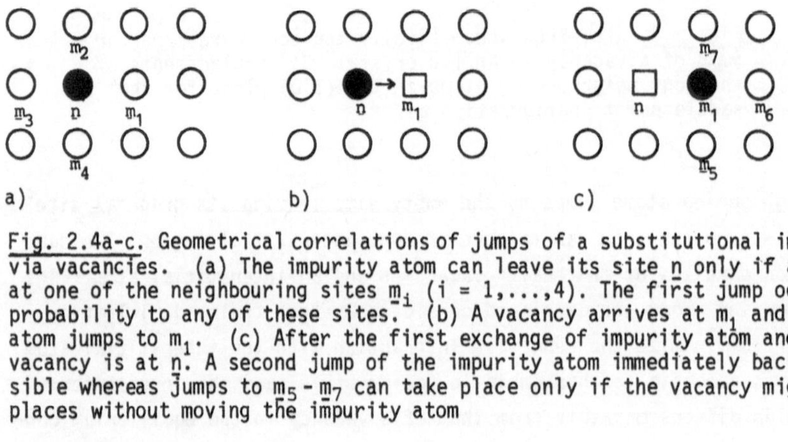

Fig. 2.4a-c. Geometrical correlations of jumps of a substitutional impurity diffusing via vacancies. (a) The impurity atom can leave its site n only if a vacancy arrives at one of the neighbouring sites m_i (i = 1,...,4). The first jump occurs with equal probability to any of these sites. (b) A vacancy arrives at m_1 and the impurity atom jumps to m_1. (c) After the first exchange of impurity atom and vacancy the vacancy is at n. A second jump of the impurity atom immediately back to n is possible whereas jumps to m_5 - m_7 can take place only if the vacancy migrates to these places without moving the impurity atom

an exchange of sites with a vacancy an impurity atom has a larger chance to jump immediately back to its former position (where the vacancy is) than to jump to a different neighbouring site which is occupied by a host atom.

For interstitials the lattice of the stable sites in general is different from the host lattice. Some possible interstitial configurations are shown in Fig. 2.5 for an fcc host lattice and in Fig. 2.6 for a bcc host lattice. The corresponding "diffusion lattices" are as follows:

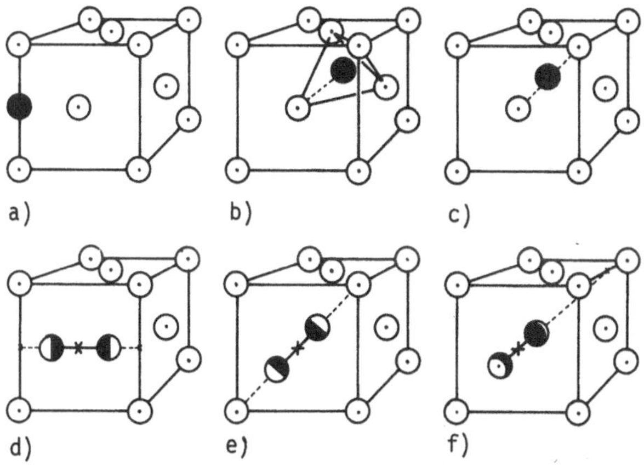

Fig. 2.5a-f. Six interstitial configurations in fcc lattices. (a) body centered or octahedral (b) tetrahedral (c) crowdion (d) <100>-split (e) <110>-split (f) <111>-split

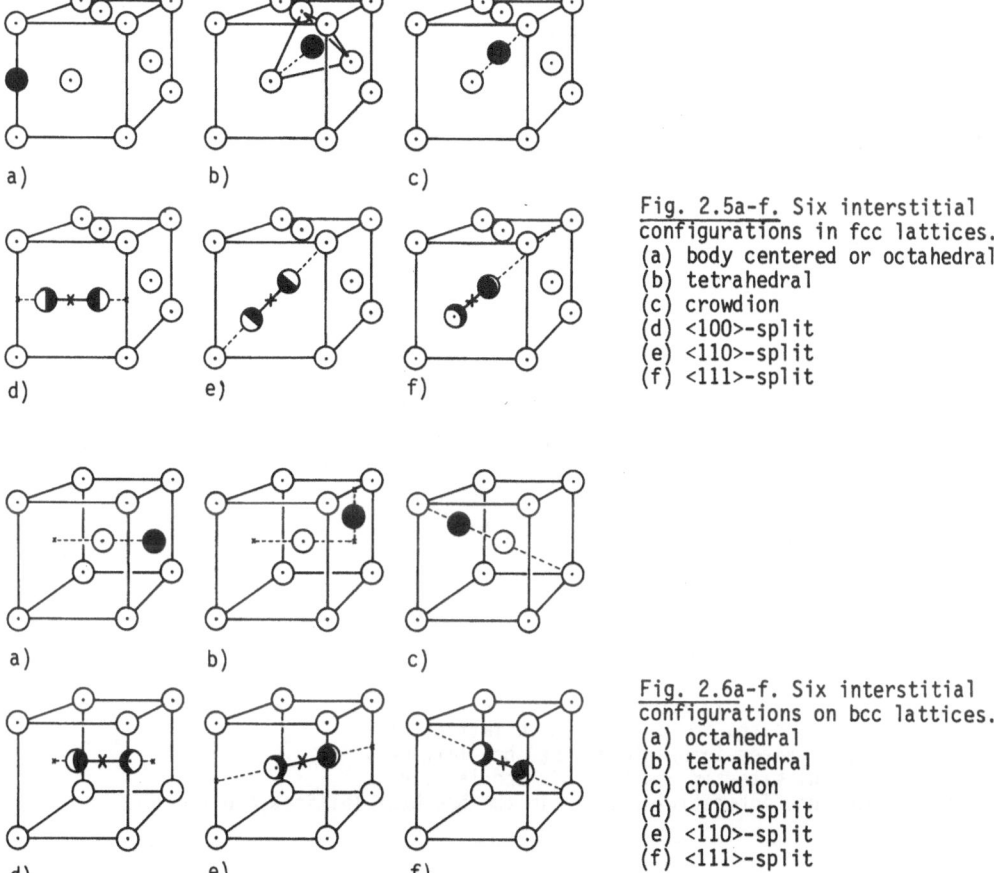

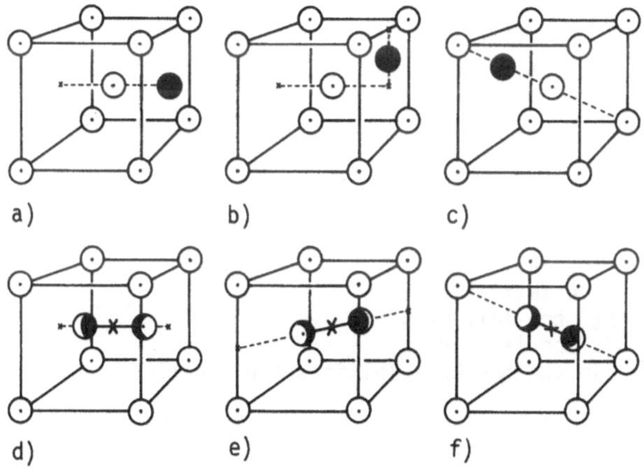

Fig. 2.6a-f. Six interstitial configurations on bcc lattices. (a) octahedral (b) tetrahedral (c) crowdion (d) <100>-split (e) <110>-split (f) <111>-split

a) fcc Host Lattice

Octahedral sites in fcc (Fig. 2.5a) form an fcc lattice with the same lattice constant a_o as the host lattice. The interstitial lattice is displaced with respect to the host lattice by $a_o/2$ along a cube axis (Fig. 2.7a). There is one octahedral site per host atom. Neutron scattering /2.1/ and channeling experiments have shown that H in Pd occupies the octahedral sites.

The tetrahedral sites (Fig. 2.5b) form a sc lattice with lattice constant $a_o/2$ which is displaced with respect to the host lattice by $\sqrt{3}\,a_o/4$ along the cube diagonal (Fig. 2.7b). There are two tetrahedral sites per host atom.

Interstitial sites in the middle of each pair of nearest-neighbour sites ("crowdion", Fig. 2.5c) form a lattice consisting of three sc sublattices with lattice constant $a_o/2$ which are displaced with respect to the host lattice along three face diagonals by $a_o/4$ (110), $a_o/4$ (011) and $a_o/4$ (101), respectively. There are six such sites per host atom.

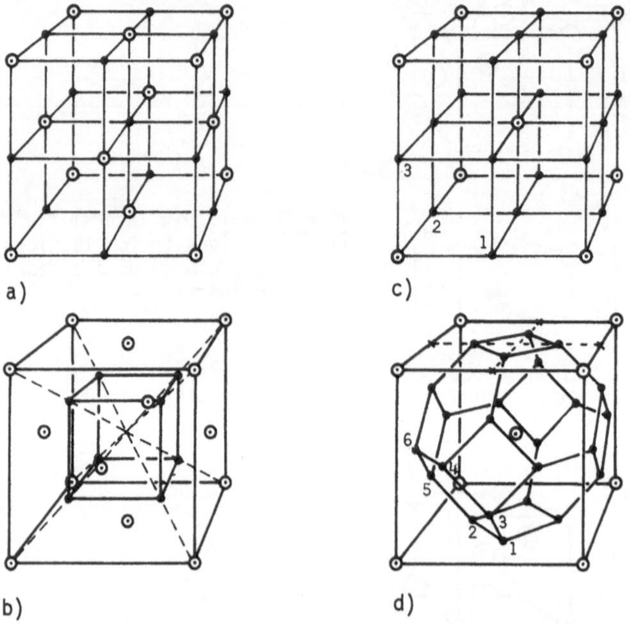

Fig. 2.7a-d. Various types of lattices of interstitial sites (full circles) in fcc and bcc lattices (open circles)
(a) octahedral sites in fcc form an fcc lattice,
(b) tetrahderal sites in fcc form an sc lattice,
(c) octahedral sites in bcc form three bcc lattices,
(d) tetrahedral sites in bcc form six bcc lattices.
The nonequivalent interstitial sites in the bcc host lattices are labelled

b) bcc Host Lattice

The situation is much more complicated in bcc lattices.

Octahedral sites (Fig. 2.6a) form a lattice consisting of three bcc sublattices with lattice constant a_o which are displaced with respect to the host lattice along the cubic axes by $a_o/2$, (Fig. 2.7c). There are three octahedral sites per host atom. The octahedral site is the stable position for C in Fe /2.10/ and O in Nb /2.11,12/.

Tetrahedral sites (Fig. 2.6b) form a lattice consisting of six bcc sublattices with lattice constant a_o which are displaced with respect to the host lattice by $a_o/4$ {(210), (120), (201), (102), (021), (012)}, respectively (Fig. 2.7d). There are six tetrahedral sites per host atom. Channeling studies /2.12/ and neutron scattering experiments /2.13/ have shown that the tetrahedral site is the stable position of D and H in Nb, V and Ta.

Interstitials in the middle of each pair of nearest neighbours ("crowdion", Fig. 2.6c) form an sc lattice with lattice constant $a_o/2$ which is displaced with respect to the host lattice along the body diagonal by $a_o/4$ (111). There are four such sites per host atom.

c) sc Host Lattice

The body center interstitial sites [e.g., $a_o/2$ (111)] form an sc lattice with the same lattice constant a_o as the host lattice. The interstitial lattice is displaced with respect to the host lattice by $a_o/2$ (111) along the cube diagonal. There is one body center interstitial site per host atom. The face center interstitial sites [e.g., $a_o/2$ (110)] form a lattice consisting of three sc sublattices with lattice constant a_o. They are displaced with respect to the host lattice along the face diagonals by $a_o/2$ (110) , $a_o/2$ (011) , $a_o/2$ (101), respectively. There are three face center interstitial sites per host atoms.

The split interstitials (Figs. 2.5,6d-f) sit on host lattice sites. The extra atom forms a dumbbell with one of the regular lattice atoms. Computer simulation studies have shown that in fcc metals the <100>-dumbbell structure (Fig. 2.5d) is the stable configuration of the self-interstitial /2.2,14/. Diffuse X-ray scattering experiments have confirmed this structure for self-interstitials in Al and Cu /2.3/ produced by electron irradiation at low temperatures. For bcc-iron the <110>-dumbbell (Fig. 2.6e) was found to be the stable configuration by computer simulation /2.15/ and experimentally by magnetic aftereffect /2.16/. Diffuse X-ray scattering shows this structure for the self-interstitial in Mo /2.17/.

Also dynamics and jumps of split interstitials have been investigated both by computer simulation /2.18/ and analytically /2.19/. For the <100>-dumbbell in fcc crystals this has revealed the peculiar jump geometry discussed above. The dumbbell moves by a combined translational - rotational jump as has been confirmed by mechanical relaxation experiments on irradiated Al /2.20/. Due to their lower than cubic symmetry the split interstitials have an orientational degree of freedom which complicates their diffusion considerably. One can formulate the theory in analogy to the case of a nonprimitive diffusion lattice which applies for instance to hydrogen diffusion in bcc metals (see Sect. 4).

2.2 Jump Frequency

Many attempts have been made to calculate the jump frequency of a defect in a solid. For a general survey and a critical evaluation of the applicability see, e.g., the book by FLYNN /2.21/, and the recent reviews by FRANKLIN /2.22/, BENNETT /2.8/, and KEHR /2.23/. Rather than trying to give a complete survey of the ideas developed so far we shall present a simple classical model to show which parameters determine the jump frequency.

We consider an impurity atom interstitially dissolved in a crystal. The potential energy of the system is $U(\underline{R} \mid \underline{R}_1 \ldots \underline{R}_N)$ where the impurity atom is at $\underline{R}$ and the N crystal atoms are at $\underline{R}_1 \ldots \underline{R}_N$. As discussed before, the stable sites of the impurity interstitial form a lattice. Neighbouring lattice sites are separated by saddle points.

At elevated temperatures the impurity interstitial can jump from one stable position, say $\underline{R}^{m}$, to a neighbouring one, $\underline{R}^{n}$, across the saddle point $\underline{R}^{mn}$ if by thermal fluctuation it has gained enough energy to overcome the barrier. We wish to calculated the frequency with which such jumps occur in thermal equilibrium.

For constant volume statistics is determined by the free energy F. The configurational part of F as a function of impurity coordinate $\underline{R}$ is given by

$$F(\underline{R}/T,V) = \ln P_o(T,V) - \beta^{-1} \ln \int_V d\underline{R}_1 \ldots d\underline{R}_N \exp\left(-\beta U(\underline{R} \mid \underline{R}_1 \ldots \underline{R}_N)\right) \tag{2.1}$$

with $\beta = 1/kT$. Here we have explicitly indicated the dependence on temperature T and volume V which we will omit in the following. The constant P_o guarantees the normalization of the configurational partition function

$$P(\underline{R}) = e^{-\beta F(\underline{R})} \ . \tag{2.2}$$

From $F(\underline{R})$, (2.1), we can calculate the stable positions and saddle points for the impurity atom.

For $T \to 0$ the configurational part of the free energy for fixed impurity coordinate reduces to the potential energy of the relaxed crystal

$$F_{conf}(\underline{R} \mid T \to 0) = U_{rel}(\underline{R} \mid \bar{\underline{R}}_1 \ldots \bar{\underline{R}}_N) \tag{2.3}$$

which is determined by

$$\frac{\partial}{\partial R_i} U(\underline{R} \mid \underline{R}_1 \ldots \underline{R}_N)\Big|_{\bar{\underline{R}}_1 \ldots \bar{\underline{R}}_N} = 0 \qquad i = 1,\ldots,N$$

and

$$\frac{\partial}{\partial R_i\, \partial R_j} U(\underline{R} \mid \underline{R}_1 \ldots \underline{R}_N)\Big|_{\bar{\underline{R}}_1 \ldots \bar{\underline{R}}_N} > 0$$

as can be shown from (2.1).

The jump frequency Γ across the saddle point involves two factors, one is the relative population of the saddle point configuration compared to the stable configuration, which is given by the ratio of two partition functions $P_s(\underline{R}^{mn})/P_e(\underline{R}^{m})$ and the second is the frequency $\bar{\nu}$ of the vibrational mode which leads the system across the saddle point along the diffusion path, i.e., the line of force joining the saddle point and the equilibrium configuration

$$\Gamma = \underline{\nu} \frac{P_s(\underline{R}^{mn})}{P_e(\underline{R}^{m})} \ . \tag{2.4}$$

The ratio of the partition functions can be calculated from the configurational part of the free energy yielding

$$\frac{P_s(\underline{R}^{mn})}{P_e(\underline{R}^{m})} = \exp\left[-\beta\left(F(\underline{R}^{mn}) - F(\underline{R}^{m})\right)\right] = \frac{\int_V d\underline{R}_1\ldots d\underline{R}_N \exp\left(-\beta\, U(\underline{R}^{mn} \mid \underline{R}_1\ldots\underline{R}_N)\right)}{\int_V d\underline{R}_1\ldots d\underline{R}_N \exp\left(-\beta\, U(\underline{R}^{m} \mid \underline{R}_1\ldots\underline{R}_N)\right)}. \tag{2.5}$$

The frequency $\bar{\nu}$ is a dynamical property of the system and much more difficult to calculate. An estimate can be given by the following arguments: Consider two neighbouring relaxed configurations of the crystal atoms, one with the impurity atom fixed in the saddle point and a second one where the impurity is located close to the saddle point displaced along the diffusion path. $\bar{\nu}$ is the frequency of the vibrational mode of the crystal atoms which leads from one configuration to the other with fixed impurity atom. In general, this mode is a linear combination of eigenmodes of the crystal atoms (it is not a single eigenmode because symmetry is not retained) and thus $\bar{\nu}$ is given by

$$\bar{\nu}^2 = \sum_{\alpha=1}^{3N} c_\alpha^2\, \nu_\alpha'^2 \;; \quad \sum_\alpha c_\alpha^2 = 1 \tag{2.6}$$

where c_α is the projection of the considered (normalized) displacements on the eigenmode α in the saddle point. The sum runs over all 3N eigenmodes of the crystal atoms with frequencies ν_α'.

In order to obtain a more explicit result for the jump frequency we make a harmonic expansion of the potential energy of the system with fixed impurity coordinate $\underline{R}$. For the denominator in (2.5) we expand around the relaxed equilibrium configuration

$$U(\underline{R}^{m} \mid \underline{R}_1\ldots\underline{R}_N) \cong E_e(\underline{R}^{m}) + \frac{1}{2}\sum_{\alpha=1}^{3N} (2\pi\nu_\alpha)^2\, y_\alpha^2$$

and for the numerator around the relaxed saddle point configuration

$$U(\underline{R}^{mn} \mid \underline{R}_1\ldots\underline{R}_N) \cong E_s(\underline{R}^{mn}) + \frac{1}{2}\sum_{\alpha=1}^{3N} (2\pi\nu_\alpha')^2\, y_\alpha^2 \,.$$

Here $E_e(\underline{R}^{m}) = U(\underline{R}^{m} \mid \bar{\underline{R}}_1\ldots\bar{\underline{R}}_N)$ is the potential energy of the relaxed equilibrium configuration, $E_s(\underline{R}^{mn})$ has the same meaning for the saddle point. y_α are the mass weighted normal coordinates of the system obtained from $\underline{x}_i = \sqrt{M}\,\underline{R}_i$ with M the mass of the crystal atoms. Performing the integrations in (2.5) yields

$$\frac{P_s(\underline{R}^{mn})}{P_e(\underline{R}^{m})} = \exp\left[-\beta\left(E_s(\underline{R}^{mn}) - E_e(\underline{R}^{m})\right)\right] \prod_{\alpha=1}^{3N}\nu_\alpha \Big/ \prod_{\alpha=1}^{3N}\nu_\alpha' \tag{2.7}$$

and we obtain for the jump frequency

$$\Gamma = \bar{\nu}\left[\prod_{\alpha=1}^{3N} \nu_\alpha \Big/ \prod_{\alpha=1}^{3N} \nu'_\alpha\right] \exp\left[-\beta\left(E_s(\underline{R}^{mn}) - E_e(\underline{R}^{m})\right)\right] \tag{2.8}$$

showing an exponential temperature dependence with an activation energy given by the difference between the energy of the relaxed saddle point configuration and the equilibrium configuration

$$E^M = E_s(\underline{R}^{mn}) - E_e(\underline{R}^{m}) \tag{2.9}$$

and a frequency factor, called "attempt frequency",

$$\nu^* = \bar{\nu} \prod_{\alpha=1}^{3N} \nu_\alpha \Big/ \prod_{\alpha=1}^{3N} \nu'_\alpha \tag{2.10}$$

which involves in a complicated way eigenfrequencies of the crystal atoms with the impurity fixed in the saddle point position and in the equilibrium position, ν'_α and ν_α, respectively.

VINEYARD /2.24/ has derived a similar expression for ν^* by considering the dynamics of an $(N+1)$ particle system without constraining the coordinates of the jumping impurity atom. Then, the system is unstable in the saddle point configuration which is a maximum of energy in the jumpdirection. The unstable mode in the $(N+1)$ particle system corresponds to the displacements we considered for the calculation of $\bar{\nu}$ in the constrained system. It is an eigenmode for the unconstrained system with an imaginary frequency. This, however, cancels against one of the eigenfrequencies, say $\tilde{\nu}'_1$, of the $(N+1)$ particle system in the saddle point thus leading to

$$\nu^* = \prod_{\alpha=1}^{3(N+1)} \nu_\alpha \Big/ \prod_{\alpha=2}^{3(N+1)} \tilde{\nu}'_\alpha \tag{2.10a}$$

with the eigenfrequencies of the $(N+1)$ particle system $\tilde{\nu}_\alpha$ and $\tilde{\nu}'_\alpha$ for the equilibrium and saddle point configuration, respectively. This expression is usually used to calculate the attempt frequency ν^* /2.8/.

The frequencies, ν, and energies, E, can depend on parameters like volume and shape of the crystal which are determined by external forces. For instance, if a crystal is strained inhomogeneously different equilibrium positions and saddle points are no longer equivalent. Then the frequencies, ν, the energies, E, and thus the jump frequencies, Γ, depend explicitly on the position of the jumping atom as we have indicated in (2.8). We shall use this in Sect. 4. The dependence on other forces like electric or magnetic fields can also be incorporated.

Equation (2.8) is valid in the framwork of classical statistical mechanics using a harmonic expansion of the potential energy around the relaxed equilibrium and saddle point configurations which for $T \neq 0$ are determined from the free energy F. This corresponds to the quasiharmonic approximation. Additional anharmonic effects can also be incorporated /2.22/.

There have been attempts to calculate the jump frequency using quantum mechanics /2.25/ which is thought to be most important for light impurities which diffuse very fast, e.g., hydrogen in bcc metals. In systems with very low barriers one should perhaps abolish the idea of stable positions and saddle points and describe the motion of the defects as a Brownian motion in a periodic potential as has been suggested for superionic conductors /2.26/.

2.3 Diffusion Equation and Green's Function

We shall assume small concentrations of mobile defects throughout the paper. Then site blocking effects are unimportant and the defects move independently. We thus can consider the motion of a single defect.

For simplicity we shall consider Bravais lattices only. But the theory can also be formulated for non-Bravais lattices (see Sect. 4).

For the probability $G^{\underline{m}\underline{n}}(t)$ to find the defect on site $\underline{m}$ at time t if it was produced on site $\underline{n}$ at time t = 0 (Green's function) the following equation holds:

$$\frac{d}{dt} G^{\underline{m}\underline{n}}(t) = \sum_{\underline{\ell}} \lambda^{\underline{m}\underline{\ell}} G^{\underline{\ell}\underline{n}}(t) - \sum_{\underline{\ell}} \lambda^{\underline{\ell}\underline{m}} G^{\underline{m}\underline{n}}(t) + \delta_{\underline{m}\underline{n}}\delta(t) \;. \tag{2.11}$$

The probability $G^{\underline{m}\underline{n}}(t)$ changes in time due to jumps from neighbouring sites $\underline{\ell}$ to $\underline{m}$ (first term on rhs) or due to jumps from $\underline{m}$ to neighbouring sites (second term) or by production of defects on site $\underline{m}$ (last term). The rates at which the jumps change the probability are proportional to a jump frequency λ times the probability to find the defect in the starting position. The initial condition is

$$G^{\underline{m}\underline{n}}(t<0) = 0 \;. \tag{2.12}$$

Since (2.11) is linear, the defect distribution for any initial condition $\rho^{\underline{n}}(0)$ and arbitrary source terms $\Pi^{\underline{n}}$ can be found by integration

$$\rho^{\underline{m}}(t) = \sum_{\underline{n}} \int_0^t dt'\, G^{\underline{m}\underline{n}}(t-t')\, \Pi^{\underline{n}}(t') + \sum_{\underline{n}} G^{\underline{m}\underline{n}}(t)\, \rho^{\underline{n}}(0) \;. \tag{2.13}$$

Equation (2.11) can be written in a more compact form

$$\frac{d}{dt} G^{\underline{m}\underline{n}}(t) + \sum_{\underline{\ell}} \Lambda^{\underline{m}\underline{\ell}} G^{\underline{\ell}\underline{n}}(t) = \delta_{\underline{m}\underline{n}}\delta(t) \tag{2.14}$$

with the jump frequency matrix

$$\Lambda^{\underline{m}\underline{n}} = \delta_{\underline{m}\underline{n}} \sum_{\underline{\ell}} \lambda^{\underline{\ell}\underline{m}} - \lambda^{\underline{m}\underline{n}} \;. \tag{2.15}$$

During the jump processes the number of particles does not change, thus Λ obeys the "particle conservation law"

$$\sum_{\underline{m}} \Lambda^{\underline{m}\underline{n}} = 0 \; . \tag{2.16}$$

In an ideal Bravais lattice the jump frequencies are independent of the starting position. Then Λ is symmetric and lattice translational invariant, i.e., it depends only on the difference of the indices $\underline{m}$ and $\underline{n}$

$$\Lambda^{\underline{m}\underline{n}} = \Lambda^{(\underline{m}-\underline{n})} = \Lambda^{\underline{n}\underline{m}} \; . \tag{2.17}$$

In the following we shall consider nearest-neighbour jumps only. This means that $\lambda^{\underline{m}\underline{n}}$ can be written as

$$\lambda^{\underline{m}\underline{n}} = \Gamma \, s_1^{(\underline{m}-\underline{n})} \tag{2.18}$$

with the nearest-neighbour step function

$$s_1^{(\underline{m}-\underline{n})} = \begin{cases} 1 & \text{if } \underline{m},\underline{n} \text{ denote nearest neighbours} \\ 0 & \text{otherwise} \end{cases} \tag{2.19}$$

and the nearest-neighbour jump frequency Γ.

Equation (2.15) then reads

$$\Lambda^{(\underline{m}-\underline{n})} = \Gamma \, [z_1 \delta_{\underline{m}\underline{n}} - s_1^{(\underline{m}-\underline{n})}] = \Gamma \, \phi^{(\underline{m}-\underline{n})} \tag{2.20}$$

with z_1 the number of nearest-neighbours.

In the ideal lattice G is also translational invariant and thus

$$G^{\underline{m}\underline{n}} = G^{(\underline{m}-\underline{n})} \; .$$

Then (2.11) can be solved by Fourier transformation

$$\tilde{G}(\underline{k},\omega) = \int_0^\infty dt \, \exp(i\omega t) \sum_{\underline{n}} \exp(-i\underline{k}\underline{R}^{\underline{n}}) \, G^{(\underline{n})}(t) \; . \tag{2.21}$$

This yields

$$\tilde{G}(\underline{k},\omega) = \frac{1}{\tilde{\Lambda}(\underline{k}) - i\omega} \tag{2.22}$$

with the Fourier transformation of the jumps frequency matrix given by

$$\tilde{\Lambda}(\underline{k}) = \sum_{\underline{m}} \Lambda^{(\underline{m})} \exp(i\underline{k}\underline{R}^{\underline{m}}) \; . \tag{2.23}$$

For Bravais lattices we have

$$\Lambda^{(\underline{m})} = \Lambda^{(-\underline{m})}$$

which yields

$$\tilde{\Lambda}(\underline{k}) = \tilde{\Lambda}(-\underline{k}) = \sum_{\underline{n}} \Gamma\, s_1^{(\underline{n})}(1 - \cos \underline{k}\underline{R}^{\underline{n}}) \geqslant 0 \ . \tag{2.23a}$$

For the three cubic Bravais lattices we obtain from (2.20,23a)

sc lattice:

$$\tilde{\Lambda}_{sc}(\underline{k}) = 6\Gamma \left[1 - \frac{1}{3} (\cos k_x a_o + \cos k_y a_o + \cos k_z a_o) \right] \tag{2.24a}$$

bcc lattice:

$$\tilde{\Lambda}_{bcc}(\underline{k}) = 8\Gamma \left(1 - \cos k_x \frac{a_o}{2} \cos k_y \frac{a_o}{2} \cos k_z \frac{a_o}{2} \right) \tag{2.24b}$$

fcc lattice:

$$\tilde{\Lambda}_{fcc}(\underline{k}) = 12\Gamma \left[1 - \frac{1}{3} \left(\cos k_x \frac{a_o}{2} \cos k_y \frac{a_o}{2} + \cos k_y \frac{a_o}{2} \cos k_z \frac{a_o}{2} + \cos k_z \frac{a_o}{2} \cos k_x \frac{a_o}{2} \right) \right] . \tag{2.24c}$$

Here a_o is the cubic lattice constant.

$G^{(\underline{n})}(t)$ can be found by the inverse transformation corresponding to (2.21). With (2.22) we obtain

$$G^{(\underline{n})}(t) = \frac{1}{2\pi} \frac{1}{V_B} \int_{-\infty}^{+\infty} d\omega \exp(-i\omega t) \int_{V_B} d\underline{k} \frac{\exp(i\underline{k}\underline{R}^{\underline{n}})}{\Lambda(\underline{k}) - i\omega} \ . \tag{2.25}$$

The $\underline{k}$-integration is extended over the first Brillouin zone of volume V_B.

Performing the ω integration one obtains

$$G^{(\underline{n})}(t) = \int_{V_B} \frac{d\underline{k}}{V_B} \exp(i\underline{k}\underline{R}^{\underline{n}}) \exp[-\tilde{\Lambda}(\underline{k})t] \tag{2.25a}$$

which shows that $\tilde{\Lambda}(\underline{k})$ is the spectrum of relaxation times.

Many authors have worked on the problem of evaluating lattice Green's functions. They mostly considered the intermediate function

$$\bar{G}^{(\underline{n})}(z) = \int_{V_B} \frac{d\underline{k}}{V_B} \frac{\exp(i\underline{k}\underline{R}^{\underline{n}})}{\phi(\underline{k}) - z} \tag{2.26}$$

for complex z.

This function has many applications in different areas of solid-state physics, e.g., ferromagnetism /2.27-31/, spin wave scattering from impurities /2.32,33/, electronic energy levels in disordered systems /2.34/, vibrational behaviour of defects in solids /2.35/, and random walk problems such as the calculation of correlation factors for diffusion /2.36/.

$\bar{G}^{(n)}(z)$ is related to the Fourier transform of the Green's function for diffusion by

$$\Gamma\, G^{(\underline{n})}(\omega) = \bar{G}^{(\underline{n})}(z = i\,\frac{\omega}{\Gamma})\,. \tag{2.27}$$

Because of the importance of the Green's function many papers are devoted to its evaluation. WATSON /2.37/ was the first to evaluate $\bar{G}$ for $z = 0$ and $\underline{R}^{\underline{n}} = 0$ for the three primitive cubic lattices in terms of the complete elliptic integrals K. Recently, several authors have shown that $\bar{G}^{(\underline{n})}(z)$ can be related to elliptical integrals in bcc /2.38/, fcc /2.39/, sc /2.40/ lattice also for $z \neq 0$ and/or $\underline{R}^{\underline{n}} \neq 0$. Also other lattices have been considered, e.g., anisotropic fcc /2.41/, the two-dimensional square /2.42/ and honeycomb /2.43/ lattices, diamond /2.44,45/, and diatomic lattices /2.46/ for varying range of the parameters $\underline{R}^{\underline{n}}$ and z.

An introduction /2.47/ to a series of papers of the group at the Tohoku University on the analytical behaviour of the lattice Green's function contains most references up to 1971.

There are a number of papers giving numerical values for $\bar{G}^{(\underline{n})}(z)$, particularly for $z = 0$ (stationary diffusion) in sc /2.48/, bcc /2.38/ and fcc /2.36/ lattices. The available numerical results for the stationary case have been compiled in an earlier report by the author /2.45/.

In our group a program is available for calculating the tensor Green's function for lattice vibrations, which describes the response of the phonon system on external forces. It can be used to calculate phonon dispersion curves and phonon spectra or scattering of phonons by defects. The scalar Green's function, (2.26), is a special case for isotropic coupling (i.e., equal coupling for transverse and longitudinal motion) to nearest neighbours. With this program the ω-dependent Green's function can be calculated as well as the stationary ($\omega = 0$) Green's function.

2.3.1 The Stationary Green's Function

Of particular interest for diffusion reactions (see Sect. 6) is the stationary Green's function

$$G_{st}^{(\underline{n})} = G_{st}(\underline{R}^{\underline{n}}) = \int_{V_B} \frac{d\underline{k}}{V_B}\, \frac{\exp(i\underline{k}\underline{R}^{\underline{n}})}{\Lambda(\underline{k})}\,. \tag{2.28}$$

This integral has to be solved numerically. But it turns out that the asymptotic expansion for $\underline{R}^{\underline{n}} \to \infty$ gives rather good values even for $\underline{R}^{\underline{n}}$ close to the origin /2.45/.

For $R \to \infty$ the numerator at the integrand in (2.28) oscillates rapidly and the main contribution to the integral comes from small $\underline{k}$.

To obtain the leading term one expands the denominator for small values of $\underline{k}$ and extends the integral to infinity.

The first nonvanishing term of the expansion of $\tilde{\Lambda}(\underline{k})$ in Bravais lattices is quadratic in $\underline{k}$

$$\tilde{\Lambda}_{(2)}(\underline{k}) \cong \frac{1}{2}\Gamma \sum_{\underline{n}} s_1^{(\underline{n})}(\underline{k},\underline{R}^{\underline{n}})^2 = \frac{1}{2}\Gamma \sum_{ij} k_i k_j \sum_{\underline{n}} R_i^{\underline{n}} R_j^{\underline{n}} s_1^{(\underline{n})} \tag{2.29}$$

and we obtain

$$G_{st}^{o}(\underline{R}^{\underline{n}}) \cong \int_{\infty} \frac{d\underline{k}}{V_B} \frac{\exp(i\underline{k}\underline{R}^{\underline{n}})}{\tilde{\Lambda}_{(2)}(\underline{k})} \tag{2.30}$$

In cubic crystals one can show that the sum over nearest neighbours in (2.29) yields

$$\sum_{\underline{n}} R_i^{\underline{n}} R_j^{\underline{n}} s_1^{(\underline{n})} = z_1 \frac{d_1^2}{3} \delta_{ij} \tag{2.31}$$

with d_1 the nearest-neighbour distance. Thus we obtain

$$\tilde{\Lambda}_{(2)}(\underline{k}) \cong D_o k^2 \tag{2.32}$$

with the diffusion constant

$$D = z_1 \Gamma d_1^2/6 \ . \tag{2.33}$$

The integral in (2.30) can now easily be solved yielding

$$G_{st}^{o}(\underline{R}^{\underline{n}}) \cong \frac{V_c}{4\pi D_o} \frac{1}{|\underline{R}^{\underline{n}}|} \tag{2.34}$$

with V_c the volume of the primitive cell. This is the solution of the stationary macroscopic diffusion equation (which is the Poisson equation)

$$D_o \, \Delta G_{st}(\underline{R}) = V_c \, \delta(\underline{R}) \ . \tag{2.35}$$

To establish that (2.34) actually is the leading term of G_{st} one has to consider the difference between (2.34) and (2.28) and show that it goes to zero faster than R^{-1}. To do this a rather tricky procedure has to be adopted which makes use of the periodicity of the integrand to get rid of the contributions from the surface of the Brillouin zone. We shall not repeat the derivation here, details can be found in /2.45/. If follows that the next term is given by

$$G_{st}^{1}(\underline{R}^{\underline{n}}) = -\frac{V_c}{4\pi D_o} \frac{d_1^2}{|\underline{R}^{\underline{n}}|^3} \frac{15}{8} B \left[\frac{3}{5} - \sum_{i=1}^{3} \left(R_i^{\underline{n}}\right)^4 \Big/ |\underline{R}^{\underline{n}}|^4 \right] \tag{2.36}$$

with the coefficient B depending on the lattice symmetry

$$B = \begin{cases} \frac{1}{3} & \text{sc} \\ -\frac{2}{9} & \text{bcc} \\ -\frac{1}{12} & \text{fcc} \end{cases} \tag{2.36a}$$

and higher order terms are of order $|\underline{R}^{\underline{n}}|^{-5}$. It turns out that the numerical values for $G_{st}^{(\underline{n})}$ agree with $G_{st}^{o}(\underline{R}^{\underline{n}})$ to about 2% even for nearest neighbours /2.45/. The angular dependence suggested by G_{st}^{1}, (2.36), is not found, in fact the agreement with the numerical values is made worse for small $\underline{R}^{\underline{n}}$ if one takes this term into account, indicating that there is a cancellation with higher order terms. The good agreement of the asymptotically leading term with the exact stationary Green's function seems to be a general behaviour for nearest-neighbour coupling also for the tensor Green's function /2.49/.

2.3.2 Asymptotic Behaviour

The same general procedure can be used to obtain the asymptotic behaviour of $G^{(\underline{n})}(t)$ and $G^{(\underline{n})}(\omega)$ for large $\underline{R}^{\underline{n}}$, starting from (2.25a,26) respectively.

The leading terms are

$$G^{(\underline{n})}(t) \cong \frac{V_c}{\sqrt{(4\pi D_o t)^3}} \exp\left[-\frac{|\underline{R}^{\underline{n}}|^2}{4D_o t}\right] , \tag{2.37}$$

the well-known solution of the macroscopic diffusion equation

$$\frac{d}{dt} G(\underline{R},t) = D_o\, \Delta G(\underline{R},t) + V_c\, \delta(\underline{R})\, \delta(t) \tag{2.38}$$

and /2.50/

$$G^{(\underline{n})}(\omega) \cong \frac{V_c}{4\pi D_o} \frac{1}{|\underline{R}^{\underline{n}}|} \exp\left[\frac{1}{\sqrt{2}}(-1+i)\sqrt{\frac{\omega}{D_o}}\,|\underline{R}^{\underline{n}}|\right] \tag{2.39}$$

showing an exponential decay of $G^{(\underline{n})}(\omega)$ for large $\underline{R}^{\underline{n}}$ if $\omega \neq 0$.

2.3.3 Spectral Representation

Since the Fourier tranform of the jump frequency matrix, (2.23a), is positive and real

$$\tilde{\phi}(\underline{k}) = \sum_{\underline{n}} s_1^{(\underline{n})} (1 - \cos \underline{k}\underline{R}^{\underline{n}}) \geq 0 \tag{2.40}$$

we can obtain the spectral representation (using the inversion symmetry to replace the exponential Fourier transformation by the cosine transformation)

$$\bar{G}^{(\underline{n})}(z) = \int_{V_B} \frac{d\underline{k}}{V_B} \cos \underline{k}\underline{R}^{\underline{n}} \int_0^\infty \frac{dx}{x-z}\, \delta\left(x - \tilde{\phi}(\underline{k})\right) = \int_0^\infty dx\, \frac{I(\underline{n},x)}{x-z} \tag{2.41}$$

where the x integration is along the real axis.

The spectral function is

$$I(\underline{n},x) = \int_{V_B} \frac{d\underline{k}}{V_B} \cos \underline{k}\underline{R}^{\underline{n}}\, \delta\left(x - \tilde{\phi}(\underline{k})\right) . \tag{2.42}$$

Since $\tilde{\phi}(\underline{k})$ has an upper bound $\tilde{\phi}_{max}$, $I(\underline{n},x)$ is different from zero only for real x in the interval $0 \leq x \leq \tilde{\phi}_{max}$.

It can be shown that $I(\underline{n},x)$ is the imaginary part of the Fourier transform of the retarded vibrational Green's function $G_V^{(\underline{n})}(\omega^2) = \bar{G}^{(\underline{n})}(z = \omega^2 + i\eta)$, $\eta \to 0_+$ for a harmonic lattice with isotropic nearest-neighbour coupling constant f = 1 and atoms of mass M = 1 /2.51/.

The analytical behaviour of this Green's function $G_V^{(n)}$ has been studied recently /2.52/ and a computer program is available to calculate $G_V^{(n)}(\omega^2)$ /2.49/. Equation (2.41) then provides a link to calculate the diffusional Green's function G by integration. With purely imaginary $z = i\frac{\omega}{\Gamma}$ we obtain from (2.41)

$$\mathrm{Re}\{G^{(\underline{n})}(\omega)\} = \frac{1}{\Gamma}\int_0^\infty dx \, \frac{x}{x^2 + (\omega/\Gamma)^2}\, I(\underline{n},x)$$
$$\mathrm{Im}\{G^{(\underline{n})}(\omega)\} = \frac{\omega}{\Gamma}\int_0^\infty dx \, \frac{1}{x^2 + (\omega/\Gamma)^2}\, I(\underline{n},x) \; . \tag{2.43}$$

These equations replace the Kramers - Kronig relations between real and imaginary parts of the Green's function, which are valid for real z.

2.3.4 Analytical Behaviour

One can obtain an analytical expression for the ω dependence of $G^{(\underline{n})}(\omega)$ for small and large ω, starting from (2.43).

Notice that for small ω mostly small x contribute to the integrals. Then one can expand the spectral function $I(\underline{n},x)$, (2.42), for small x. Because of the δ function only small k contribute and we can expand the cosine. The integral in (2.42) then yields /2.51,52/

$$I(\underline{n},x) \cong a_o\sqrt{x} + a_1(\underline{n})\sqrt{x^3} + a_2(\underline{n})\sqrt{x^5} \tag{2.44}$$

with

$$a_o = \frac{2\pi}{V_B(z_1 d_1^2/6)^{3/2}} \; ; \quad a_1(\underline{n}) = \frac{-\pi\,|\underline{B}^{\underline{n}}|^2}{V_B(z_1 d_1^2/6)^{5/2}} \; ; \quad a_2(\underline{n}) = \frac{\pi\,|\underline{B}^{\underline{n}}|^4}{32V_B(z_1 d_1^2/6)^{7/2}} \; . \tag{2.45}$$

With this expansion inserted into (2.43) one sees that the two integrands diverge for small x. This means that $G^{(\underline{n})}(\omega)$ cannot be expanded into a power series of ω, but that there are singular terms. It turns out that a power series in terms of $\sqrt{\omega}$ evolves. The leading terms are /2.51/

$$\mathrm{Re}\{G^{(\underline{n})}(\omega)\} = G_{st}^{(\underline{n})} - \frac{\pi a_o}{\sqrt{2}\Gamma}\left(\frac{\omega}{\Gamma}\right)^{1/2} - \frac{\pi a_1(\underline{n})}{\sqrt{2}\Gamma}\left(\frac{\omega}{\Gamma}\right)^{3/2} + O(\omega^2)$$
$$\mathrm{Im}\{G^{(\underline{n})}(\omega)\} = \frac{\pi a_o}{\sqrt{2}\Gamma}\left(\frac{\omega}{\Gamma}\right)^{1/2} + O(\omega) \; . \tag{2.46}$$

The terms to order $\sqrt{\omega}$ agree asymptotically with (2.39), expanded for small ω.

For large ω, i.e., $\frac{\omega}{\Gamma} \gg \tilde{\phi}_{max}$, one can expand the integrand in (2.43) into a power series of $(\omega/\Gamma)^2$ and integrate term by term. Thus one obtains

$$G^{(\underline{n})}(\omega) = \frac{1}{\Gamma} \sum_{s=0}^{\infty} \left[b_{2s+1}^{(\underline{n})} \left(\frac{\Gamma}{\omega}\right)^2 + i b_{2s}^{(\underline{n})} \cdot \frac{\Gamma}{\omega} \right] \left(\frac{\Gamma}{\omega}\right)^{2s} . \tag{2.47}$$

Here the coefficients

$$b_s^{(\underline{n})} = \int dx \; x^s \; I(\underline{n},x) \tag{2.48}$$

are the moment of order s of the spectral function. These moments all exist, since $I(\underline{n},x) = 0$ for $x \geq \tilde{\phi}_{max}$.

2.3.5 Numerical Results

Starting from (2.43) we have calculated $G^{(\underline{n})}(\omega)$ for a few points in an fcc lattice using the program of SCHOBER et al. /2.49/. The results are shown in Fig. 2.8a-e.

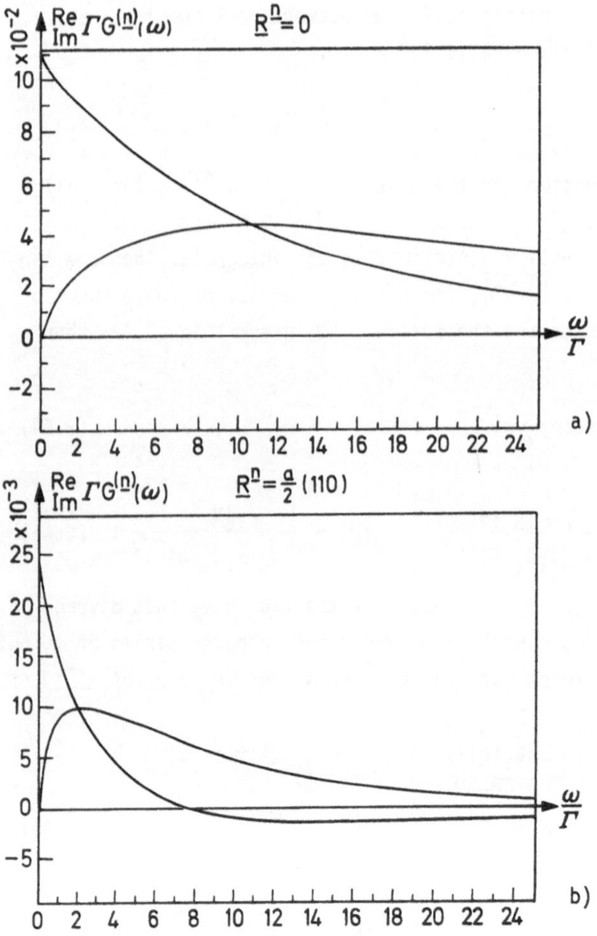

Fig. 2.8a,b. Diffusional Green's function $G^{(\underline{n})}(\omega)$ in fcc crystals as a function of ω.

a) $\underline{R}^{\underline{n}} = 0$

b) $\underline{R}^{\underline{n}} = \frac{a_o}{2}(110)$

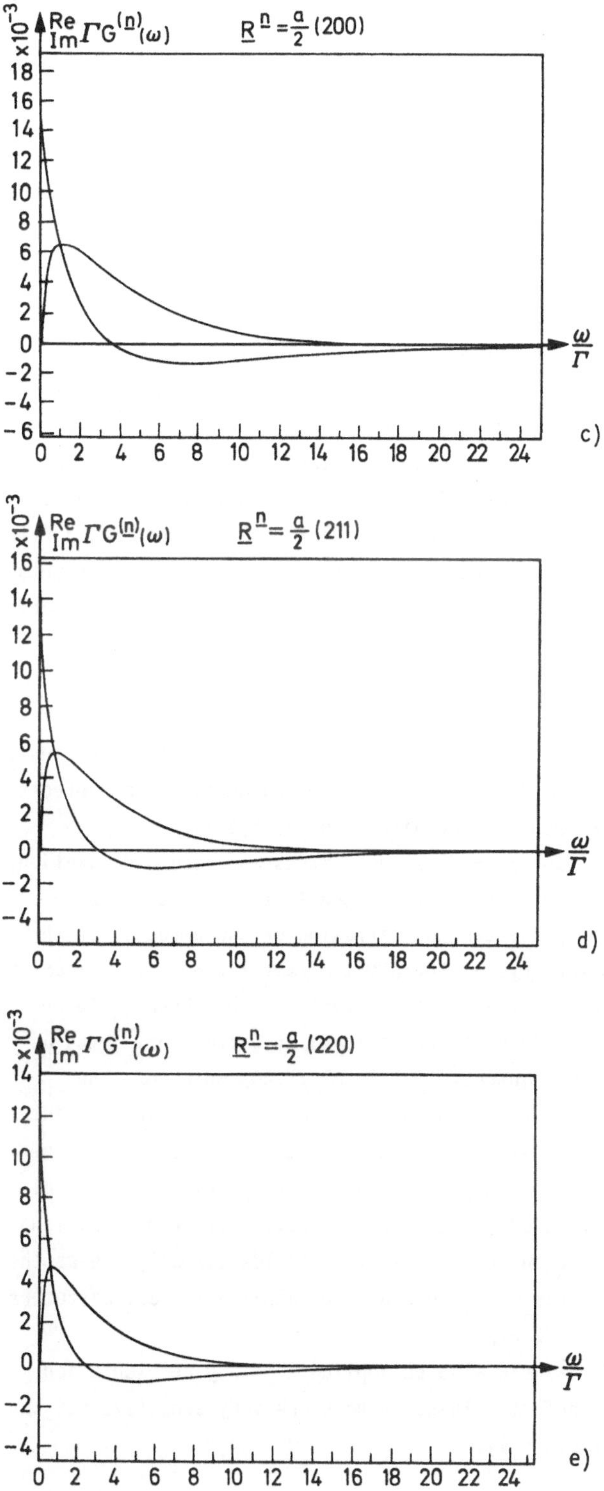

Fig. 2.8c-e. Diffusional Green's function $G^{(\underline{n})}(\omega)$ in fcc crystals as a function of ω.

c) $\underline{R}^{\underline{n}} = \frac{a}{2}$ (200)

d) $\underline{R}^{\underline{n}} = \frac{a}{2}$ (211)

e) $\underline{R}^{\underline{n}} = \frac{a}{2}$ (220)

For small and large ω the behaviour discussed above is obtained. For $\underline{R}^{\underline{n}} \neq 0$ both real and imaginary part of $G^{(\underline{n})}(\omega)$ oscillate. For other cubic lattices qualitatively the same results are expected.

2.4 Experimental Methods

The classical method to determine diffusion coefficients is the so-called *tracer technique* /2.53/. Here the depth distribution of radioactive tracers initially put on the surface of a crystal is measured after holding the crystal at a fixed temperature for a certain time. An analysis based on the macroscopic diffusion equation (2.38) with the solution (2.37) integrated over the source distribution on the surface according to (2.13) then yields the diffusion constant. This method can only be applied if radioactive tracers are available and if the defects to be studied are thermally stable at the temperature at which they diffuse at a substantial rate. New sectioning techniques have made the method very sensitive. Diffusion constants as low as $D \cong 10^{-18} cm^2/s$ have been measured /2.54/. Diffusion coefficients in many metals have been determined by this method both for self-diffusion (mostly via vacancies) and impurity diffusion (interstitially or via vacancies) /2.53/. Detailed analysis of the temperature, pressure, and isotope dependence of diffusion coefficients combined with measurements of other properties of high temperature intrinsic defects, as, e.g., formation in thermodynamic equilibrium and annealing of defects, can provide information about the mechanism of diffusion /2.55,56/.

By *mechanical relaxation* techniques /2.57/, such as stress or strain relaxation, and internal friction, the response of defects to external stresses is measured. This yields information about jump processes and diffusion of the defects. If the defects have lower symmetry than the crystal the *SNOEK effect* /2.58/, i.e., local redistribution among different orientations of the defects can be observed in response to homogeneous stress fields. This can be achieved by diffusional jumps as has, e.g., been observed for the self-interstitial in Al /2.20/ which has the <100>-dumbbell structure and for C in Fe /2.59/ which occupies the octahedral interstitial site. In single crystals the symmetry of the defect configuration and the jump direction can be obtained from the response to differently oriented uniaxial stresses /2.57/. The *Gorski effect* /2.60/, i.e., redistribution of defects by long-range migration in response to inhomogeneous dilatation fields directly yields the diffusion constant. Unfortunately this method can only be applied to fast diffusers ($D \geqslant 10^{-8} cm^2/s$) such as H in metals /2.61/.

Magnetic relaxation methods /2.62/ can also be applied to study reorientation and long-range diffusion of point defects. These methods are very sensitive but can be applied to ferromagnetic materials only.

Microscopic methods which involve wavelengths comparable to or less than the jump distance require for their interpretation the knowledge of the lattice Green's function discussed in Sect. 2.3. These include neutron scattering, Mössbauer effect and nuclear magnetic resonance.

Incoherent quasielastic neutron scattering /2.63/ can be used to measure diffusion coefficients because due to the diffusive motion the incoming neutron beam is energetically broadened. The linewidth gives information about the jump frequency and jump geometry. The information about the system behaviour is, e.g., contained in the incoherent scattering law $S_{inc}(\underline{k},\omega)$ with $\hbar\underline{k}$ the momentum transfer and $\hbar\omega$ the energy transfer of the neutron to the target. $S_{inc}(\underline{k},\omega)$ can be expressed by the self-correlation function which for simple jump models is directly related to the lattice Green's function. For Bravais lattices one obtains

$$S_{inc}(\underline{k},\omega) = \frac{1}{\pi} \mathrm{Re}\{\tilde{G}(\underline{k},\omega)\} = \frac{1}{\pi} \frac{\tilde{\Lambda}(\underline{k})}{\omega^2 + [\tilde{\Lambda}(\underline{k})]^2} , \qquad (2.49)$$

$S_{inc}(\underline{k},\omega)$ is a Lorentzian with the width $2\tilde{\Lambda}(\underline{k})$. For small k the macroscopic behaviour is measured and the width is directly proportional to the diffusion constants as can be seen from the expansion of $\tilde{\Lambda}(\underline{k})$, (2.32)

$$2\tilde{\Lambda}(\underline{k}) = 2D_o k^2 \quad \text{for } ka_o \ll 1 .$$

For $\underline{k}$ at the edge of the Brillouin zone the lattice structure enters and $\tilde{\Lambda}(\underline{k})$ is given by the mean residence time $\tau = (z_1\Gamma)^{-1}$ on one lattice site. $\tilde{\Lambda}(\underline{k})$ is a periodic function in the reciprocal lattice and assumes its maximum at the edge of the Brillouin zone. For the three cubic lattices we have from (2.24)

$$\tilde{\Lambda}_{max} = \begin{cases} 12\Gamma & \text{at } k = \frac{\pi}{a_o}(111) \quad \text{sc} \\ 16\Gamma & \text{at } k = \frac{2\pi}{a_o}(111) \quad \text{bcc} \\ 16\Gamma & \text{at } k = \frac{2\pi}{a_o}(110) \quad \text{fcc} . \end{cases} \qquad (2.50)$$

Since the shape of the Brillouin zone is determined by the symmetry of the diffusion lattice the analysis of the $\underline{k}$ dependence of the width of the quasi-elastic line yields information about the equilibrium sites of the diffusing defects. With neutron scattering SKÖLD and NELIN /2.64/ were able to show that H in Pd occupies octahedral sites. The results on H diffusion in the bcc metals Nb /2.65/, V /2.66/ and Ta /2.67/ were not consistent with a simple jump model. Recently, new results on these metals /2.13/ could be interpreted consistently using a more complicated model including selected jumps to more distant neighbours. The H atoms were found to occupy tetrahedral sites. H diffusion has been very widely studied /2.68/ with incoherent neutron scattering because of the large incoherent scattering length of H, but also self-diffusion in Na (via vacancies) has been studied /2.69/.

Mössbauer effect has also been used to study diffusion. Essentially the same formalism as for incoherent neutron scattering can be applied to calculate the diffusion broadening of the Mössbauer line /2.70/. Since the Mössbauer atoms usually migrate by means of vacancies correlation effects have to be taken into account /2.51,71/. Due to the large k-vector of the Mössbauer quanta the discrete jump vectors give rise to an anisotropy of the diffusional broadening in single crystals. The method has been applied, e.g., to self-diffusion in iron /2.72/ and to diffusion of Fe in copper /2.73/ and gold /2.74/. Experimental results and possible interpretations have recently been reviewed by JANOT /2.75/.

Nuclear magnetic resonance has been widely used to study diffusion in solids. The fluctuating interaction due to the motion of the atoms contributes to the measured relaxation times T_1 (spin-lattice relaxation time), T_2 (spin relaxation time), and $T_{1\rho}$ (spin-lattice relaxation time in the rotating frame). To relate these relaxation times to the atomic jump frequencies a knowledge of the lattice Green's function is necessary /2.76,77/. The method has been applied, e.g., to H diffusion in metals /2.78/ and to self-diffusion in Li /2.79/.

3. Interaction of Defects in Metals

We shall restrict our discussion to metals where the elastic interaction between defects is dominant because charges are screened by the conduction electrons already within distances comparable to the interatomic distance. In ionic crystals and semiconductors defects can be charged and thus, in addition, Coulomb interaction and electric polarization (induced dipole) interaction have to be considered.

3.1 Static Displacements Around Defects

If one introduces a defect into a crystal forces are exerted on the neighbouring host lattice atoms which lead to long-range displacements from the regular lattice sites. In the close vicinity of the defect the displacements can be so large that also the coupling between the host atoms is changed. In the harmonic approximation the new equilibrium configuration of the crystal is given by /3.1/

$$\sum_{\underline{n},j} (\phi_{ij}^{\underline{mn}} + \varphi_{ij}^{\underline{mn}})\, s_j^{\underline{n}} = F_i^{\underline{m}} . \tag{3.1}$$

Here $\phi_{ij}^{\underline{mn}}$ is the coupling matrix of the ideal crystal, $\varphi_{ij}^{\underline{mn}}$ the change of the coupling matrix; $s_j^{\underline{n}}$ are the displacements of the host atoms from the regular lattice sites and F_i^m the forces due to the defect. Alternatively, the displacements can also be interpreted in terms of "Kanzaki forces" K_i^m which are defined to produce the same displacements in the ideal harmonic crystal as the defect does in the real crystal /3.2/. They contain the effect of the coupling changes (anharmonicities) and of the forces; the relationship is (using an obvious matrix notation)

$$\underline{\underline{\phi}}\,\underline{s} = \underline{K} \; ; \quad \underline{K} = \underline{F} - \underline{\underline{\varphi}}\,\underline{s} = (\underline{\underline{1}} - \underline{\underline{t}}\,\underline{\underline{G}})\,\underline{F} \tag{3.2}$$

with the t matrix corresponding to the coupling changes

$$\underline{\underline{t}} = (1 + \underline{\underline{\varphi}}\,\underline{\underline{G}})^{-1}\,\underline{\underline{\varphi}} . \tag{3.3}$$

Here the static Green's function of the ideal lattice G has been introduced which is the inverse of the ideal coupling matrix

$$G_{ij}^{\underline{m}\underline{n}} = \int\limits_{V_B} \frac{d\underline{k}}{V_B} \exp[i\underline{k}(\underline{R}^{\underline{m}} - \underline{R}^{\underline{n}})] \left(\frac{1}{\tilde{\underline{\underline{\phi}}}(\underline{k})}\right)_{ij} . \tag{3.4}$$

The integral over the Brillouin zone V_B has to be calculated numerically. $\tilde{\phi}_{ij}(\underline{k})$ is the Fourier transform of the coupling matrix $\phi_{ij}^{\underline{m}\underline{n}} = \phi_{ij}^{(\underline{m}-\underline{n})}$ which can, e.g., be determined from Born - von-Karman fits of phonon dispersion curves. The displacements are given by

$$s_i^{\underline{m}} = \sum_{\underline{n},j} G_{ij}^{\underline{m}\underline{n}} K_j^{\underline{n}} . \tag{3.5}$$

For large distances $\underline{R} = (\underline{R}^{\underline{m}} - \underline{R}^{\underline{n}})$ one can use the asymptotic form of G, the elastic Green's function $G_{ij}(\underline{R})$ which depends only on the elastic constants of the medium. It is obtained from (3.4) if $\tilde{\phi}_{ij}(\underline{k})$ is replaced by its small k limit $\sim k^2$ and the integration extended to infinity (see Sect. 2.3.1)

$$G_{ij}(\underline{R}) = \frac{1}{(2\pi)^3} \int\limits_{\infty} \frac{d\underline{k}}{k^2} \exp(i\underline{k}\underline{R})\, \tilde{g}_{ij}(\hat{\underline{k}}) = \frac{1}{R}\, g_{ij}(\hat{\underline{R}}) . \tag{3.6}$$

Here $\tilde{g}_{ij}$ and g_{ij} are functions of the directions $\hat{\underline{k}}$ and $\hat{\underline{R}}$ of $\underline{k}$ and $\underline{R}$ only. Due to the $1/k^2$ dependence in Fourier space, $G_{ij}(\underline{R}) \sim 1/R$. The angular dependence can be calculated analytically only in certain cases: for cubic crystals only for isotropic or weakly anisotropic cases and for $c_{12} + c_{44} = 0$ /3.3/ (c_{ij} = elastic constants in Voigt's notation).

The asymptotic behaviour of the displacement field $s_i^{\underline{m}}$, (3.5), can be obtained by a multipole expansion. Since the total force vanishes ($\sum_{\underline{n}} K_i^{\underline{n}} = 0$) the first non-vanishing term is determined by the first moment of the forces, the so-called dipole tensor

$$P_{ij} = \sum_{\underline{n}} R_i^{\underline{n}} K_j^{\underline{n}} \tag{3.7}$$

namely

$$s_i(\underline{R}) = -\sum_{jk} \partial_{R_k} G_{ij}(\underline{R})\, P_{jk} \sim 1/R^2 \ . \tag{3.8}$$

The displacements vary with distance $\sim 1/R^2$, the angular variation is determined by the ansiotropy of the medium (via G_{ij}) and of the defect (via P_{jk}). For an isotropic defect ($P_{ij} = P_o \delta_{ij}$) in an isotropic medium we obtain

$$\underline{s}(\underline{R}) = \frac{P_o}{4\pi c_{11}} \cdot \frac{\hat{\underline{R}}}{R^2} \, . \tag{3.9}$$

HEALD /3.4/ has recently reviewed the results of lattice calculations of displacement fields. Whereas the Green's function approaches the asymptotic 1/R dependence very soon (already at the nearest-neighbour distance only a small deviation is found /3.5/), the displacements seem to approach the asymptotic $1/R^2$ dependence much slower, the deviations extend as far as 8 lattice constants in Al /3.6/ and 10 lattice constants in Mo /3.7/. However, the dipole tensor fully characterizes the long-range displacement field of a defect. For instance, the volume change due to a defect which can be measured by the change of the lattice constant is for a cubic crystal given by /3.8/.

$$\delta V = \frac{1}{c_{11} + 2c_{12}} \mathrm{Tr}\{\underline{\underline{P}}\} \ . \tag{3.10}$$

Also the interaction with external stresses (see Sect. 4) and the long-range elastic interaction of defects (see next section) can be expressed in terms of the dipole tensor.

3.2 Point Defect Interaction

If one introduces a second defect B into a crystal containing a defect A the forces of defect A do work against the displacements of defect B and vice vera. This is the reason for the elastic interaction. As long as the forces and coupling changes do not overlap one can use the Kanzaki forces of the individual defects to describe the interaction energy. For two defects separated by $\underline{R}^{\underline{p}}$ we obtain

$$\begin{aligned} E_{AB}(\underline{R}^{\underline{p}}) &= -\frac{1}{2} \sum_{\underline{m},i} \left[K_i^A(\underline{m})\, s_i^B(\underline{m}) + K_i^B(\underline{m})\, s_i^A(\underline{m}) \right] \\ &= -\sum_{\substack{\underline{m}\underline{n} \\ ij}} K_i^A(\underline{m})\, G_{i\ j}^{\underline{m}\ \underline{n}+\underline{p}}\, K_j^B(\underline{n}+\underline{p}) \ . \end{aligned} \tag{3.11}$$

In addition, we obtain an induced interaction due to the coupling changes in the vicinity of the defects

$$E_{AB}^{ind}(\underline{R}^{p}) = -\frac{1}{2}\sum_{m,i}\left[\delta K_i^A(\underline{m})\ s_i^B(\underline{m}) + \delta K_i^B(\underline{m})\ s_i^A(\underline{m})\right]. \tag{3.12}$$

Here /3.9/

$$\delta\underline{K}^A = \underline{\underline{t}}^A\underline{s}^B\ ;\quad \delta\underline{K}^B = \underline{\underline{t}}^B\underline{s}^A \tag{3.13}$$

are the additional Kanzaki forces in the vicinity of one defect induced by the displacements due to the other one. The connection of the t matrix with the force constant changes is given in (3.3).

For $\underline{R}^{p}$ large compared to the range of the forces (⩾ a few lattice constants) we can again use the continuum elastic Green's function, (3.6), and a multipole expansion of the forces. The leading term of the direct interaction, (3.11), is given by the dipole-dipole interaction

$$E_d(\underline{R}) = \sum_{ik,st} P_{is}^A\left(\partial_{R_s}\partial_{R_t}\,G_{ik}(\underline{R})\right)P_{kt}^B = -\frac{\alpha}{R^3}\,f(\Omega) \tag{3.14}$$

where $\underline{\underline{P}}^{A,B}$ is the dipole force tensor, (3.7), of defect A,B. This interaction is also called first-order size interaction in the English literature.

Due to the complicated structure of the Green's function in general the interaction cannot be given in closed form. We shall briefly discuss the main features:

1) $E_d(\underline{R})$ falls off with distance $\sim R^{-3}$ since the second derivative of the Green's function, (3.6), is involved. This is in analogy to the interaction of electrical dipoles.

2) The dipole interaction averaged over the solid angle at fixed distance is zero: $\int d\Omega\, f(\Omega) = 0$. This is also true for higher multipole interactions. This implies that there are always attractive and repulsive directions irrespective of the nature of the defects. Thus the long-range part of the interaction always causes a tendency for clustering reactions: When the defects become mobile they will preferentially jump to sites with low potential energy (i.e., attractive interaction) and eventually react with each other.

3) As a consequence of 2) the dipole interaction vanishes for isotropic defects ($P_{ij}^{A,B} = P_o^{A,B}\delta_{ij}$) in an isotropic medium, since no direction is preferred. Then higher order multipoles which drop off faster determine the interaction.

4) The dipole interaction is nonzero if the medium and/or at least one of the defects are anisotropic. For cubic crystals we obtain to lowest order in the anisotropy parameter $d = c_{11} - c_{12} - 2c_{44}$ for isotropic defects (pure dilatation or compression centers) the well-known ESHELBY formula /3.8/

$$E_d(\underline{R}) = -\frac{15}{8\pi}\,d\left(\frac{5}{3c_{11}+2c_{12}+4c_{44}}\right)^2\frac{P_o^A P_o^B}{R^3}\left(\frac{3}{5} - \sum_i \frac{R_i^4}{R^4}\right). \tag{3.15}$$

For most metals $d < 0$ and for identical defects (e.g., two interstitials with $P_o > 0$) the interaction is attractive in <100>- and repulsive in <110>- and <111>-directions.

Higher order terms in the anisotropy d have been calculated /3.10/. In Fig. 3.1 the results for two dilatation centers to third order in d are compared to numerical values /3.11/ for Cu whose anisotropy is quite large ($d/c_{44} = -1{,}90$). One sees that the angular variation becomes more pronounced for the higher order terms and approaches the exact results.

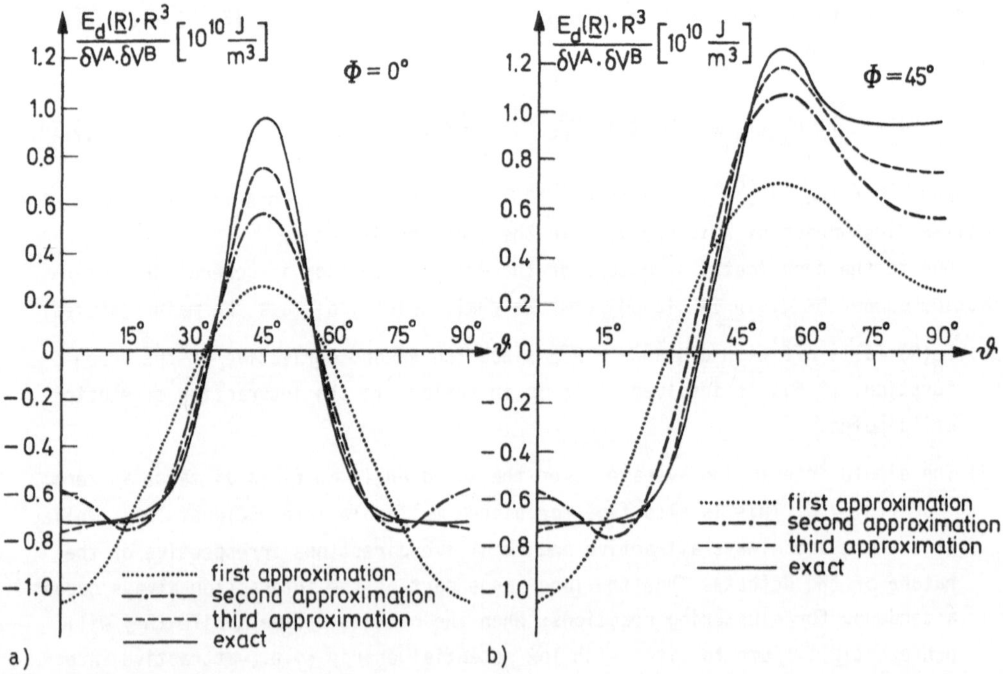

Fig. 3.1a,b. Angular dependent part of the interaction energy of two dilatation centers in Cu. Shown are the results of a perturbation expansion with respect to d/c_{44} up to third order /3.10/ and the exact results /3.11/.

a) (010)-plane (azimuth $\phi = 0$; polar angle $0 \leq \vartheta \leq 90^o$),

b) (110)-plane (azimuth $\phi = 45^o$; polar angle $0 \leq \vartheta \leq 90^o$)

5) The strength of the interaction is determined by the strength of the defects (dipole force tensor). Thus self-interstitials which have a much stronger displacement field than vacancies will interact more strongly with each other and with other defects.

Corrections to the dipole interaction are higher multipole contributions /3.12/ which fall off at least $\sim 1/R^5$, corrections to the static Green's function due to dispersion of the phonon energies /3.13/ also falling off $\sim 1/R^5$, and the induced interaction (3.12) which is also called inhomogeneity interaction. This can also be evaluated in the continuum limit. One obtains /3.9/

$$E_{AB}^{ind}(\underline{R}) = -\frac{1}{2}\left[\underline{\underline{P}}^A\, \underline{\underline{G}}''\, \underline{\underline{\alpha}}^B\, \underline{\underline{G}}''\, \underline{\underline{P}}^A + \underline{\underline{P}}^B\, \underline{\underline{G}}''\, \underline{\underline{\alpha}}^A\, \underline{\underline{G}}''\, \underline{\underline{P}}^B\right] = \frac{h(\Omega)}{R^6}\ . \tag{3.16}$$

We have abbreviated the second derivative of G by $\underline{G}''$ and fourth-rank tensors are indicated by thick underlining. The tensor $\underline{\underline{\alpha}}$ is the elastic polarizability of the defect which is given by the second moment of the t matrix, (3.3)

$$\alpha_{ijk\ell} = -\sum_{\underline{m},\underline{n}} R_i^{\underline{m}}\, t_{jk}^{\underline{mn}}\, R_\ell^{\underline{n}}\ . \tag{3.17}$$

It determines, e.g., the diaelastic change of the elastic constants $\Delta\underline{\underline{C}} = -\rho\underline{\underline{\alpha}}$ where ρ is the defect density /3.14/.

The structure of the induced interaction is as follows: $\underline{\underline{G}}''\underline{\underline{P}}^A$ is the strain $\underline{\underline{\varepsilon}}^A$ produced by defect A. This polarizes defect B giving rise to an induced dipole tensor $\delta\underline{\underline{P}}^B = \underline{\underline{\alpha}}^B\underline{\underline{G}}''P^A$ which then interacts with the displacement field of defect A in the usual way.

Since the explicit form of E^{ind} is very complicated, we just want to mention some general features. It falls off with distance $\sim 1/R^6$ analogous to the van der Waals interaction of induced electrical dipoles. The different elements of the elastic polarizability $\underline{\underline{\alpha}}$ can be positive or negative depending on the change of the coupling constants $\underline{\underline{\varphi}}$. They determine the sign and the angular dependence of E^{ind}. For self--interstitials in Al and Cu the shear eigenvalues of $\underline{\underline{\alpha}}$ are positive as is known from the large negative change of the elastic constants /3.15/. In this case the induced interaction is attractive for all directions. But even in this extreme case E^{ind} is comparable to the dipole interaction only for short distances $R \lesssim 2$ lattice constants /3.9/.

3.3 Interaction of Point Defects with Dislocations

The elastic interaction of point defects and defect agglomerates with dislocations is very important technologically since it provides the basic mechanism for creep /3.16/ and hardening /3.17/ of metals.

The concept of the dipole tensor to describe the long-range interaction can also be used for defect clusters. For instance, for a dislocation loop the dipole tensor is given by /3.18/

$$P_{ij} = \sum_{k\ell} C_{ijk\ell}\, b_k\, A_\ell \tag{3.18}$$

where $\underline{b}$ is the Burgers' vector and $\underline{A}$ the normal to the loop plane with $|\underline{A}|$ being the area of the loop. The Burgers' vector $\underline{b}$ is defined to be parallel to $\underline{A}$ for interstitial loops and antiparallel for vacancy loops. Then the volume change is simply given by $\delta V_{loop} = \underline{b}\cdot\underline{A}$. It represents the volume of the added interstitial platelet or missing vacancy platelet. Thus also the long-range interaction with point defects can be formulated in the same way as in the previous section. However, the asymptotic form $\sim R^{-3}$ is reached only for distances larger than the loop radius.

Very large loops are better described as an assembly of dislocation lines with which the mobile defects interact separately, because in a large fraction of space the interference of the strain fields of the different parts of the loop leading to an interaction $\sim R^{-3}$ is absent. Also a tangled dislocation network for not too high dislocation densitites can be described by single dislocation lines.

The first-order size interaction between a point defect A with a dipole tensor $\underline{\underline{P}}^A$ and a dislocation line is in elastic continuum theory given by

$$W_{dis\ A} = - P^A_{ij}\ \varepsilon^{dis}_{ij} \ . \tag{3.19}$$

(In this section we use Einstein's summation convention, i.e., over doubly occuring indices is to be summed). Here

$$\varepsilon^{dis}_{ij} = \frac{1}{2}\left(\partial_{R_j} s^{dis}_i + \partial_{R_i} s^{dis}_j \right) \tag{3.20}$$

is the strain field due to the dislocation line at the site of the point defect and $\underline{s}^{dis}$ the displacement field.

One can in general express ε^{dis}_{ij} as a line integral along the dislocation line as has been shown for isotropic /3.19/ and anisotropic /3.20/ crystals. The general expression for the derivative of the displacement field is

$$\partial_j s_i(\underline{R}) = \oint d\ell'_q \int d\underline{k}\ \exp[-i\underline{k}(\underline{R}-\underline{R}')]\ \frac{-i}{k}\Big[\hat{k}_t b_i \varepsilon_{tjq} + \hat{k}_j \tilde{g}_{im}(\hat{\underline{k}})\ C_{mnrs}\ \hat{k}_n \hat{k}_p b_r \varepsilon_{spq} \Big] . \tag{3.21}$$

Here $C_{ijk\ell}$ are the components of the elastic tensor, $\hat{\underline{k}} = \underline{k}/k$ is the unit vector in the direction of $\underline{k}$, $\tilde{g}(\hat{\underline{k}})$ is the angular part of the Fourier transform of the elastic Green's function defined in (3.6) and ε_{ijk} is the totally antisymmetric tensor of third rank whose elements are different form zero only if all indices are different. We have $\varepsilon_{123} = 1$, and every permutation of two indices yields a factor of (-1); thus the nonzero elements are

$$\varepsilon_{123} = \varepsilon_{231} = \varepsilon_{312} = 1 \ ; \quad \varepsilon_{132} = \varepsilon_{321} = \varepsilon_{231} = -1 \ . \tag{3.22}$$

Equation (3.22) can be evaluated explicitly only for certain geometries, e.g., for straight dislocations lying in highly symmetrical directions of the crystal. Let us shortly discuss the main features of the interaction of straight dislocations with point defects:

1) One can always express the strain field and thus the interaction with point de-defects as a one-dimensional integral in the plane perpendicular to the dislocation line /3.21/.

2) Using cylindrical coordinates $\underline{R} = (z,r,\theta)$ with z parallel to the dislocation line (see Fig. 3.2) the functional form of the strain field is (i,j = 1,2,3 cartesian coordinates)

$$\varepsilon_{ij}^{dis}(\underline{R}) = \frac{f_{ij}^{dis}(\theta)}{r} \tag{3.23}$$

with

$$f_{ij}^{dis}(\theta) = - f_{ij}^{dis}(\theta + \pi) \ . \tag{3.24}$$

This means that the interaction

$$E_{dis\ A} = \frac{\omega(\theta)}{r} \ ; \quad \omega(\theta) = f_{ij}^{dis}(\theta)\ P_{ij}^{A} \tag{3.25}$$

always has attractive and repulsive parts. $\left(\int_0^{2\pi} d\theta\ \omega(\theta) = 0 \right)$

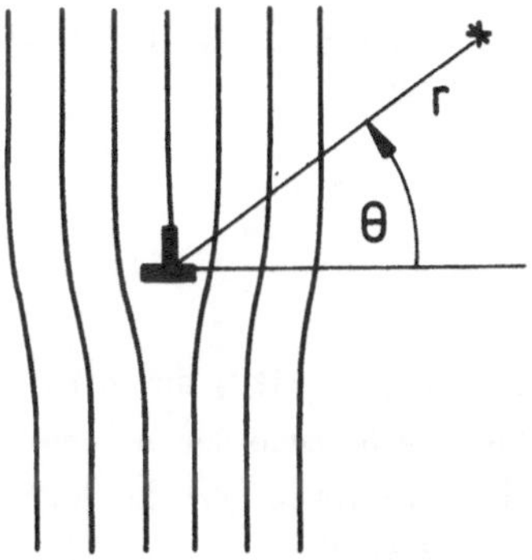

Fig. 3.2. Schematic picture of an edge dislocation (⊥), i.e., the line at which an extra lattice plane terminates. Cylindrical coordinates R = (z,r,θ) are used to describe the interaction with a point defect (*); the z axis is along the dislocation line which is perpendicular to the plane of the drawing

3) Particularly simple explicit expressions are obtained for the strain and stress fields of dislocations in isotropic materials /3.22/. Isotropic defects (dilatation or compression centers) with diagonal dipole tensor ($P_{ij}^{A} = P_{o}^{A}\delta_{ij}$) only interact with edge dislocations in isotropic materials since a screw dislocation produces pure shear strains. The interaction is given by

$$E_{dis\ A}(\underline{R}) = \pi\ b_{edge}\ c_{44}\ \delta V_{\infty}^{A}\ \frac{\sin\theta}{r} \ . \tag{3.26}$$

Here b_{edge} is the edge component of the Burgers' vector of the dislocation (⊥ to the dislocation line) and

$$\delta V_\infty^A = \frac{1}{3}\frac{1+\nu}{1-\nu}\,\delta V^A = \frac{1+\nu}{1-\nu}\,\frac{P_o^A}{c_{11}+2c_{12}} \tag{3.27}$$

is the volume change due to the defect A in an infinite medium (or local volume change) /3.8/ with Poisson's ratio

$$\nu = \frac{c_{12}}{c_{11}+c_{12}} \,. \tag{3.28}$$

4) For the general solution of the elastic equations it is necessary to find the zeros of a sixth-order polynomial /3.23/ but for particular geometries this polynomial can be factorized. This is always the case if the z axis (parallel to the dislocation line) is perpendicular to a reflection plane or is an axis of even-fold symmetry /3.22/. This is true for the following cases which can be solved explicitly /3.22/: dislocation in the basal plane of hexagonal crystals, dislocation along the <110>-direction in fcc crystals, dislocations along the <110>-direction in bcc crystals. For more complicated but still tractable cases see the references in /3.22/.

5) As in the case of point defects there are other contributions to the interaction of point defects with dislocations which have been discussed by BULLOUGH and NEWMAN /3.24/. They fall off faster than 1/R, e.g., the second-order size interaction (due to higher order elasticity theory) and the inhomogeneity interaction (due to elastic polarizability of defects) fall off $\sim 1/R^2$.

3.4 Short-Range Interaction

The short-range interaction which determines the structure, stability, and binding energy of defect agglomerates cannot be handled analytically because due to the strong overlap of defect forces the simple picture of Kanzaki forces breaks down.

But computer simulations have been made by introducing defects into a small crystallite containing a few thousand atoms which interact via an empirical potential. The stable configurations of defects are found by minimizing the total energy of the crystallite. Also binding energies of complex defects, energies of metastable configurations, activation energies for migration and reorientation, and the structure of saddle point configurations can be found by these calculations /3.25,26/. A detailed account of the results for structural defects is contained in the recent reviews by DEDERICHS et al /3.27/ and by the author /3.28/.

The most important results are:

1) Self-interstitials are unstable in the neighbourhood of a vacancy. The instability region contains about 100 atomic sites as has been determined by computer

simulation /3.29/ and by irradition experiments to very large doses /3.30/. The "binding energy" of this Frenkel pair is the sum of the formation energies of the interstitial and vacancy, in metals this is of the order of 5eV.

2) Self-interstitials have a strong tendency to cluster with binding energies of the order of 1eV per defect /3.26/. Experimentally this was found in annealing experiments of irradiated metals /3.31/.

3) Interstitials can very effectively be trapped at undersized impurity atoms /3.26/. The pair forms a so-called mixed dumbbell, i.e., a dumbbell with one constituent being the impurity atom. The binding energies are also of the order of 1eV. The binding energy to oversized impurities is only a few tenths of 1eV /3.27,32/.

4) Vacancies also form clusters as is known from annealing studies of irradiated and quenched metals /3.33/. These clusters however are less tightly bound than interstitial clusters. Divacancies have binding energies of a few tenths of an eV, for larger agglomerates the binding energy gradually increases until it reaches the formation energy of a vacancy of the order of 1 eV for large dislocation loops. However, since vacancies are much less mobile than interstitials (the migration energy is of the order of 1 eV), the dissociation energy which is the sum of binding and migration energy is always 1eV.

5) Vacancies can be bound to interstitial and substitutional impurity atoms /3.34/, the binding energy never exceeds a few tenths of 1 eV. For instance, it is ~ 0.25eV for Ge in Cu /3.35/ and $\gtrsim 0.4$eV for H in Cu /3.36/.

6) Impurity atoms also can have a positive binding to each other. If this is the case they tend to form precipitates as is, e.g., known for Cu and Ag in Al. No quantitative number for the binding energy of these clusters can be given.

In Sects. 5 - 7 we only consider formation of stable complexes by diffusion of defects and discuss the influence of the lattice structure and of the long range interaction between the reactants on the rate of formation. From the above discussion it is clear that this theory has a wide range of application. Only at high temperatures when kT is of the order of the binding energy, must dissociation be taken into account.

4. Diffusion in Force Fields

In this section we discuss the influence of force fields on the diffusion of point defects. The forces can originate from external conditions like application of stress or electric fields or from the interaction of point defects with each other or with dislocations or grain boundaries. We are particularly emphasizing the effect of slowly varying force fields on the macroscopic quantities like diffusion coefficients and current because in Sect. 7 we discuss the influence of the long-range interaction between point defects on the reaction rate for cluster formation.

Macroscopically in ideal crystals the current density $\underline{j}$ of mobile particles is proportional to the gradient of the particle density ρ

$$j_i(\underline{R},t) = -\sum_j D_{ij}\, \partial_{R_j} \text{grad}\, \rho(\underline{R},t) \ . \tag{4.1}$$

The diffusion tensor D_{ij} is diagonal in cubic crystals, i.e., $D_{ij} = D_o \delta_{ij}$ with D_o the diffusion constant. Together with the continuity equation

$$\frac{d}{dt}\rho(\underline{R},t) + \text{div}\, \underline{j}(\underline{R},t) = \Pi(\underline{R},t) \tag{4.2}$$

where $\Pi(\underline{R},t)$ describes the production of particles, (3.1) yields for cubic crystals the macroscopic diffusion equation which for cubic crystals reads

$$\frac{d}{dt}\rho(\underline{R},t) = D_o\, \Delta\rho + \Pi(\underline{R},t) \ . \tag{4.3}$$

As has been discussed in Sect. 2.3 this can be obtained as the asymptotic form of the lattice diffusion equation (2.11) for large $\underline{R}$.

In the presence of external fields, like stress fields of electric fields, one obtains, in addition to the above diffusion current, a drift current which is set proportional to the gradient of the interaction energy $E(\underline{R})$ of the particle in the external field. For cubic crystals the total current is

$$j_i(\underline{R},t) = D_o\, \partial_{R_i} \rho(\underline{R},t) - D_o\, \beta\, \rho(\underline{R},t)\, \partial_{R_i} E(\underline{R},t) \tag{4.4}$$

Here the Einstein relation $B = D_o\beta$ for mobility and diffusion constant has been used.

It has been shown recently /4.1/ that (4.4) is not quite correct for point defects in solids since several effects due to the structure of the defects and the geometry of the jumps have not been considered.

i) The geometry of the jumps can considerably influence the symmetry of diffusion. For certain defects one obtains, e.g., one- or two-dimensional diffusion.

ii) As is obvious from the discussion of the jump frequencies in sect. 2.2 one has to distinguish between the interaction energy in the equilibrium configuration $E_e(\underline{R})$ and in the saddle point configuration $E_s(\underline{R})$. It turns out that one has to identify $E(\underline{R})$ in (3.4) with $E_e(\underline{R})$. However in addition, the diffusion constant becomes R dependent

$$D(\underline{R}) = D_o\, \exp\left[-\beta\left(E_s(\underline{R}) - E_e(\underline{R})\right)\right] \ . \tag{4.5}$$

This is a consequence of detailed balance and implies that for stationary conditions (not necessary thermal equilibrium) the total current (4.4) becomes independent of the interaction in the equilibrium configuration, $E_e(\underline{R})$. This means that only the change of the saddle point energy $E_s(\underline{R})$ can influence the diffusion rate which is very important for reaction rates (see Sect. 7).

iii) External fields can effectively lower the cubic symmetry of the crystal so that the diffusion tensor becomes ansiotropic. As we shall see below the anisotropy of the saddle point configuration, e.g., in elastic stress fields the anisotropy of the dipole tensor, determines the symmetry of the diffusion tensor. In the simple case of a homogeneous deformation of a cubic crystal with a strain field $\varepsilon_{k\ell}$ one obtains additional contributions to the diffusion tensor proportional to $\varepsilon_{k\ell}$

$$D_{ij} = D_o\delta_{ij} + \sum_{k\ell} d_{ijk\ell}\,\varepsilon_{k\ell} \tag{4.6}$$

as has been pointed out by FLYNN /4.2/. Since the "elasto diffusion tensor" is determined by the symmetry of the saddle point configuration a measurement of the diffusion tensor in single crystals with differently oriented uniaxial stresses yields information about the saddle point configuration.

Some of these complications have previously been discussed. KOEHLER /4.3/ considered the anisotropy of the saddle point for vacancy jumps. KRONMOLLER et al. /4.4/ also discussed the effects of the saddle point interaction but obtained an isotropic diffusion tensor. SAVINO /4.5/ discussed the consequences of the saddle point anisotropy for the interpretation of irradiation creep. INGLE and CROCKER /4.6/ have studied the anisotropic diffusion of vacancies in bcc crystals by computer simulation.

4.1 Derivation of Continuum Theory from Lattice Theory

We shall derive the continuum equation for diffusion from lattice theory following rather closely the treatment of /4.1/.

4.1.1 Cubic Defects

To illustrate the effect of external fields we first consider defects with cubic symmetry, as an example we treat H in Pd which occupies octahedral sites (see Figs. 2.5a and 2.7a).

We start from the microscopic equation for the density[1] $\rho_{\underline{n}}$ at site $\underline{n}$

[1] We consider *one* particle only: $\sum_{\underline{m}} \rho_{\underline{m}}(t) = 1$, and formulate the equations for occupation probabilities. To obtain the results for many identical independent particles (low concentration) one has to multiply with the average density N/V.

$$\frac{d}{dt}\rho_{\underline{n}} = \sum_{\underline{m}} \left[\lambda^{\underline{nm}} \rho_{\underline{m}}(t) - \lambda^{\underline{mn}} \rho_{\underline{n}}(t)\right] + \Pi_{\underline{n}}(t) \ . \tag{4.7}$$

The first terms represent the changes due to jumps, $\pi_{\underline{n}}(t)$ is the production of particles per unit time at site $\underline{n}$.

In Sect. 2.2 we have derived an expression for the jump frequencies λ in terms of an effective attempt frequency ν and the energy difference between the saddle point configuration and the equilibrium configuration $\Delta E = E(s) - E(A)$. Due to the presence of an external field these quantities depend on the position $\underline{n}$ of the mobile defect. In practice the change of the energies is much more important than of the attempt frequency /4.7/.

Thus we shall use the following simplified expression

$$\lambda^{\underline{mn}} = \nu\, s_1^{(\underline{m}-\underline{n})} \exp(-\beta E^M) \exp[-\beta(E_s^{(\underline{m},\underline{n})} - E_e^{(\underline{n})})] \tag{4.8}$$

with the attempt frequency assumed to be the same for all sites. The energies are counted from the equilibrium energy in the ideal crystal which is set equal to zero. E^M is the saddle point energy in the ideal crystal. The interaction energy with the external field of the defect in the saddle point between site $\underline{n}$ and $\underline{m}$ is denoted by $E_s^{(\underline{m},\underline{n})}$ and in the equilibrium positions $\underline{n}$ and $\underline{m}$ it is $E_e^{(\underline{n})}$ and $E_e^{(\underline{m})}$, respectively (see Fig. 4.1).

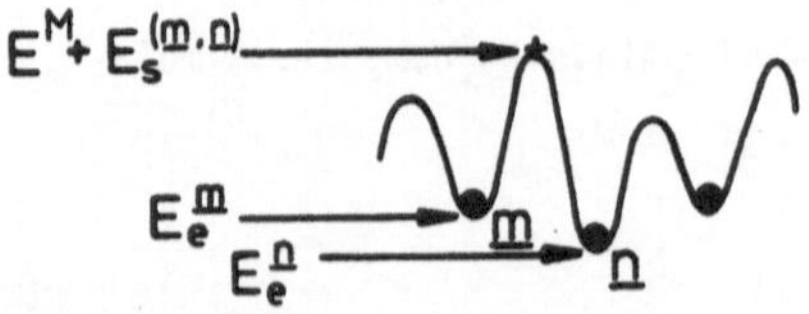

Fig. 4.1. Sketch of the energy profile for a defect (•) in a distorted crystal. $E_e^{(\underline{m})}$ is the energy in the equilibrium site $\underline{m}$, $E^M + E_s^{(\underline{m},\underline{n})}$ the energy in the saddle point $(\underline{m},\underline{n})$ between the equilibrium sites $\underline{m}$ and $\underline{n}$

In *thermal equilibrium* $\Pi_{\underline{m}} = 0 = \frac{d}{dt}\rho_{\underline{m}}$ the particle density is given by

$$\rho_{\underline{m}} = \frac{\exp(-\beta E_e^{(\underline{m})})}{\sum_{\underline{m}'} \exp(-\beta E_e^{(\underline{m}')})} \ . \tag{4.9}$$

This is independent of the interaction E_s in the saddle point. However the time to reach the equilibrium strongly depends on E_s.

For the theory of reaction rates (in Sect. 7) the *stationary state* is very important which is obtained for time-independent production. We then have

$$\sum_{\underline{m}} \left[\lambda^{\underline{nm}} \rho_{\underline{m}} - \lambda^{\underline{mn}} \rho_{\underline{n}}\right] + \Pi_{\underline{n}} = 0 \ . \tag{4.10}$$

By introducing a renormalized density

$$W_{\underline{m}} = \rho_{\underline{m}} \exp(\beta E_e^{(\underline{m})}) \tag{4.11}$$

which according to (4.9) describes the deviation from thermal equilibrium. Equation (4.10) can be written in the form

$$\sum_{\underline{m}} \tilde{\lambda}^{\underline{nm}}(W_{\underline{m}} - W_{\underline{n}}) + \Pi_{\underline{m}} = 0 \tag{4.12}$$

with the symmetrical jump frequency

$$\tilde{\lambda}^{\underline{mn}} = \tilde{\lambda}^{\underline{nm}} = \nu \exp[-\beta(E^M + E_s^{(\underline{m},\underline{n})})] \tag{4.13}$$

which depends only on the saddle point interaction but not on the interaction in the equilibrium positions. Since for given production terms $\Pi_{\underline{m}}$ and interactions $E_s^{(\underline{m},\underline{n})}$, $W_{\underline{m}}$ is determined uniquely from (4.12), it is independent of the interaction E_e at the equilibrium sites. Also the microscopic currents

$$j_{\overleftarrow{\underline{mn}}} = \lambda^{\underline{mn}}\rho_{\underline{n}} = \tilde{\lambda}^{\underline{mn}}\, W_{\underline{n}} \tag{4.14}$$

are independent of E_e and so is any macroscopic current which is the sum of microscopic $j_{\overleftarrow{\underline{mn}}}$. Only the densities $\rho_{\underline{m}}$ do depend on the interaction in the equilibrium positions according to the relation (4.11).

The stationary diffusion equation (4.12) can be obtained from a variational principle with the following functional /4.8/

$$L\{W_{\underline{m}}\} = \frac{1}{4}\sum_{\underline{mn}} \tilde{\lambda}^{\underline{mn}}(W_{\underline{n}} - W_{\underline{m}})^2 + \sum_{\underline{m}} \Pi_{\underline{m}}\, W_{\underline{m}} \ . \tag{4.15}$$

The extremal condition $\delta L = 0$ yields directly (4.12) and the stationary value L_{ex} is given by

$$L_{ex} = -\frac{1}{2}\sum_{\underline{m}} \pi_{\underline{m}}\, W_{\underline{m}} \ . \tag{4.16}$$

We now specify the external field to be slowly varying elastic strain field $\varepsilon_{ij}(\underline{R})$, e.g., a homogeneous deformation of the crystal or the long-ranging strain field of a dislocation. In this case the interaction energies can be described by the dipole tensors P^e and P^s for equilibrium and saddle point configuration, respectively,

$$E_e^{(\underline{m})} = \sum_{ij} P_{ij}^e\, \varepsilon_{ij}(\underline{R}^{\underline{m}}) \ ; \quad E_s^{(\underline{m},\underline{n})} = \sum_{ij} P_{ij}^s(\underline{m},\underline{n})\, \varepsilon_{ij}(\underline{R}^{\underline{m},\underline{n}}) \ . \tag{4.17}$$

For cubic defects $\underline{P}^e$ is isotropic ($P_{ij}^e = P_o^e \delta_{ij}$). However, the saddle point dipole tensor $\underline{P}^s$ in general is anisotropic and depends on the direction of the jump as is illustrated in Fig. 4.2. The defect exerts strong forces on the neighbouring atoms

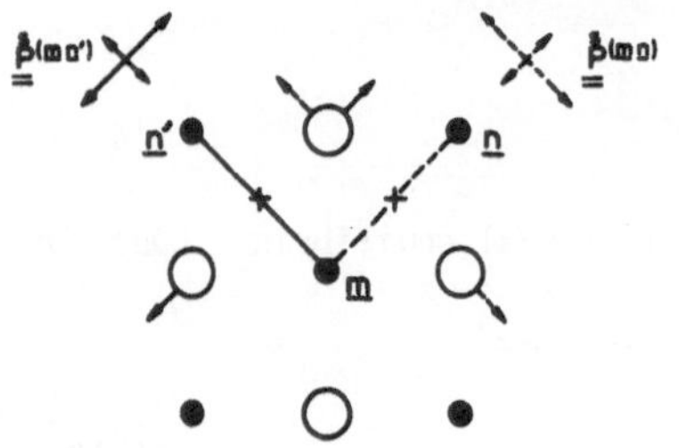

Fig. 4.2. Dipole tensor P^s in the saddle point (+) between two octahedral sites (•) in an fcc crystal. At the saddle point between sites $\underline{m}$ and $\underline{n}$ the defect exerts strong forces (dashed arrows) on the neighbouring host atoms (○), thus the dipole tensor P^s has a large component in this direction. For a jump between $\underline{m}$ and $\underline{n}'$ the dipole tensor is rotated by 90^0

so that the dipole tensor has a large component in the direction of these neighbours and two smaller ones in perpendicular directions. One can see that the dipole tensor $P^s(\underline{m},\underline{n}')$ for the jump $\underline{m}\text{—}\underline{n}'$ (full lines in Fig. 4.2) is obtained from $P^s(\underline{m},\underline{n})$ for a jump $\underline{m}\text{—}\underline{n}$ (dashed lines in Fig. 4.2) by a rotation by 90^0.

The transition to continuum theory is most conveniently performed with the functional $L\{W_{\underline{m}}\}$, (4.15). Assuming $W_{\underline{m}} = W(\underline{R}^{\underline{m}})$ to be slowly varying we can expand $W_{\underline{n}}$ in a Taylor series

$$W_{\underline{n}} \cong W + \sum_i (R_i^{\underline{n}} - R_i^{\underline{m}})\, \partial_{R_i^{\underline{m}}} W(\underline{R}^{\underline{m}}) + \ldots \quad . \tag{4.18}$$

Then instead of summing in (4.15) over $\underline{m}$ and $\underline{n}$ we first sum over the difference $\underline{m} - \underline{n} = \underline{h}$, i.e., over the z_1 nearest neighbours of $\underline{m}$ and then transform the sum over $\underline{m}$ into an integral over R. This yields

$$L\{W(\underline{R})\} = \int d\underline{R} \left| \frac{1}{2} \sum_{ij} \tilde{D}_{ij}(\underline{R})\, \partial_{R_i} W(\underline{R})\, \partial_{R_j} W(\underline{R}) - \Pi(\underline{R}) W(\underline{R}) \right| \tag{4.19}$$

with

$$\begin{aligned} \tilde{D}_{ij}(\underline{R}) &= \frac{1}{2} \sum_{\underline{h}} \lambda^\circ(\underline{h})\, R_i^{\underline{h}} R_j^{\underline{h}} \exp\left[\beta \sum_{k\ell} \varepsilon_{k\ell}(\underline{R})\, P_{k\ell}^s(\underline{h})\right] \\ &= D_o \frac{3}{z_1} \sum_{\underline{h}} \hat{R}_i^{\underline{h}} \hat{R}_j^{\underline{h}} \exp\left[\beta \sum_{k\ell} \varepsilon_{k\ell}(\underline{R})\, P_{k\ell}^s(h)\right] \end{aligned} \tag{4.20}$$

for cubic crystals.

Here $\lambda^\circ(\underline{h}) = s_1^{(\underline{h})}\, \nu \exp(-\beta E^M)$ is the jump rate in the ideal lattice. $\hat{\underline{R}}^{\underline{h}}$ are unit vectors in the different jump directions. In deriving (4.20) we have replaced $\varepsilon(\underline{R}^{\underline{m},\underline{n}})$ by $\varepsilon(\underline{R}^{\underline{m}})$. However the orientation dependence of $P^s(\underline{m},\underline{n}) = P^s(\underline{h})$ has been taken into account exactly, since due to the angular dependence $P^s(\underline{h})$ has an essential singularity at $\underline{h} = 0$ and cannot be expanded in a Taylor series around $\underline{h} = 0$. This means that

$$E_s^{(\underline{h})}(\underline{R}) = \sum_{k\ell} \varepsilon_{k\ell}(\underline{R})\, P_{k\ell}^s(\underline{h}) \tag{4.21}$$

explicitly depends on the direction of the jump. For cubic crystals this gives rise to the anisotropy of $D_{ij}(\underline{R})$ according to (4.20).

From (4.19) we obtain the stationary diffusion equation for the renormalized density $W(\underline{R})$

$$\sum_{ij} \partial_{R_i} [\tilde{D}_{ij}(\underline{R})\ \partial_{R_j} W(\underline{R})] + \Pi(\underline{R}) = 0 \tag{4.22}$$

which together with the appropriate boundary condition, e.g., $W(\underline{R}) = 0$ for $R \to \infty$ determines $W(\underline{R})$ uniquely. As in lattice theory $W(\underline{R})$ is independent of the interaction in the equilibrium positions and so is the current

$$j_i(\underline{R}) = - \sum_j \tilde{D}_{ij}(\underline{R})\ \partial_{R_j} W(\underline{R}) \ . \tag{4.23}$$

The interaction energy $E_e(\underline{R})$ can only enter via the boundary condition if the density $\rho(\underline{R}_S)$ is finite at the boundary S. However in many cases $E_e(\underline{R})$ can be assumed to vanish at the boundary so that $W(\underline{R}_S) = \rho(\underline{R}_S)$ and $W(\underline{R})$ is independent of E_e.

By introducing the density $\rho(\underline{R})$ into (4.23) we can decompose the current into a diffusion current and a drift current

$$j_i(\underline{R}) = j_i^{\mathrm{Diff}} + j_i^{\mathrm{Drift}} = - \sum_j D_{ij}(\underline{R}) \left[\partial_{R_j} \rho(\underline{R}) - \beta\rho(\underline{R})\ \partial_{R_j} E_e(\underline{R}) \right] \tag{4.24}$$

with the $\underline{R}$-dependent diffusion constant

$$D_{ij}(\underline{R}) = \tilde{D}_{ij}(\underline{R}) \exp[\beta E_e(\underline{R})] = D_o \frac{3}{z_1} \sum_{\underline{h}=NN} R_i^{\underline{h}}\ R_j^{\underline{h}} \exp\left[-\beta\left(E_s^{(\underline{h})}(\underline{R}) - E_e(\underline{R})\right)\right] \ . \tag{4.25}$$

By comparison with the phenomenological equation (4.4) we notice the effects due to the lattice structure: the two different interaction energies E_e and E_s, and the $\underline{R}$ dependence and anisotropy of D_{ij}.

4.1.2 Noncubic Defects

The situation is more complicated for noncubic defects. For instance, tetrahedral sites and octahedral sites in bcc crystals have tetragonal symmetry (D_{2d} and D_{4h}, respectively) (see Fig. 2.9). For the octahedral sites there are three nonequivalent sites with preferred orientation in x, y and z direction. Sites with the same orientation form a bcc lattice, the total interstitial lattice is nonprimitive. By denoting the Bravais lattice by vectors $\underline{M}$, $\underline{N}$ and the s sites within a cell by $\alpha = 1,\dots,s$ we obtain the following equation of motion

$$\frac{d}{dt} \rho_\alpha^{\underline{M}}(t) = \sum_{N\beta} \left[\lambda_{\alpha\beta}^{\underline{MN}}\ \rho_\beta^{\underline{N}}(t) - \lambda_{\beta\alpha}^{\underline{NM}}\ \rho_\alpha^{\underline{M}}(t) \right] + \Pi_\alpha^{\underline{M}}(t) \tag{4.26}$$

where $\lambda_{\alpha\beta}^{\underline{MN}}$ is the transition rate from position ($\underline{N}\beta$) to ($\underline{M}\alpha$). Another example for noncubic defects are the self-interstitials in the form of dumbbells, e.g., the <100>-dumbbell in Al with tetragonal (D_{4h}) symmetry as shown in Fig. 2.5d. This dumbbell can have 3 different orientations on one site in the x, y and z direction. The <110>-dumbbell in bcc crystals, Fig. 2.6e, can have 6 different orientations. Thus we have to define partial densities for each orientation $\mu = 1...s$ on site $\underline{m}$, which yields an equation of motion similar to (4.26)

$$\frac{d}{dt}\rho_\mu^{\underline{m}}(t) = \sum_{\underline{n}\nu}\left[\lambda_{\mu\nu}^{\underline{mn}}\rho_\nu^{\underline{n}}(t) - \lambda_{\nu\mu}^{\underline{nm}}\rho_\mu^{\underline{m}}(t)\right] + \Pi_\mu^{\underline{m}}(t) . \tag{4.27}$$

Due to the orientational degree of freedom, special jump vectors can be preferred thus leading to diffusion in less than three dimensions. Examples in fcc crystals are (i) the <110>-dumbbell (so-called "crowdion") which is supposed to move only along the <110>-chain and (ii) the di-interstitial consisting of two parallel <100>--dumbbells on nearest-neighbour sites. According to computer simulations the elementary jump of this defect consists of two subsequent jumps of one of the <100>--dumbbells, confining the defect to planar diffusion on the plane perpendicular to the dumbbell axes /4.9/.

Most defects, however, perform three-dimensional diffusion as, e.g., the <100>--dumbbell in fcc crystals /4.9-11/. Then it can be shown /4.1/ that for long-range migration in ideal crystals an effective jump frequency matrix with cubic symmetry is obtained even if the elementary jump frequencies $\lambda_{\mu\nu}^{\underline{mn}}$ of a given orientation do not have cubic symmetry. The reason is that after only a few jumps the orientations have been mixed locally so that for the long-range migration one effectively starts with an orientational averaged distribution. The effective jump frequency matrix in the ideal crystal can then be defined by

$$\lambda_{eff}(\underline{h}) = \frac{1}{s}\sum_{\mu\nu}\lambda^{\circ}{}_{\mu}^{\underline{m}}{}_{\nu}^{\underline{m}+\underline{h}} . \tag{4.28}$$

It is independent of the cell index $\underline{m}$.

This leads to the following equation of motion for the long-range diffusion in terms of the total density in the unit cell $\bar{\rho}(\underline{R}^{\underline{m}}) = \sum_\mu \rho_\mu^{\underline{m}}$

$$\frac{d}{dt}\bar{\rho}(\underline{R}^{\underline{m}},t) = D\,\Delta\rho(\underline{R}^{\underline{m}},t) \tag{4.29}$$

with

$$D = \frac{1}{2}\sum_{\underline{h}} R_i^{\underline{h}} R_j^{\underline{h}} \lambda_{eff}(\underline{h}) . \tag{4.30}$$

To obtain an expression for the diffusion coefficient in distorted crystals one starts again with the functional which yields the stationary case ($\frac{d}{dt} = 0$) of (4.27).

$$L\{W_\mu^{\underline{m}}\} = \frac{1}{4}\sum_{\substack{\underline{mn}\\ \mu\nu}}\tilde{\lambda}_{\mu\nu}^{\underline{mn}}(W_\nu^{\underline{n}} - W_\mu^{\underline{m}})^2 - \sum_{\substack{\underline{m}\\ \mu}}\Pi_\mu^{\underline{m}}\,W_\mu^{\underline{m}} \tag{4.31}$$

with

$$\tilde{\lambda}^{\underline{m}\underline{n}}_{\mu\nu} = \tilde{\lambda}^{\underline{n}\underline{m}}_{\nu\mu} = s_1^{(\underline{m}-\underline{n})}\nu \exp\left[-\beta\left(E^M + E_s \,{}^{\underline{m}}_{\mu}\,{}^{\underline{n}}_{\nu}\right)\right] . \tag{4.32}$$

Again we can assume that after only a few jumps a local equilibrium has been reached such that the renormalized partial density $W^{\underline{m}}_{\mu}$ does not depend on orientation and is slowly varying.

Then we can formulate the Lagrangian in the continuum limit in terms of the total renormalized density on one site or in the unit cell

$$\overline{W}(\underline{R}) = \sum_{\mu} W^{\underline{m}}_{\mu} \tag{4.33}$$

using the same procedure as in Sect. 4.1.1 (see Ref. /4.1/). We thus obtain

$$L\{\overline{W}(\underline{R})\} = \frac{1}{s}\int d\underline{R}\left[\frac{1}{2}\sum_{ij}\tilde{D}_{ij}(\underline{R})\ \partial_{R_j}\overline{W}(\underline{R})\ \partial_{R_i}\overline{W}(\underline{R}) - \overline{\Pi}(\underline{R})\ \overline{W}(\underline{R})\right] \tag{4.34}$$

with the $\underline{R}$-dependent diffusion tensor

$$\tilde{D}_{ij}(\underline{R}) = \frac{1}{2}\sum_{\underline{h}} R_i^{\underline{h}}\ R_j^{\underline{h}}\left[\frac{1}{s}\sum_{\mu\nu}\tilde{\lambda}^{\underline{m}\ \underline{m}+\underline{h}}_{\mu\ \nu}\right] . \tag{4.35}$$

Due to the presence of the external field the effective jump frequencies

$$\tilde{\lambda}^{(\underline{h})}_{\text{eff}}(\underline{R}^{\underline{m}}) = \frac{1}{s}\sum_{\mu\nu}\tilde{\lambda}^{\underline{m}\ \underline{m}+\underline{h}}_{\mu\ \nu} \tag{4.36}$$

now depend on the site index $\underline{m}$.

A similar result is obtained for the case of more than one site per unit cell, e.g., for tetrahedral and octahedral interstitials in bcc crystals.

As in the preceding section the tensor $\tilde{D}_{ij}(\underline{R})$ depends only on saddle point properties, in particular on the interaction $E_s \,{}^{\underline{m}}_{\mu}\,{}^{\underline{n}}_{\nu}$.

For an external strain field $\varepsilon_{k\ell}(\underline{R})$ one obtains

$$\tilde{D}_{ij}(\underline{R}) = \frac{1}{2}\sum_{\underline{h}} R_i^{\underline{h}}\ R_j^{\underline{h}}\ \lambda^{\circ(\underline{h})}_{\text{eff}} \exp\left[\beta\sum_{k\ell}\varepsilon_{k\ell}(\underline{R})\ P^s_{k\ell}(\underline{h})\right] \tag{4.37}$$

where we have assumed that the dipole tensor only depends on the jump direction $\underline{h}$ irrespective of the initial and final orientation of the defect. This is, e.g., the case for the <100>-dumbbell in fcc as well as for the tetrahedral and octahedral interstitials in bcc crystals.

This is the same result as for cubic defects, (4.20), which shows that the anisotropy of the equilibrium configuration does not influence the anisotropy of the diffusion tensor in cubic crystals. This is only determined by the anisotropy of the dipole tensor in the saddle point configuration.

For the total density in the cell at $\underline{R}^{\underline{m}}$ we obtain

$$\overline{\rho}(\underline{R}^{\underline{m}}) = \sum_{\mu}\rho^{\underline{m}}_{\mu} = \sum_{\mu} W^{\underline{m}}_{\mu}\exp(-\beta E_e \,{}^{\underline{m}}_{\mu}) = \overline{W}(\underline{R}^{\underline{m}})\ \exp[-\beta\ E_e^{\text{eff}}(\underline{R}^{\underline{m}})] . \tag{4.38}$$

For the last equation and the definition of $E_e^{eff}(\underline{R}^{\underline{m}})$ we have used the independence of $W_\mu^{\underline{m}}$ from the orientational degree of freedom.

The equation of motion for $\bar{\rho}$ resulting from (4.34) reads

$$\sum_{ij} \partial_{R_i}\left[\tilde{D}_{ij}(\underline{R})\, \partial_{R_j}\left(\bar{\rho}(\underline{R})\, \exp[\beta\, E_e^{eff}(\underline{R})]\right)\right] + \bar{\Pi}(\underline{R}) = 0 \ . \tag{4.39}$$

By comparison with (4.22) we see that the effective interaction E_e^{eff} defined by

$$\exp[-\beta\, E_e^{eff}(\underline{R})] = \sum_\mu \exp(-\beta\, E_\mu^{\underline{m}}) \tag{4.40}$$

replaces the interaction E_e in the equilibrium configuration for cubic defects. The diffusion tensor is then

$$D_{ij}(\underline{R}) = \frac{1}{2}\sum_{\underline{h}} R_i^{\underline{h}}\, R_j^{\underline{h}}\, \lambda_{eff}^{o\,(\underline{h})}\, \exp\left[-\beta\left(E_s^{(\underline{h})}(\underline{R}) - E_e^{eff}(\underline{R})\right)\right] \ . \tag{4.41}$$

4.2 Diffusion in a Homogeneously Deformed Crystal

The simplest deformation of a crystal is a homogeneous strain field $\varepsilon_{k\ell}$. Then all interactions are constant throughout the crystal and D_{ij} is independent of $\underline{R}$. Thus for the density $\rho(\underline{R})$ we obtain the usual stationary diffusion equation

$$\sum_{ij} D_{ij}\, \partial_{x_i}\partial_{x_j}\rho(\underline{R}) + \Pi(\underline{R}) = 0 \ .$$

Since experimentally feasible strains are quite small, typically $\varepsilon \cong 10^{-4}$, one can expand D_{ij} linearly in ε

$$D_{ij} = D_{ij}^{o} + \sum_{k\ell} d_{ijk\ell}\, \varepsilon_{k\ell} \ ; \quad D_{ij}^{o} = D_o\delta_{ij} \ \text{for cubic crystals.} \tag{4.42}$$

The "elasto diffusion tensor" $d_{ijk\ell}$ describes the influence of the homogeneous strain on the diffusion. In cubic crystals it has the symmetry of the elastic constants

$$d_{ijk\ell} = d_{jik\ell} = d_{ij\ell k} = d_{k\ell ij} \tag{4.43}$$

and only three independent constants remain /4.1/

$$d^{(1)} = d_{11} + 2d_{12} \ ; \quad d^{(2)} = d_{11} - d_{12} \ ; \quad d^{(4)} = 2d_{44} \ . \tag{4.44}$$

(Here Voigt's notation for fourth-rank tensors is used.) They correspond to the three independent elastic constants for cubic crystals, the compression modulus $K = c_{11} + 2c_{12}$ and the two shear moduli $c' = \frac{1}{2}(c_{11} - c_{12})$ and c_{44}.

Expanding the diffusion tensor D_{ij}, (4.41), with E_e^{eff}, (4.40), we obtain

$$d_{ijk\ell} = \frac{\beta}{2}\sum_{\underline{h}} R_i^{\underline{h}}\, R_j^{\underline{h}}\, \lambda^{o}(\underline{h})[P_{k\ell}^{s}(\underline{h}) - P_o^{e}\delta_{k\ell}] \ ; \quad P_o^{e} = \frac{1}{3}\,\mathrm{Tr}\{\underline{\underline{P}}^{e}\} \ . \tag{4.45}$$

Since for noncubic defects we have to average over all possible orientations only the trace of $\underline{\underline{P}}^e$ enters.

For nearest-neighbour jumps $\lambda^o(\underline{h})$ we can extract the diffusion constant D_o in the ideal crystal and obtain

$$d_{ijk\ell} = 3D_o\beta\left\langle \hat{R}_i^{\underline{h}}\, \hat{R}_j^{\underline{h}}\, [P^s_{k\ell}(\underline{h}) - P^e_o\, \delta_{k\ell}]\right\rangle \tag{4.46}$$

where $\hat{R}^{\underline{h}}$ are unit vectors in the jump directions and $\langle\ \rangle$ means an average over all jump directions. With this formula the invariants $d^{(1)}$, $d^{(2)}$ and $d^{(4)}$ can easily be calculated. It turns out that only the elasto diffusion constant $d^{(1)}$, coupling to a uniform compression, depends on the dipole tensor of the equilibrium configuration. Since a compression does not break cubic symmetry, the change ΔD of the diffusion constant D_o is given by

$$\Delta D = \frac{1}{3} D_o\, Tr\{(P^s - P^e)\}\, Tr\{\varepsilon\} = -\beta\, p(\delta V^s - \delta V^e) \tag{4.47}$$

where p is the hydrostatic pressure, and $\delta V^s - \delta V^e$ is the activation volume for diffusion. This result is well known /4.12/.

By measuring all three constants $d^{(1)}$, $d^{(2)}$ and $d^{(4)}$ one can get useful information about the symmetry of the saddle point configuration. For the most important symmetries $\underline{\underline{P}}^s$ has the following form

$$\underline{\underline{P}} = \begin{bmatrix} P_{11} & 0 & 0 \\ 0 & P_{11} & 0 \\ 0 & 0 & P_{11} \end{bmatrix}_{\text{cubic}} \qquad \begin{bmatrix} P_{11} & 0 & 0 \\ 0 & P_{22} & 0 \\ 0 & 0 & P_{22} \end{bmatrix}_{\text{tetragonal}}$$

$$\underline{\underline{P}} = \begin{bmatrix} P_{11} & P_{12} & P_{12} \\ P_{12} & P_{11} & P_{12} \\ P_{12} & P_{12} & P_{11} \end{bmatrix}_{\text{trigonal}} \qquad \begin{bmatrix} P_{11} & P_{12} & 0 \\ P_{12} & P_{11} & 0 \\ 0 & 0 & P_{33} \end{bmatrix}_{\text{110-orthorhombic}} \tag{4.48}$$

Depending on the symmetry of the saddle point, the elasto diffusion constants are proportional to combinations of elements of the dipole tensors as listed in the following table:

Table 1 Elasto diffusion constants for different saddle point symmetries

symmetry	cubic	tetragonal	trigonal	orthorhombic
$d^{(1)}$	$P^s_{11} - P^e_o$	$\frac{1}{3}(P^s_{11} + 2P^s_{22}) - P^e_o$	$P^s_{11} - P^e_o$	$\frac{1}{3}(2P^s_{11} + P^s_{33}) - P^e_o$
$d^{(2)}$	0	$P^s_{11} - P^s_{22}$	0	$P^s_{11} - P^s_{33}$
$d^{(4)}$	0	0	P^s_{12}	P^s_{12}

4.3 Discussion

The diffusion coefficients in external fields can in principle be measured with the same methods as discussed in Sect. 2. However, to obtain full information about the saddle point, single crystals have to be used and external strains of different symmetries have to be applied. The highly symmetric defects listed in Table 1 can be distinguished uniquely by a full set of measurements: The zeros in the table determine the symmetry of the saddle point and the nonzero elements yield the elements of the dipole tensor in the saddle point, if the dipole tensor in the equilibrium configuration is known from other experiments, e.g., from Huang scattering /4.13,14/ or mechanical relaxation /4.15/. In polycrystals an average over the different components of the elasto diffusion tensor is measured.

To obtain an estimate for the change of the diffusion constant we assume an external strain $\varepsilon = 10^{-4}$ and a value $P^s = 10eV$ for the relevant component of the dipole force tensor. Equation (4.46) then yields a relative change of diffusion constant $\Delta D/D_o \cong 4\%$ at room temperature. Changes of diffusion constants in the percent range are hard to measure because of the limited accuracy of the determination of diffusion constants (see references in Sect. 2.4).

On the other hand, internal strains due to other lattice defects can be much larger for short distances. Thus the local change of the diffusion constant can give an important contribution to the reaction rate as we will discuss in Sect. 7. The stress-induced anisotropy is especially important for irradiation creep as has recently been discussed by SAVINO /4.5/.

5. Phenomenological Theory for Reactions of Point Defects

In the following sections we will discuss the theory of diffusion-limited reactions of point defects. We will always assume that one of the reaction partners is mobile while the other partner - which we call "sink" - is immobile. This is no loss of generality since the reaction of two mobile defects can be described by using the sum of the diffusivities as has, e.g., been shown by WAITE /5.1/. This is plausible since the reaction rate which is proportional to the encounter probability in an homogeneous system is determined by the relative motion of the partners.

In this section we first present the results of a phenomenological theory to define the parameters used to describe reactions. The motion of defects is treated as isotropic continuum diffusion and the interaction between mobile defects and sinks is described by a phenomenological reaction radius R_a: Whenever a defect approaches a sink closer than R_a, an irreversible reaction is assumed to take place instantaneously. Outside of the reaction spheres $V_a = (4\pi/3)\cdot R_a^3$ no interaction is taken into account, i.e., the defects are assumed to diffuse freely.

We shall not consider time-dependent problems since the influence of lattice structure and long-range interaction on the reaction radius can most simply be demonstrated for the stationary case. It can be shown /5.2/ that a large portion of the time dependence of diffusion-limited reactions can very well be described by the long time behaviour which is entirely determined by the parameters calculated from stationary theory.

In Sect. 6 we consider the influence of the jump diffusion on a discrete lattice (but no long-range interaction) and show that the concept of the reaction radius with the use of continuum diffusion remains valid. It is a very good approximation even if the reaction volume contains just one lattice site /5.3/. We shall derive explicit expressions for the reaction radius in terms of the Green's function for stationary diffusion on a lattice which has been discussed in Sect. 2.3.

In Sect. 7 we take into account the long-range interaction between point defect and sink. We shall show that the drift of the mobile defect leads to a temperature dependence of the effective reaction radius. With the help of a variation principle we also give approximate results for realistic anisotropic interactions /5.4/.

As explicit examples we will consider the recombination of a Frenkel pair, i.e., the reaction of a mobile interstitial with an immobile vacancy, and the reactions of point defects (vacancies or interstitials) with dislocations.

5.1 Stationary Diffusion and Boundary Condition

In this section we describe the motion of defects by the macroscopic diffusion equation and the interaction between mobile defects and sinks by a phenomenological reaction radius R_a. Since we are concerned with stationary problems only, the relevant equation for the stationary defect distribution $\rho(\underline{R})$ is

$$\frac{d}{dt}\rho = 0 = D_o \Delta \rho(\underline{R}) + \Pi(\underline{R}) \tag{5.1}$$

with D_o the diffusion constant and $\Pi(\underline{R})$ the number of defects produced per unit time at $\underline{R}$. Equation (5.1) has to be solved subject to boundary conditions and production terms appropriate for the physical situation.

The boundary condition at the sink depends on the details of the microscopic configurations of the complex formed and can strictly be justified from microscopic considerations only. Macroscopically the most general boundary condition is /5.5/

$$D_o \left.\frac{d\rho}{dR}\right|_{R_a} = a_o\left(\eta\rho(R_a) - \eta_e\rho_e\right) \quad \text{at } R_a \tag{5.2}$$

which relates the current into the sink with the concentration difference across the boundary. a_o is the lattice constant, η and η_e are jump rates across the surface from the outside and inside, respectively, and ρ_e is the defect concentration inside the sink. The first term on the rhs of (5.2) is the number of particles (per unit time and area) which crosses the sink surface to form complexes, the second term describes the re-emission of defects because complexes dissociate. We shall assume in the following that dissociation is not possible and neglect the second term. This is justified if the binding energy exceeds the thermal energy kT. If, in addition, the jump rate η is large ($\eta a_o^2/D_o \gg 1$), i.e., if all defects which arrive at the sink by diffusion can immediately react, (5.2) simplifies to the absorption boundary condition

$$\rho(R_a) = 0 \tag{5.3}$$

and the reaction is called diffusion-limited. For $\eta = 0$ the surface is impenetrable, i.e., $d\rho/dR = 0$. Any intermediate case describes some hinderence of the penetration. In the limit of small η ($\eta a_o^2/D_o \ll 1$) the reaction is called rate limited because the number of reacting particles is then determined by the jump rate across the surface and not by the number of defects arriving by diffusion. One can show /5.6/ that a finite η can be accounted for by using a smaller effective reaction radius (with possible temperature dependence)

$$R'_a = R_a\left(1 + \frac{D_o}{a_o\,\eta\,R_a}\right)^{-1} \tag{5.4}$$

with the boundary condition (5.3).

In the following we will use the boundary condition (5.3) for simplicity. Then the number of reacting defects per unit time is given by the total flux through the surface of the reaction volume which for a single sink is given by

$$J = 4\pi R_a^2 D_o \left.\frac{d\rho}{dR}\right|_{R_a} . \tag{5.5}$$

We shall consider the following problems:

1) The reaction probability of a defect which is produced at a finite distance from the sink at R = 0, i.e.,

$$\Pi(\underline{R}) = \frac{\delta(R - R_p)}{4\pi R_p^2} ; \quad \rho(R \to \infty) = 0 . \tag{5.6}$$

This case is relevant, e.g., for the recombination of Frenkel pairs in metals produced by high energy irradiation /5.7/ and for the recombination of an electron-hole pair produced by photoeffect in a semiconductor or ionic crystal /5.8/.

2) The reaction rate per unit time of mobile defects with a sink at R = 0 when the defect density is fixed to a constant value at infinity, i.e.,

$$\Pi(\underline{R}) = 0 , \quad \rho(R \to \infty) = \rho_\infty . \tag{5.7}$$

This boundary condition is called the "independent sink approximation". It can, e.g., be used to describe the steady-state growth of precipitates from a homogeneous solution of defects for low concentrations of nuclei. Also a large portion of the time dependence of defect reactions can be described with this reaction rate if the density of sinks is small /5.2/.

3) For finite sink densities the competition between sinks has to be taken into account. A particularly simple and mathematically tractable model is the so-called "spherical cell approximation" which assigns a reaction cell to each sink by imposing a boundary condition at finite distances, e.g.,

$$\left.\frac{d\rho}{dR}\right|_{R_B} = 0 \; ; \quad \Pi(\underline{R}) = P \quad \text{for } R_a \leqslant R \leqslant R_B \; . \tag{5.8}$$

By relating the steady-state flux into the sink to the average steady state density $\bar{\rho}$ of mobile particles in the cell an effective reaction radius can be defined. For $R_B \to \infty$, the situation becomes equivalent to the independent sink approximation discussed in 2).

Another possibility is to fix the density of mobile particles at the boundary and then calculate the reaction rate

$$\rho(R_B) = \rho_B \; ; \quad \Pi(\underline{R}) = 0 \; . \tag{5.8a}$$

By relating ρ_B to the average defect density again an effective reaction radius can be defined. Obviously for $R_B \to \infty$ and $\rho_B = \rho_\infty$ the situation becomes identical to 2).

A boundary condition at finite distances such as (5.8) or (5.8a) has to be used for reactions in less than three dimensions, e.g., for the two-dimensional problem of the reaction of point defects with straight dislocations. For less than three dimensions the situations discussed under 1) and 2) yield results which are independent of the size of the sink. The reaction probability calculated with the boundary condition (5.6) turns out to be unity and the reaction rate for the boundary condition (5.7) becomes infinite.

The spherical cell approximation has actually been used to calculate the sink density dependence of the reaction rate /5.2,9/. The assignment of a cell of definite size to each sink implies a very strict correlation for the distribution of sink sites, namely a regular lattice. Thus the resulting sink density dependence of the reaction rate cannot be expected to be valid in general. Indeed, recent calculations of the sink density dependence of the reaction rate for random distributions of sink centers /5.10-12/ yield a different dependence. We shall briefly discuss the dependence of the reaction rate on the sink density in Sect. 5.5.

Equation (5.1) is formally Poisson's equation and there is a close analogy between stationary diffusion and electrostatics. The corresponding quantities are listed in Table 2. This analogy has very efficiently been used to solve problems of

Table 2 Analogy between Stationary Diffusion und Electrostatics

Diffusion	Electrostatics
density $\rho(\underline{R})$	electrostatic potential $V(\underline{R})$
production rate $\Pi(\underline{R})$	charge density $n(\underline{R})$
current $\underline{j}(\underline{R})$	electrical field $\underline{E}(\underline{R})$
surface of reaction volume	equipotential surface $V = 0$
reaction rate J	total induced charge on grounded conductor Q
effective reaction radius R_a	capacitance C of grounded conductor

diffusion-limited reactions. For instance, the reaction rate for sinks of various shape has been calculated /5.2/ and the problem of the dependence of the reaction rate on the sink density /5.10,11/ has been treated.

5.2 Reaction Probability of a Single Defect with a Single Sink

We want to calculate the probability that a mobile defect reacts with a sink, if the initial separation is R_p. As an example we consider a Frenkel pair (interstitial - vacancy pair). The interaction is described by a phenomenological reaction radius R_a (see Fig. 5.1). Rather than integrating the time-dependent flux of interstitials

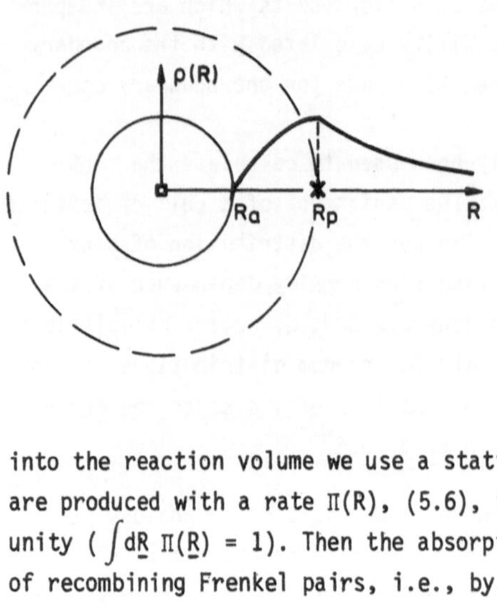

Fig. 5.1. Stationary distribution of interstitials (*) produced at a distance R_p from a vacancy (□). The reaction sphere with radius R_a is indicated by the full circle, equivalent production sites are on the dotted circle

into the reaction volume we use a stationary approach: We suppose that Frenkel pairs are produced with a rate $\Pi(R)$, (5.6), independent of time. $\Pi(R)$ is normalized to unity ($\int d\underline{R}\ \Pi(\underline{R}) = 1$). Then the absorption probability A is given by the fraction of recombining Frenkel pairs, i.e., by the flux of interstitials per unit time into the recombination volume surrounding the vacancy. The stationary distribution of interstitials is governed by the steady-state diffusion equation (5.1). Since Π is different from zero only on the surface of the sphere with radius R_p the flux through any spherical surface with $R < R_p$ is constant and equal to the absorption probability A (since Π is normalized). For $R > R_p$ the flux is equal to another constant B, the escape probability

$$4\pi R^2\, j_R(R) = -4\pi D_o R^2\, \partial_R \rho(R) = \begin{cases} A & R_a \leq R \leq R_p \\ B & R_p \leq R\,. \end{cases} \tag{5.9}$$

This equation can be integrated to yield the stationary interstitial distribution which obeys the boundary conditions (5.3,6) at $R = R_a$ and $R \to \infty$, respectively,

$$\begin{aligned} \rho_i(R) &= \frac{A}{4\pi D_o R_a}\left(1 - \frac{R_a}{R}\right) \qquad & R_a \leq R \leq R_p \\ \rho_a(R) &= \frac{B}{4\pi D_o R} & R_p \leq R\;. \end{aligned} \tag{5.10}$$

The two constants A and B are determined by the continuity conditions at R_p

$$\rho_i(R_p) = \rho_a(R_p) \tag{5.11a}$$

$$4\pi R_p^2\big(j_a(R_p) - j_i(R_p)\big) = 1\;. \tag{5.11b}$$

This yields for the absorption probability

$$A = \frac{R_a}{R_p} \tag{5.12a}$$

and for the escape probability

$$B = 1 - A = 1 - \frac{R_a}{R_p}\;. \tag{5.12b}$$

We shall use (5.12a) to define an effective reaction radius when we consider jump diffusion on a discrete lattice in Sect. 6.

5.3 Independent Sink Approximation

To calculate the reaction rate for a homogeneous distribution of sinks is a complicated many-particle problem since in general one has to take into account the competition of sinks. In a very dilute system, however, when the average distance between the sinks is much larger than the absorption radius R_a, one can neglect the competition between sinks and consider the absorption at each sink separately /5.13/. This is called the "independent sink approximation".

The solution of the steady-state diffusion equation (5.1) with the boundary conditions (5.3,7), i.e., for a uniform defect density at infinity yields the following distribution of defects around a sink at the origin

$$\rho(R) = \rho_\infty(1 - \frac{R_a}{R}) \qquad (5.13)$$

which is plotted in Fig. 5.2. The number of particles reacting with the sink per unit time, calculated according to (5.5) yields

$$J = K\cdot\rho_\infty = 4\pi D_o R_a \rho_\infty \qquad (5.14)$$

which defines the rate constant in the independent sink limit

$$K = 4\pi D_o R_a \ . \qquad (5.15)$$

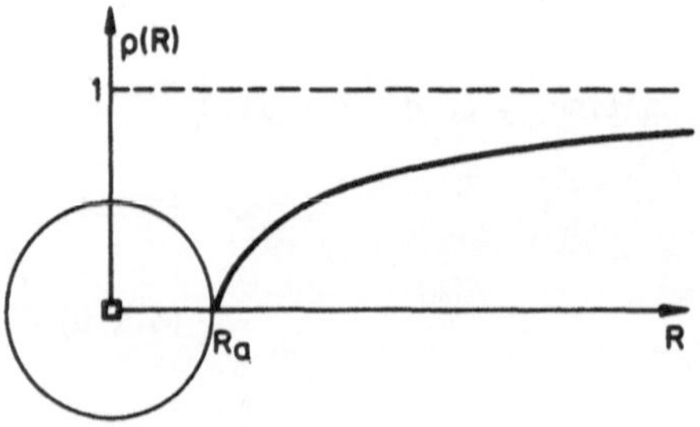

Fig. 5.2. Stationary distribution of defects around a single sink (□) in the dilute limit. The reaction sphere with radius R_a is indicated by the circle. At infinity the defect density is normalized to unity

On a macroscopic scale the equation for the average defect density $\bar{\rho}$ becomes

$$\frac{\partial}{\partial t}\,\bar{\rho}(\underline{R},t) = D_o\,\Delta\bar{\rho}(\underline{R},t) - K\rho_s\bar{\rho}(\underline{R},t) \qquad (5.16)$$

where ρ_s is the sink density.

We shall use (5.15) to discuss the influence of long-range interaction fields on the rate constant in Sect. 7.

5.4 Reaction of Point Defects with Straight Dislocations

We shall use the boundary condition (5.8) to calculate the reaction rate of mobile defects with straight dislocations to the lowest order in the dislocation density. The reaction at the dislocation line is described by the absorption boundary condition (5.3) at the cylindrical core of the dislocation line. The external radius R_B is connected to the dislocation density ρ_D. It can, e.g., be identified with the average radius of a cylindrical cell assigned to the dislocations which is given by /5.2/

$$\pi R_B^2 \rho_D = 1 . \tag{5.17}$$

For two dimensions the stationary diffusion equation reads

$$D_o \frac{1}{r} \partial_r r \rho(r) + p = 0 \quad \text{for } R_a \leq r \leq R_B \tag{5.18}$$

with the solution satisfying the boundary conditions (5.3,8)

$$\rho(r) = \frac{p}{2D_o} R_B^2 \left\{ \ln \frac{r}{R_a} + \frac{1}{2}\left(\frac{R_a}{R_B}\right)^2 \left[1 - \left(\frac{r}{R_a}\right)^2 \right] \right\} . \tag{5.19}$$

The total flux per unit time into a cylinder of unit length is given by the total production inside the cylindrical cell

$$J = 2\pi R_a D_o \partial_r \rho(r) \Big|_{R_a} = \pi R_B^2 p \left[1 - \left(\frac{R_a}{R_B}\right)^2 \right] . \tag{5.20}$$

Relating this to the average stationary defect density

$$\rho = 2\pi \int_{R_a}^{R_B} dr\ r\ \rho(r) \Big/ \left(2\pi \int_{R_a}^{R_B} dr\ r\right)$$

one obtains for the reaction rate K_D to lowest order in the dislocation density ρ_D

$$K_D = \frac{J}{\rho} = \frac{2\pi D_o}{\ln(R_B/R_a)} . \tag{5.21}$$

HAM /5.2/ has derived this result for the rate constant K_D more rigorously using an eigenfunction expansion of the defect distribution for cylindrical symmetry.

The same result for the rate constant K_D to lowest order in the dislocation density is obtained with the boundary condition (5.8a) /5.14/. In this case the reacting defects are provided by an infinite source outside the cell. Although the stationary defect distributions in the cell $r \leq R_B$ are quite different for the two boundary conditions, as can be seen in Fig. 5.3, the gradient at the dislocation core $r = R_a$ is the same, thus yielding the same rate constant.

For this two-dimensional case the rate constant K_D is not independent of the sink density even in the case of low sink densities ($\pi R_a^2 \rho_D \ll 1$). As can be seen in Fig. 5.3 however, for large R_B/R_a the defect density $\rho(r)$, (5.19), is slowly varying at large r and thus for all practical purposes the concept of independently acting sinks is a good approximation at low dislocation densities. We shall use the result, (5.21), as a reference when we discuss the influence of a long-range interaction between dislocations and point defects on the reaction rate (Sect. 7).

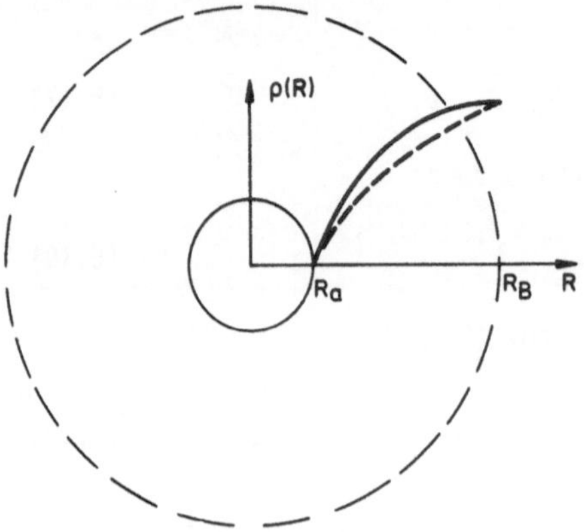

Fig. 5.3. Stationary defect distribution in a cylindrical cell around a straight dislocation. The absorption cylinder at the core with radius R_a is indicated by the full circle, the cell with radius R_B by the dotted circle.
——— for boundary condition (5.8) with homogeneous production inside the cell.
– – – for boundary condition (5.8a) with defect density fixed at R_B

5.5 Finite Sink Densities

5.5.1 Spherical Cell Approximation

A simple estimate for the influence of larger sink densities ρ_s on the rate constant K can be obtained by the spherical cell approximation. Each sink is placed into a spherical cell whose radius is determined by

$$\frac{4\pi}{3} R_B^3 \rho_s = 1 \; , \tag{5.22}$$

i.e. the entire sample volume is covered with cells.

Within this cell one then solves the stationary diffusion equation

$$D_o \Delta \rho(R) + p = 0 \qquad R_a \leq r \leq R_B \tag{5.23}$$

with the boundary conditions (5.3,8) yielding

$$\rho(R) = \frac{p}{3D_o} \frac{R_B^3}{R_a} \left\{ 1 - \frac{R_a}{R} + \frac{1}{2}\left(\frac{R_a}{R_B}\right)^3 \left[1 - \left(\frac{R}{R_a}\right)^2 \right] \right\} . \tag{5.24}$$

Because of the impenetrable outer surface of the cell implied by the boundary condition (5.8) the absorption rate per unit time is equal to the total production within the cell

$$J = 4\pi D_o R_a^2 \, \partial_R \, \rho(R)\Big|_{R_a} = \frac{4\pi}{3} R_B^3 \, p \left[1 - \left(\frac{R_a}{R_B}\right)^3 \right] . \tag{5.25}$$

However, one can define an effective rate constant by relating the absorption flux to the average steady-state density $\overline{\rho}$

$$K(\rho_s) = J/\overline{\rho} \; . \tag{5.26}$$

This yields a rate constant which depends on the radius of the cell and via (5.22) on the sink density ρ_s. The first two terms of an expansion in terms of the sink density yield

$$K(\rho_s) = 4\pi D_o R_a \left[1 + \frac{9}{5} \left(\frac{4\pi}{3} R_a^3 \rho_s\right)^{1/3} + O(\rho_s^{2/3})\right] . \qquad (5.27)$$

A similar result has been obtained by STRIEDER and ARIS /5.9/ using a variational principle set up by HAM /5.2/. They obtained an upper bound of 11.3 for the factor in front of the second term in the bracket of (5.27) (instead of 9/5 in our result).

It is important to note that for larger sink densities the rate constant increases beyond the independent sink value. The additional terms represent corrections due to the overlap of the defect concentration gradients around the individual sink. The dependence of the correction term on the cube root of the sink density in (5.27) is characteristic of the cell approximation which implies very strict correlations in the distribution of sink centers. As we will see in the next section for a random sink distribution the first correction term is proportional to the square root of the sink density.

5.5.2 Random Distribution of Sinks

Recently the dependence of the reaction rate on the sink concentration for randomly distributed sinks has very thoroughly been treated by FELDERHOF and DEUTCH /5.10/, and independently by BRAILSFORD /5.11/. In order to establish the first correction term to the independent sink approximation we can follow a much simpler approach which was, e.g., used by PEAK et al /5.12/.

Consider a single sink embedded in an absorbing medium containing sinks with density ρ_s. The absorption rate $a(\rho_s)$ (per unit time) is supposed to be known and parameterized by a density-dependent reaction rate

$$a(\rho_s) = K(\rho_s)\cdot\rho_s . \qquad (5.28)$$

A steady-state defect distribution is established by a homogeneous production which replaces all reacting particles

$$p = a(\rho_s)\cdot\rho_\infty \qquad (5.29)$$

where ρ_∞ is the sample average defect density which is reached far away from each sink. The stationary defect density which obeys the equation

$$0 = D_o \Delta \rho(R) - a(\rho_s)\, \rho(R) + p \qquad (5.30)$$

and satisfies the absorption boundary condition, (5.3), at the edge of the considered sink is given by

$$\rho(R) = \rho_\infty \left(1 - \frac{R_a}{r} \exp[-\beta(r - R_a)]\right) ; \quad \beta = \left(\frac{a(\rho_s)}{D_o}\right)^{1/2} . \qquad (5.31)$$

The exponential describes the shielding due to the average absorption by the other sinks.

The total absorption rate (per unit time and unit volume) is given by the flux of defects into the considered sink multiplied with the density of the sinks since each sink acts in the same way; it is of course equal to the production rate, (5.29). This yields

$$a(\rho_s)\cdot\rho_\infty = 4\pi R_a^2\rho_s D_o\,\partial_R\,\rho(R)\Big|_{R_a} = 4\pi R_a D_o\rho_s\rho_\infty\left[1+\left(\frac{a(\rho_s)\cdot R_a^2}{D_o}\right)^{1/2}\right] . \qquad (5.32)$$

This equation for $a(\rho_s) = K(\rho_s)\cdot\rho_s$ has been used by BRAILSFORD et al. /5.15/ to define the loss rate of defects in an effective medium. We shall restrict ourselves to the first two terms of an expansion of $K(\rho_s)$ for small sink densities

$$K(\rho_s) \cong 4\pi D_o R_a\left[1 + (4\pi R_a^3\rho_s)^{1/2} + O(\rho_s)\right] . \qquad (5.33)$$

The first term is the "independent sink approximation", obtained in (5.15), whereas the next terms are corrections due to the competition of sinks for higher sink densities. The $\rho_s^{1/2}$ dependence is characteristic for a random sink distribution.

FELDERHOF and DEUTCH /5.10/, and BRAILSFORD /5.11/ have developed a systematic expansion of $K(\rho_s)$ using the electrostatic analogon for the stationary diffusion problem (see Table 2). They take into account induced charges and dipole layers on one sink due to the presence of the other sinks. The more general boundary condition (5.2), describing incomplete absorption, can also be incorporated into the theory /5.10/. All authors agree to the first two terms of (5.33), for higher order terms differences appear owing to the different approximations used.

6. Lattice Theory for the Reaction Probability

With a phenomenological reaction radius R_a to describe the interaction of a mobile defect with a sink and using continuum theory to describe the motion of the defect we have in Sect. 5.2 obtained the expression

$$A(R_p) = \frac{R_a}{R_p} \qquad (6.1)$$

for the probability that the defect reacts with the sink if it starts its diffusion path at a distance R_p. An analysis of experimental data on diffusion annealing of irradiation-produced defects in metals, like vacancies and interstitials, generally yields reaction radii of the order of a few lattice constants and due to the small initial distances rather short diffusion paths /6.1/. This raises the question whether the use of continuum theory is adequate. Using Monte Carlo methods several authors

/6.2-4/ have shown that it is indeed possible to fit the parameters of a continuum theory of diffusion-limited reactions to the results of random walk models on discrete lattices. In this section we shall systematically examine the dependence of the reaction probability on the shape and size of microscopic reaction regions describing the motion of the defect by a random walk on discrete lattice. The presentation will follow rather closely the treatment of our earlier paper on this subject /6.5/.

It turns out that (1.1) is the asymptotic form of the exact expression for $A(R_p)$ and in general is a very good approximation.

We shall consider steady-state situations only which are established by time-independent production of defects.

In this section we do not take into account any disturbance of the defect jump rates due to external or internal fields, and consider nearest-neighbour jumps only (see Sect. 2). The defect distribution $w^{\underline{m}}$, i.e., the mean number of defects on site $\underline{m}$ is determined by the equation

$$0 = \sum_{\underline{n}} \lambda^{(\underline{m}-\underline{n})}(w^{\underline{n}} - w^{\underline{m}}) + \Pi^{\underline{m}} . \tag{6.2}$$

To describe a reaction we shall assume a reaction region around the sink consisting of lattice sites $\{\nu\}$. Whenever a defect jumps to one of these sites a reaction takes place instantaneously. Thus the boundary condition at the sink reads

$$w^{\underline{\mu}} = 0 \quad \text{for any site } \mu \in \{\nu\} . \tag{6.3}$$

The simplest reaction region consists of just one site, e.g., $\underline{\mu} = 0$.

If we introduce absorption rates $a^{\underline{m}}$ into (6.2) which are determined to satisfy the boundary condition (6.3) we can assume ideal random walk on the entire lattice. The stationary diffusion equation then reads

$$0 = \sum_{\underline{n}} \lambda^{(\underline{m}-\underline{n})}(w^{\underline{n}} - w^{\underline{m}}) + \Pi^{\underline{m}} - a^{\underline{m}} . \tag{6.4}$$

For our nearest-neighbour jump model the absorption rates $a^{\underline{m}}$ are different from zero only on surface sites of the reaction region, i.e., sites which can be reached by nearest-neighbour jumps from outside. The inner sites are shielded by these surface sites and the boundary condition is satisfied automatically. $a^{\underline{\mu}}$ is the rate (per unit time) with which a defect reaches the site $\underline{\mu}$ of the reaction region and thus the reaction probability A is given by the sum of all $a^{\underline{\mu}}$ divided by the number of defects produced per unit time. We shall use normalized production (one defect per unit time), i.e.,

$$\sum_{\underline{m}} \Pi^{\underline{m}} = 1 . \tag{6.5}$$

Then the reaction probability is given by

$$A = \sum_{\underline{\mu}} a^{\underline{\mu}} \tag{6.6}$$

where the sum runs over all surface sites of the reaction region $\{\nu\}$.

6.1 General Expression for the Reaction Probability

Equation (6.4) can be solved using the Green's function for stationary diffusion $G_{st}^{(n)}$ discussed in Sect. 2.3.1

$$w^{\underline{m}} = \sum_{\underline{n}} G_{st}^{(\underline{m}-\underline{n})} (\Pi^{\underline{n}} - a^{\underline{n}}) \ . \tag{6.7}$$

From the boundary condition (6.3) we obtain a system of linear equations for the absorption rates $a^{\underline{\mu}}$

$$\sum_{\underline{\mu}} G_{st}^{(\underline{\nu}-\underline{\mu})} a^{\underline{\mu}} = \sum_{\underline{m}} G_{st}^{(\underline{\nu}-\underline{m})} \Pi^{\underline{m}} \tag{6.8}$$

where the sum on the lhs runs over all surface sites of the reaction region and the sum on the rhs runs over all possible production sites outside of the reaction. The absorption rates $a^{\underline{\mu}}$ can thus be determined by a finite matrix inversion

$$a^{\underline{\mu}} = \sum_{\underline{\nu}} (\tilde{g}^{-1})^{\underline{\mu}\underline{\nu}} \sum_{\underline{m}} G_{st}^{(\underline{\nu}-\underline{m})} \Pi^{\underline{m}} \tag{6.9}$$

where $\tilde{g}$ is the projection of the stationary Green's function to the surface sites of the reaction region. The reaction probability is given by

$$A = \sum_{\underline{\mu}} a^{\underline{\mu}} = \sum_{\underline{\mu}\underline{\nu}} (\tilde{g}^{-1})^{\underline{\mu}\underline{\nu}} \sum_{\underline{m}} G_{st}^{(\underline{\nu}-\underline{m})} \Pi^{\underline{m}} \ . \tag{6.10}$$

For the simplest reaction region containing only the site $\mu = 0$ and production of defects on site $\underline{R}^{\underline{p}}$ ($\Pi^{\underline{m}} = \delta_{\underline{m},\underline{p}}$) only one absorption rate $a^{\underline{o}}$ has to be determined and the absorption probability is given by

$$A(R^{\underline{p}}) = a^{\underline{o}} = \frac{G_{st}^{(\underline{p})}}{G_{st}^{(o)}} \quad ; \quad \{\nu\} = \{\underline{0}\} \ . \tag{6.11}$$

In general the dimension of the matrix $\tilde{g}$ is equal to the number of surface sites of the reaction region. However, by symmetry arguments one can reduce the number of unknown $a^{\underline{\mu}}$ and thus the dimension of the matrix to be inverted. For instance, if the diffusion lattice and the reaction region both have cubic symmetry the reaction probability is the same for production of defects on any of the equivalent sites of a "shell" which are obtained from the site $\underline{R}^{\underline{p}}$ by application of one of the 48 cubic symmetry operations $\underline{\underline{S}}$

$$\underline{R}^{\underline{p}s} = \underline{\underline{S}}\,\underline{R}^{\underline{p}} \ . \tag{6.12}$$

If the point symmetry of the reaction region and of a lattice point are different only the common symmetry operations can be used to define a shell of equivalent sites.

For symmetrized production, i.e., equal production rate on all sites of a shell, the absorption rates $a^{\underline{\nu}}$ on equivalent sites of a surface shell are equal. This reduces the number of $a^{\underline{\nu}}$ to be determined by about an order of magnitude.

As an example we consider a cubic reaction region in a cubic lattice containing the site $\underline{R} = 0$ and the z_1 sites of the first neighbour shell (see Fig. 6.1). The

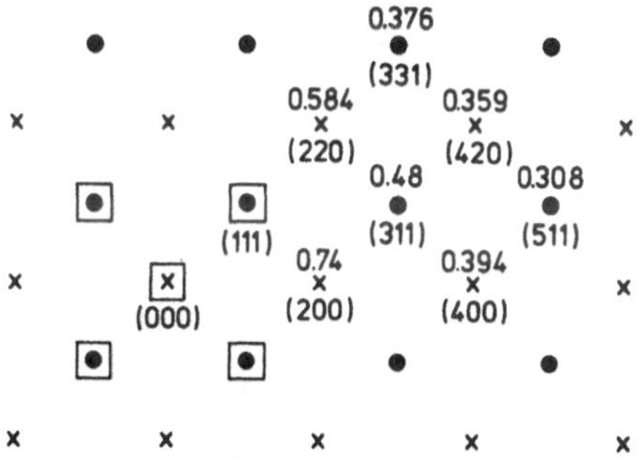

Fig. 6.1. Reaction probability $A(\underline{R}^{\underline{p}})$ in a bcc lattice for a recombination region containing the origin and the first neighbour shell (□). Shown is the (001)-plane (×) and the plane above (•). The numbers for $A(\underline{R}^{\underline{p}})$ are entered above each lattice site $\underline{p}$

surface of this region consists of the z_1 sites of the first neighbour shell which shield the site $\underline{R} = 0$ completely. The symmetrized (and normalized) production on all sites equivalent to $\underline{R}^{\underline{p}}$ is

$$\Pi_s^{\underline{m}} = \frac{1}{48} \sum_s \delta_{\underline{m},\underline{p}s}$$

where the sum is over the 48 cubic symmetry operations. The inhomogeneity on the rhs of (6.8) then reads

$$I^{\underline{\nu}} = \sum_{\underline{m}} G_{st}^{(\underline{\nu}-\underline{m})}\, \Pi_s^{\underline{m}} = \frac{1}{48} \sum_s G_{st}^{(\underline{\nu}-\underline{p}s)} \ .$$

Using the invariance of the Green's function for a cubic lattice against cubic symmetry operations, i.e.,

$$G_{st}^{(\underline{m}s-\underline{n}s)} = G_{st}^{(\underline{m}-\underline{n})}$$

for any cubic symmetry operation S, we can transform the $I^{\underline{\nu}}$ further

$$I^{\underline{\nu}} = \frac{1}{48} \sum_{s} G_{st}^{(\underline{\nu}_s - \underline{p})} .$$

The sites $\underline{\nu}_s$ are the z_1 sites of the first neighbour shell. They are generated $48/z_1$ times by the 48 cubic symmetry operations and thus we can finally write

$$I^{\underline{\nu}} = \frac{1}{z_1} \sum_{\underline{\nu} \in N.N.} G_{st}^{(\underline{\nu} - \underline{p})}$$

which shows that the inhomogeneity is independent of the specific site $\underline{\nu}$ considered. Due to the symmetrization of the production over a cubic shell all sites $\underline{\mu}$ of the first neighbour shell have the same environment. Thus the absorption terms $a^{\underline{\mu}}$ are all equal and only one constant, say $a^{(1)}$, has to be determined. Equation (6.8) reduces to

$$a^{(1)} \sum_{\underline{\mu}} G_{st}^{(\underline{\nu} - \underline{\mu})} = \frac{1}{z_1} \sum_{\underline{\nu}} G_{st}^{(\underline{\nu} - \underline{p})} . \tag{6.13}$$

From the difference equation for the stationary Green's function for nearest neighbour jumps (jump rate Γ)

$$z_1 \, \Gamma \, G_{st}^{(\underline{m})} - \Gamma \sum_{\underline{n} \in N.N.} G_{st}^{(\underline{n} - \underline{m})} = \delta_{m,0} \tag{6.14}$$

we obtain for $\underline{m} \neq 0$

$$\frac{1}{z_1} \sum_{\underline{n} \in N.N.} G_{st}^{(\underline{n} - \underline{m})} = G_{st}^{(\underline{m})} .$$

Using this in (6.13) the equation for $a^{(1)}$ reads

$$z_1 a^{(1)} G_{st}^{(1)} = G_{st}^{(\underline{p})}$$

and the total absorption probability is given by

$$A(\underline{R}^{\underline{p}}) = z_1 a^{(1)} = \frac{G_{st}^{(\underline{p})}}{G_{st}^{(1)}} \qquad \text{for } \{\nu\} = \{\underline{0}, \text{ first neighbour shell}\} \tag{6.15}$$

with $G_{st}^{(1)}$ the stationary Green's function for a site of the first neighbour shell.

By these symmetry considerations the dimension of the matrix $\tilde{g}$ can be reduced to the number of surface shells, because only one representative site of each shell needs to be considered. In our example with only one surface shell $\tilde{g}$ becomes proportional to a unity matrix, i.e., it can be reduced to a scalar. Larger reaction regions contain more than one surface shell. In the three cubic lattices (fcc, bcc, sc) the number of surface shells (and thus the dimension of the matrix $\tilde{g}$) is less than four for reaction regions containing up to about 100 lattice sites which are formed by successively adding further neighbour shells. Details of the reduction method can be found in our earlier report /6.6/.

6.2 Asymptotic Form

For large distances $\underline{R}^{\underline{p}}$ of the production sites from the sink one can replace $G_{st}^{(\underline{\nu}-\underline{m})}$ in (6.8-10) by the asymptotic form $G_{st}^{o}(\underline{R}^{\underline{m}})$, given in (2.34)

$$G_{st}^{(\underline{\nu}-\underline{m})} \rightarrow G_{st}^{o}(\underline{R}^{\underline{m}}) = \frac{V_c}{4\pi D_o |\underline{R}^{\underline{m}}|} \;. \tag{6.16}$$

We thus obtain the asymptotic form for the reaction probability

$$A(\underline{R}^{\underline{p}}) \sim \frac{R_a}{|\underline{R}^{\underline{p}}|} \tag{6.17}$$

with

$$R_a = \sum_{\underline{\mu}\underline{\nu}} (\tilde{g}^{-1})^{\mu\nu} \frac{V_c}{4\pi D_o} \;. \tag{6.18}$$

We call R_a the "effective reaction radius" because a reaction sphere of radius R_a yields the same reaction probability when continuum theory of diffusion is used (see Sect. 5.2).

6.3 Variational Principle for Calculating R_a

The final expression for the effective reaction radius, (6.18), still involves the inversion of the matrix $\tilde{g}$. One can use a variational principle to obtain a simpler expression for R_a. Using the asymptotic form, (6.16), for the stationary Green's function on the rhs of (6.8) this reads

$$\sum_{\underline{\mu}} G_{st}^{(\underline{\nu}-\underline{\mu})} a^{\underline{\mu}} = G_{st}^{o}(\underline{R}^{\underline{p}}) \;. \tag{6.19}$$

This equation, which determines the asymptotic reaction probability and the value of R_a can be obtained from a variational principle

$$\frac{\delta H\{a^{\underline{\nu}}\}}{\delta a^{\underline{\nu}}} = 0 \tag{6.20}$$

with the functional

$$H\{a^{\underline{\nu}}\} = \frac{1}{2} \sum_{\underline{\mu}\underline{\nu}} a^{\underline{\nu}} G_{st}^{(\underline{\nu}-\underline{\mu})} a^{\underline{\mu}} - \sum_{\underline{\mu}} G_{st}^{o}(\underline{R}^{\underline{p}}) a^{\underline{\mu}} \;. \tag{6.20a}$$

The sums run over all surface sites of the reaction region. The value of H for the exact solution of (6.19) is given by

$$H_o = -\frac{1}{2} G_{st}^{o}(\underline{R}^{\underline{p}}) A(\underline{R}^{\underline{p}}) \;.$$

Since $G_{st}^{(\underline{\mu}-\underline{\nu})}$ is a positive definite matrix, any approximation for the absorption rates will yield a value of H larger than the exact value and thus a lower limit for the asymptotic reaction probability.

The simplest ansatz is to set all absorption rates equal to a constant independent of the shell

$$a^{\underline{\mu}} = \alpha \tag{6.21}$$

and determine α from the variation procedure, (6.20). For reaction regions with only one surface shell this yields the exact result (if one uses symmetrized production).

If we insert (6.21) into (6.19) and solve the equation for α resulting from (6.20) we obtain

$$\alpha = \frac{z_s}{\sum\limits_{\underline{\mu}\underline{\nu}} G_{st}^{(\underline{\mu}-\underline{\nu})}} G_{st}^{o}(\underline{R}^{\underline{p}}) \tag{6.22}$$

where z_s is the number of surface sites.

The asymptotic reaction probability becomes

$$A^{var}(\underline{R}^{\underline{p}}) = z_s\alpha = \frac{z_s^2}{\sum\limits_{\underline{\mu}\underline{\nu}} G_{st}^{(\underline{\mu}-\underline{\nu})}} G_{st}^{o}(\underline{R}^{\underline{p}}) \tag{6.23}$$

and the effective reaction radius

$$R_a^{var} = \frac{z_s^2}{\sum\limits_{\underline{\mu}\underline{\nu}} G_{st}^{(\underline{\mu}-\underline{\nu})}} \frac{V_c}{4\pi D_o} \tag{6.24}$$

which involves only a summation over elements of the stationary Green's function and no matrix inversion.

For a reaction region in a bcc lattice containing the first and second neighbour shell (which are both surface shells) the exact result for R_a according to (6.18) and the variational result, (6.24), are

$$\begin{aligned} R_a &= 0.9195\ a_o \\ R_a^{var} &= 0.9190\ a_o\ , \end{aligned} \tag{6.25}$$

respectively. The result of the variational calculation is almost exact although the absorption terms on the two shells differ by about 13% in this case.

A further approximation called $R_a^{var,as}$ is obtained if one uses the asymptotic values $G_{st}^{o}(\underline{R}^{\underline{\mu}-\underline{\nu}})$ to calculate the sum in (6.24). Since $G_{st}^{o}(\underline{R} = 0)$ according to (6.16) is infinite we calculate the value for $\underline{R} = 0$ from the difference equation (6.14) for the stationary Green's function using the asymptotic value for the nearest-neighbour Green's function. Due to the very good agreement of the Green's function with the asymptotic expansion no serious error is made.

6.4 Results

6.4.1 Exact Results for the bcc Lattice

We have systematically studied the reaction probability $A(\underline{R}^p)$ for various reaction regions in cubic lattices (bcc, fcc, sc). In Fig. 6.1 the exact results for one example are shown. Plotted is the (001)-plane of a bcc lattice and the plane above. The reaction region considered consists of 9 sites, one shielded center at $\underline{R} = 0$ and 8 nearest-neighbour sites (of which only 4 can be seen). The exact value of $A(\underline{R}^p)$ according to (6.15) is entered above each production site, e.g., for the site $\underline{R}^p = a_o/2$ (420) it is $A = 0.359$. The effective reaction radius calculated with (6.18) is $R_a = 0.81\ a_o$ (a_o = cubic lattice constant) which is approximately equal to the nearest-neighbour distance $R_1 = 0.866\ a_o$ in the bcc lattice which is the linear extension of the considered reaction region. In contrast, the volumes of the reaction region containing 9 lattice sites $V_{lat} = 9V_c$ and of the effective reaction sphere $V_a = (4\pi/3)R_a^3 = 4.5V_c$ differ greatly. This shows that the linear extension of the reaction region is the significant parameter when comparing with continuum theory.

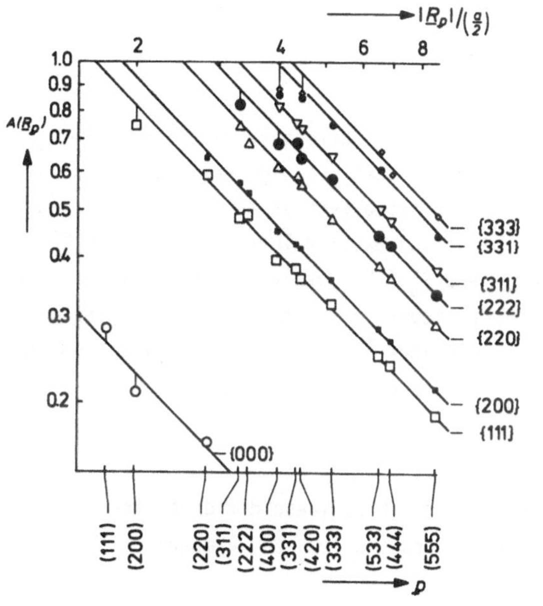

Fig. 6.2. Reaction probability $A(\underline{R}^p)$ vs $|\underline{R}^p|$ in a double logarithmic plot for compact reaction regions in a bcc lattice. (The labelling is explained in the text.) The asymptotic form for $A(\underline{R}^p)$, (6.17), is also shown (straight lines)

In Fig. 6.2 $A(\underline{R}^p)$ is plotted in a double logarithmic plot against the production distance $|\underline{R}^p|$ for a number of reaction regions in the bcc lattice. The different symbols denote different regions which are built up by successively adding further neighbour shells starting from one reaction site at $\underline{R} = 0$. We call such regions "compact". For instance, the second region labelled {111} is obtained by adding the first neighbour {111} shell to the site $\underline{R} = 0$. This is the region considered before for Fig. 6.1. Adding the {200}-shell yields the region labelled {200}, etc. The largest region considered contains 91 sites.

The two largest regions, labelled {331} and {333} are not compact in the strict sense, because there are possible production sites with $|\underline{R}^p| < R_{max}$. We chose them to minimize computational work because they contain only 3 surface shells whereas strictly compact regions of similar size contain 4 surface shells.

6.4.2 Asymptotic Results

The asymptotic form for $A(\underline{R}^p)$, (6.17), shows up as a straight line in the double logarithmic plot of Fig. 6.2, one line for each reaction region. Obviously it is a good approximation for $A(\underline{R}^p)$ even down to small production distances. If the production site is a nearest-neighbour to a surface site of the reaction region a maximal deviation of 10% is found (except for the largest "non-compact" regions where 20% deviation is found if $|\underline{R}^p| < R_{max}$).

For each reaction region the equation $A(R_a) = 1$ defines the effective reaction radius R_a. They are plotted in Fig. 6.3 together with the approximations obtained

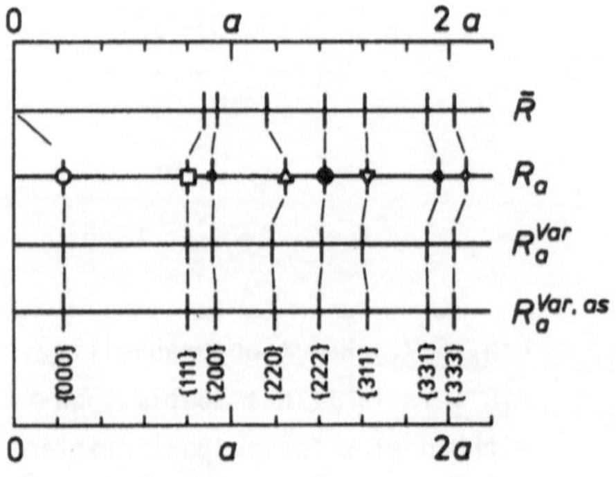

Fig. 6.3. Effective reaction radii R_a for compact reaction regions in bcc lattice (same symbols as in Fig. 6.2). Shown are the exact values R_a, (6.18), R_a^{var} and $R_a^{var,as}$ obtained from the variational procedure of Sect. 6.3 and the average distance $\bar{R}$ of the surface sites from the center of the reaction region

from the variational principle discussed in Sect. 6.3. These differ by less than 5% from the exact values for all reaction regions considered. For comparison also the average distance of the surface sites

$$\bar{R} = \frac{1}{z_s} \sum z_\sigma R_\sigma$$

(z_s = total number of surface sites, z_σ = number of sites in shell σ, R_σ = distance of shell σ from center of region) is plotted. As can be seen $\bar{R}$ can be used as a rather good estimate for the reaction radius R_a. For compact regions one can also use the distance R_{max} of the outermost surface shell.

The reason for the accuracy of the asymptotic form of $A(\underline{R}^p)$ is the close agreement between the exact values of the Green's function $G_{st}^{(\underline{n})}$ for stationary diffusion and its asymptotic expansion even for distances as small as the nearest-neighbour distance (see discussion in Sect. 2.3.1). We thus have shown that the knowledge of the continuum Green's function, (2.34), is sufficient to determine the absorption probability for microscopic reaction regions in a random walk model with a very good accuracy.

6.4.3 Other Cubic Lattices

We have also investigated the reaction probability $A(\underline{R}^p)$ in other cubic lattices for several reaction regions containing up to 79 and 87 sites for fcc and sc lattices, respectively. The results are shown in Figs. 6.4 and 6.5 in a similar plot as Fig. 6.2. Since for fcc and sc lattices fewer exact numerical values for the stationary Green's function are available we have used the asymptotic values $G_{st}^{o}(\underline{R}^n)$ for $n^2 \geq$ 24 and 15 in the fcc and the sc lattice, respectively. However, for these distances the difference to the exact values is far less than 1%, see [Ref. 6.6, Table 2 and 3].

The same accuracy of the asymptotic form of $A(\underline{R}^p)$ is found as in the bcc lattice and the approximations for the effective reaction radius R_a discussed above can be used with the same success.

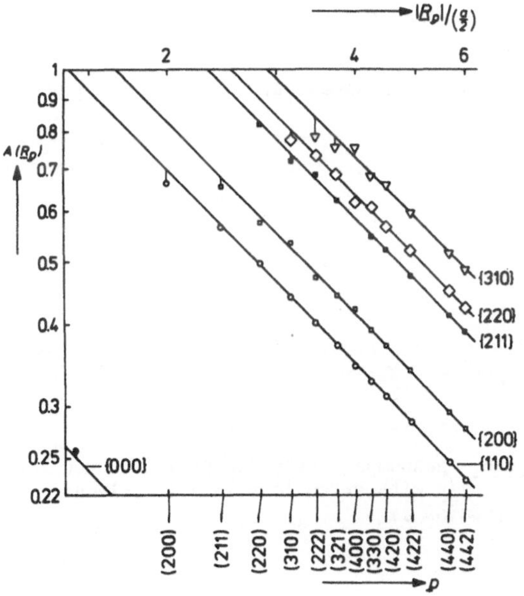

Fig. 6.4. Reaction probability $A(\underline{R}^p)$ for compact reaction regions in an fcc lattice. The plot and notation are the same as for Fig. 6.2

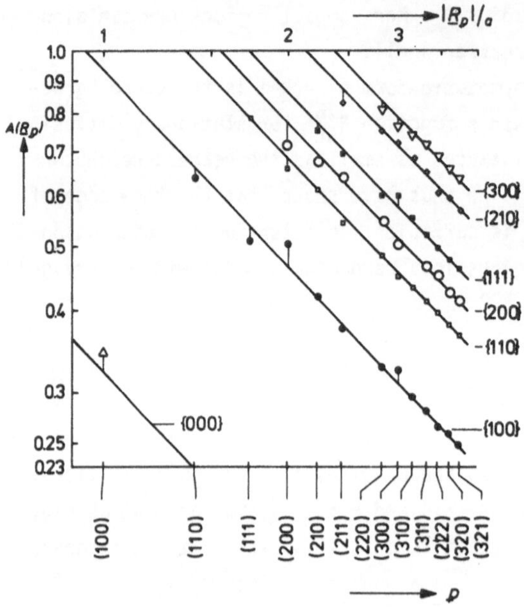

Fig. 6.5. Reaction probability $A(\underline{R}^p)$ for compact reaction regions in an sc lattice. The plot and notation are the same as in Fig. 6.2

6.4.4 Noncompact Reaction Regions

To study the influence of the form of the reaction region we have also investigated noncompact regions in the bcc lattice, namely stars with extensions in the close packed <111>-directions. Region a extends to the {222}-shell (Fig. 6.6) and region b to the {333}-shell. In these cases there are possible production sites with $|\underline{R}^p| < R_{max}$, the distance of the outermost surface shell. The values obtained for $A(\underline{R}^p)$ are plotted in Fig. 6.7 in the same way as in Fig. 6.2 for the compact regions but on a different scale. For $|\underline{R}^p| > R_{max}$ the asymptotic form for $A(\underline{R}^p)$ be-

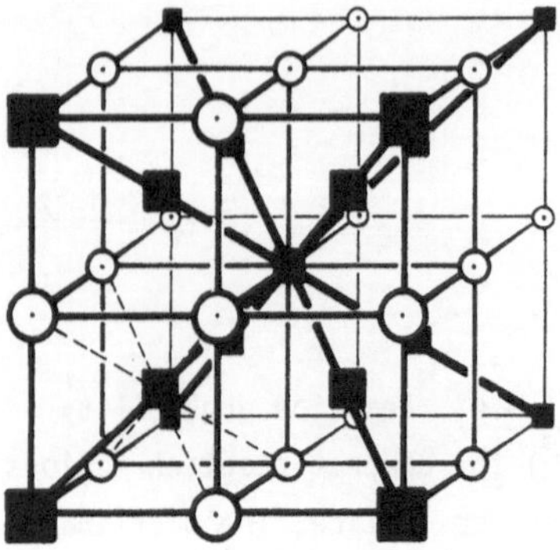

Fig. 6.6. Noncompact reaction region in bcc lattice: a star with extensions in all <111>-directions (region a)

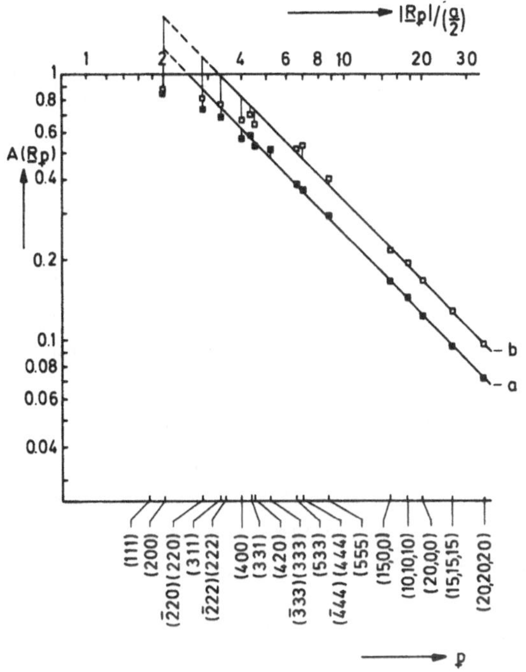

Fig. 6.7. Reaction probability $A(\underline{R}^p)$ for noncompact reaction regions in bcc lattice. Same plot as in Fig. 6.2 but with different scale. The regions a and b are explained in the text

comes a reasonable approximation again and the various approximations for the effective reaction radius are plotted in Fig. 6.8. One can see that the agreement is also within 10%. For production sites close to the reaction region the asymptotic form is less accurate; here the discrete nature of the lattice and the open structure of the reaction region have a decisive influence. This effect is even stronger for linear reaction regions containing sites of only one <111>-direction /6.7/.

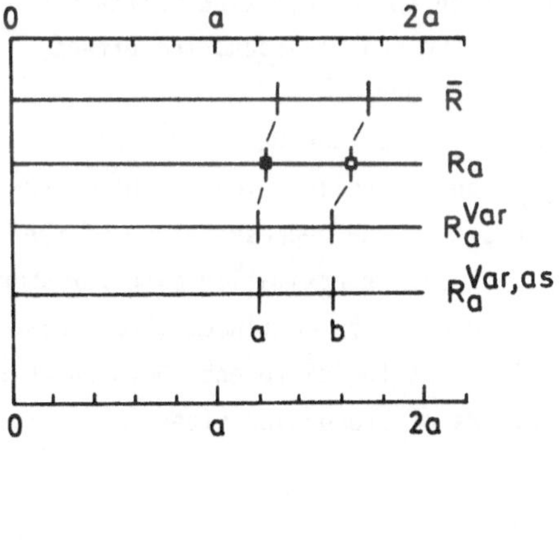

Fig. 6.8. Effective reaction radii R_a for noncompact regions in bcc (same symbols as in Fig. 6.7, same plot as in Fig. 6.3)

6.5 Discussion and Conclusion

We have shown in this section that it is possible to treat reactions considering the jump diffusion of a defect on a discrete lattice. Explicit results can be obtained if values for the Green's function are available which is the case for a number of lattices (see Sect. 2). Due to the fact that for nearest-neighbour jump models the asymptotic expansion (continuum Green's function) discussed in Sect. 2.3.1 is a surprisingly good approximation for the lattice Green's function the reaction probability can very well be approximated by the asymptotic form, (6.1). Thus the effective reaction radius R_a is the relevant parameter by which a microscopic reaction region is to be represented. Further, we have shown, that the calculation of R_a can be simplified by a variational procedure and a satisfactory approximation is obtained when the continuum Green's function is used in the final result. As a rule of thumb one can use $R_a \cong \bar{R}$ with $\bar{R}$ the average distance of the surface sites (sites of the reaction region which can be reached by nearest-neighbour jumps from outside) from the center of the reaction region. For compact regions this reduces approximately to $R_a \cong R_{max}$ with R_{max} the distance of the outermost surface shell.

We have not considered any disturbance of the jump rates of the defect due to interaction with the sink and thus have used the ideal Green's function discussed in Sect. 2. If the jumps are disturbed a different Green's function has to be used to describe the motion of the defect. For short-range interaction, i.e., local disturbance of the jumps one can calculate the disturbed Green's function by the usual scattering formalism.

A simple example has explicitly been treated in this way /6.8/: The barrier height for the jump into the reaction region (which consists of one site only) is assumed to be different from the barrier in the rest of the lattice. Since the disturbance is local the asymptotic form of the reaction probability remains a good approximation but the effective reaction radius becomes temperature dependent. In principle also longer-ranged disturbances can be treated but then the calculation of the disturbed Green's function soon becomes very cumbersome. We shall discuss the effect of long-range interaction in Sect. 7 using continuum theory.

From the results of this section the following conclusion can be drawn: One can use continuum theory of diffusion to describe reactions in solids even if the microscopic reaction region contains only a few lattice sites. The representation of the reaction region by an effective reaction radius is a good approximation also for short diffusion paths. Thus one can treat time-dependent problems by continuum theory also since the differential reaction rate can be calculated as the difference between reaction probabilities for slightly different distributions of production sites.

7. Influence of Long-Range Potentials on the Rate Constant

In this section we shall consider the change of the rate constant K due to a long-range interaction potential between mobile defect and sink. In Sect. 3 we have discussed the elastic interaction which is most important in metals. We shall use general potentials as far as possible and in particular consider power potentials of the form

$$E(\underline{R}) = -\frac{\alpha}{R^n} f(\Omega) \tag{7.1}$$

which vanish for $R \to \infty$. The angular function $f(\Omega)$ can vary arbitrarily, e.g., can be positive for some angles and negative for others, while we consider α a positive constant which describes the strength of the interaction.

As in the previous sections we shall consider stationary diffusion only. We shall restrict our treatment to cubic defects moving in cubic crystals. For simplicity we shall also assume that the defect has cubic symmetry in the saddle point. This is approximately the case for the <100>-split interstitial in fcc metals /7.1/. With this assumption we neglect the anisotropy of the diffusion constant in (4.25) due to the nonequivalence of different saddle points in strain fields. In Sect. 8 we shall briefly discuss applications where this anisotropy can be important.

With these simplifications the stationary diffusion equation reads ($\partial_{\underline{R}}$ = gradient operator)

$$\partial_{\underline{R}}[\tilde{D}(\underline{R})\, \partial_{\underline{R}} W(\underline{R})] + \Pi(\underline{R}) = 0 \tag{7.2}$$

with the renormalized density

$$W(\underline{R}) = \rho(\underline{R}) \exp[\beta E_e(\underline{R})] \tag{7.3}$$

and the position-dependent diffusion constant

$$\tilde{D}(\underline{R}) = D_o \exp[-\beta E_s(\underline{R})] \ , \tag{7.4}$$

D_o is the diffusion constant in the ideal lattice.

We have distinguished between the interaction $E_e(\underline{R})$ in the equilibrium configuration of the mobile defect and $E_s(\underline{R})$ in the saddle point configuration, which in general are different (see Sect. 4).

As in Sect. 5 we shall use the absorption boundary condition

$$\rho(R_a) = W(R_a) = 0 \tag{7.5}$$

at the distance R_a from the sink.

The subject of this section is to show that the interaction modifies the reaction rate in such a way that the "naked" reaction radius, R_a, entering the rate constant

$K = 4\pi D_o R_a$, (5.15), is replaced by an effective reaction radius R_{eff} which is determined by the interaction potential in the saddle point configuration. This means that for certain applications one can replace the actual sink with its associated interaction potential by an effective sink with radius R_{eff} and use free diffusion.

The change of the rate constant due to a drift field has been treated previously by several authors. For instance, DEBYE /7.2/ has considered the influence of the Coulomb interaction on reactions in fluids. HAM /7.3/ was the first one to introduce an effective reaction radius for reactions in solids including nonspherical potentials. He considered the capture of defects by straight dislocations and showed also that time-dependent problems can be described using the concept of an effective reaction radius.

Recently the incorporation of long-range potentials into the rate of diffusion--limited reactions has become relevant again for the description of swelling and creep of reactor materials under irradiation. The subjects have been reviewed recently by BRAILSFORD and BULLOUGH /7.4/, and HEALD and BULLOUGH /7.5/. The important point is that certain sinks (dislocations and dislocation loops) preferently absorb interstitials rather than vacancies which can be explained by the larger displacement field of interstitials resulting in a stronger interaction with the sinks. In previous papers /7.6,7/ we have shown that the interaction in the saddle point configuration of the mobile defect determines the rate constant and we have calculated the temperature dependence of the effective reaction radius. Only for special cases, e.g., spherically symmetric interaction potentials, exact solutions can be found, because then (7.2) reduces to an ordinary differential equation. However, two general remarks can be made:

i) $W(\underline{R})$ is determined by the diffusion equation (7.2) and by boundary conditions. Thus $W(\underline{R})$ and the rate constant only depend on the interaction in the saddle point configuration [which explicitly enters via $\tilde{D}(\underline{R})$, (7.4)] if the boundary conditions do not contain the interaction potential in the equilibrium configuration. This is, e.g., the case for the independent sink approximation described by (5.7). Here we have $\Pi(\underline{R}) = 0$, and the external boundary is at infinity where the interaction potentials vanish. Also, the absorption boundary condition at the sink, (7.5), is independent of the interaction potential.

ii) The interaction potential enters only in the combination $\beta E_s(\underline{R})$ into the diffusion equation (7.2). Thus we expect a decisive influence of the potential only if it is comparable to the thermal energy at the edge of the reaction sphere. As long as $\beta|E_s(\underline{R})| \ll 1$ for all $R \geqslant R_a$ we can use perturbation theory to account for the drift current. For $\beta|E_s(R_a)| \gg 1$, however, we expect that the rate constant is determined by the interaction potential alone. We shall be mostly concerned with this limit.

In the following we first present the formulation of variational principles which allow one to obtain approximate results for the rate constant, and then consider specific examples.

7.1 Variational Principles

Real interaction potentials are usually angular dependent and have a complicated structure (see Sect. 3). Thus we cannot hope to solve the diffusion equation (7.2) and determine the rate constant exactly. But we can use variational principles to obtain approximate values for the rate constant. Such variational principles have been applied earlier to other problems of diffusion and reactions /7.8,9/. We shall briefly discuss the appropriate functionals for our problem, and show how upper and lower bounds for the rate constant can be obtained.

With the simplifications mentioned above the functional L, (4.19), which yields the stationary diffusion equation (7.2), reads

$$L\{W(\underline{R})\} = \int_V d\underline{R} \left[\frac{1}{2} \tilde{D}(\underline{R}) \left(\partial_R W(\underline{R}) \right)^2 - \Pi(\underline{R})\, W(\underline{R}) \right] . \tag{7.6}$$

Because of the boundary condition (7.5) at the sink the integration extends over $R \geqslant R_a$ only.

The functional (7.6) has a minimum L_{ex} for the exact solution of the diffusion equation, the value L_{ex} depends on the physical situation, i.e., on the external boundary condition and source distribution.

We shall first assume that there are no sources inside the volume considered, i.e., $\Pi(\underline{R}) = 0$. We show that an upper bound for the rate constant is obtained when the density is fixed at the external boundary and a lower bound is obtained when the current is fixed there.

7.1.1 Upper Bound for the Rate Constant

With $\Pi(\underline{R}) = 0$ the functional L reads

$$L\{W(\underline{R})\} = \frac{1}{2} \int_V d\underline{R}\, \tilde{D}(\underline{R}) \left(\partial_R W(\underline{R}) \right)^2 . \tag{7.7}$$

To show how the minimum L_{ex} is related to the rate constant we use Gauss' theorem. We obtain from (7.7)

$$L\{W(\underline{R})\} = \frac{1}{2} \int_V d\underline{R}\, \partial_{\underline{R}} \left(\tilde{D}(\underline{R})\, \partial_{\underline{R}} W(\underline{R}) \right) W(\underline{R}) + \frac{1}{2} \int_{S_{R_a}+S_{R_B}} \left(d\underline{F} \,, \tilde{D}(\underline{R})\, \partial_{\underline{R}} W(\underline{R}) \right) W(\underline{R}) . \tag{7.8}$$

The volume integral vanishes for the exact solution of the stationary diffusion equation with $\Pi(\underline{R}) = 0$

$$\partial_R\left(\tilde{D}(\underline{R})\,\partial_R W(\underline{R})\right) = 0 \ . \tag{7.9}$$

The surface integral over the surface S_{R_a} at the sink also vanishes because of the boundary condition (7.5). Thus the minimum L_{ex} is given by the integral over the external surface S_{R_B}

$$L_{ex} = \frac{1}{2} \int\limits_{S_{R_B}} \left(d\underline{F}\,,\, \tilde{D}(\underline{R})\,\partial_{\underline{R}} W(\underline{R})\right) W(\underline{R}) \ . \tag{7.10}$$

We shall consider the independent particle approximation only, i.e., we let the radius R_B of the external surface go to infinity. Then the interaction between sink and defect can be neglected at the external surface and we obtain a spherically symmetric boundary condition for $W(\underline{R})$

$$W(R_B) = \rho(R_B) \qquad R_B \to \infty \ . \tag{7.11}$$

The minimum L_{ex} is related to the total flux of defects through the external surface

$$J(R_B) = \int\limits_{S_{R_B}} \left(d\underline{F}\,,\, \tilde{D}(\underline{R})\,\partial_{\underline{R}} W(\underline{R})\right) = \int\limits_{S_{R_B}} \left(d\underline{F}\,,\, D_o\,\partial_{\underline{R}}\,\rho(\underline{R})\right) \tag{7.12}$$

by

$$L_{ex} = \frac{1}{2} W(R_B)\; J(R_B) \qquad R_B \to \infty \ . \tag{7.13}$$

Since there are no sources inside the volume V, the total flux through any surface enclosing the sink is constant and equal to the flux J into the sink. In analogy to Sect. 5.3 we define the effective rate constant by

$$J = W(R_B)\cdot K_{eff} \ , \tag{7.14}$$

i.e., K_{eff} is the flux into the sink if the density is unity at the external surface. Thus we finally obtain

$$L_{ex} = \frac{1}{2} W^2(R_B)\cdot K_{eff} \ . \tag{7.15}$$

Since L_{ex} is a minimum we obtain an upper limit for the rate constant by

$$K_{eff} = \frac{2L_{ex}}{W^2(R_B)} < \frac{2}{W^2(R_B)} \cdot L\{W(\underline{R})\} \tag{7.16}$$

where $W(\underline{R})$ is an arbitrary trial function which obeys the boundary conditions (7.5,11) but does not necessarily solve the diffusion equation.

This variation principle is analogous to the one used by HAM /7.10/ to obtain the lowest eigenvalue which determines the long-time behaviour of the defect distribution in nonstationary situations. It has been shown by HAM that this lowest eigenvalue is closely related to the rate constant calculated from stationary theory.

7.1.2 Lower Bound for the Rate Constant

The functional L can be rewritten as a functional of the current density

$$\underline{j}(\underline{R}) = -\tilde{D}(\underline{R})\,\partial_{\underline{R}} W(\underline{R}) \tag{7.17}$$

namely

$$M\{\underline{j}(\underline{R})\} = \frac{1}{2}\int d\underline{R}\;\tilde{D}^{-1}(\underline{R})\;[\underline{j}(\underline{R})]^2 \,. \tag{7.18}$$

This has to be minimized subject to the auxiliary condition

$$\partial_{\underline{R}}\underline{j}(\underline{R}) = 0 \tag{7.19}$$

which together with (7.17) gives the diffusion equation (7.9). With a Lagrangian multiplier $B(\underline{R})$ we incorporate (7.19) into the functional and consider

$$M'\{\underline{j}(\underline{R})\} = \frac{1}{2}\int_V d\underline{R}\left[\tilde{D}^{-1}(\underline{R})[\underline{j}(\underline{R})]^2 - 2B(\underline{R})\,\partial_{\underline{R}}\,\underline{j}(\underline{R})\right] . \tag{7.20}$$

The exact solution $\underline{j}_o(\underline{R})$ will make the first variation of M' vanish. If we insert a trial function $\underline{j}(\underline{R}) = \underline{j}_o(\underline{R}) + \underline{\varepsilon}(\underline{R})$ the terms linear in $\underline{\varepsilon}(\underline{R})$ are

$$M_1' = \int_V d\underline{R}\left[\tilde{D}^{-1}(\underline{R})\;\underline{j}_o(\underline{R})\;\underline{\varepsilon}(\underline{R}) - B(\underline{R})\;\partial_{\underline{R}}\underline{\varepsilon}(\underline{R})\right] . \tag{7.21}$$

The second term is transformed by Gauss' theorem yielding

$$M_1' = \int_V d\underline{R}\left[\tilde{D}^{-1}(\underline{R})\;\underline{j}_o(\underline{R}) + \partial_{\underline{R}}B(\underline{R})\right]\underline{\varepsilon}(\underline{R}) - \int_{S_{R_a}+S_{R_B}}\left(d\underline{F}\,,\,\underline{\varepsilon}(\underline{R})\right)B(\underline{R}) \,. \tag{7.22}$$

To make this term vanish the following equations have to be satisfied

$$\begin{aligned} &\underline{j}_o(\underline{R}) = -\tilde{D}(\underline{R})\,\partial_{\underline{R}}\,B(\underline{R}) && \text{in } V \\ &B(\underline{R}) = 0 && \text{on } S_{R_a} \\ &\underline{\varepsilon}(\underline{R}) = 0 && \text{on } S_{R_B}\;. \end{aligned} \tag{7.23}$$

If we make the identification

$$B(\underline{R}) = W(\underline{R}) \tag{7.24}$$

(7.23a) becomes identical to (7.17), and (7.23b) to the boundary condition (7.5). Equation (7.23c) expresses the fact that we have to use a normalized flux at the external boundary to minimize M.

Using (7.23,24) in (7.18) we obtain the minimum value of the functional $M\{\underline{j}\}$ which after applying Gauss'theorem, reads

$$M_{ex} = \frac{1}{2} W(R_B)\, J(R_B) \qquad R_B \to \infty \, . \tag{7.25}$$

This is identical to (7.13) if we use identical normalization, i.e., if

$$\underline{j}(R_B) = -\tilde{D}(\underline{R})\, \partial_{\underline{R}} W(\underline{R})\Big|_{R_B} \, . \tag{7.26}$$

Because of the absence of sources in V we can again identify the total flux $J(R_B)$ through the external surface with the total flux into the sink. Using the definition, (7.14), for the rate constant we obtain

$$M_{ex} = \frac{1}{2} \frac{J^2(R_B)}{K_{eff}} \, . \tag{7.27}$$

Thus a lower limit for K_{eff} is given by

$$K_{eff} = \frac{J^2(R_B)}{2M_{ex}} \geqslant \frac{J^2(R_B)}{2} \, [M\{j\}]^{-1} \tag{7.28}$$

for any trial current density $\underline{j}(\underline{R})$ which satisfies (7.19) and yields a total flux $J(R_B)$ through the external boundary.

7.1.3 Discussion

The difference between the two variational principles is that L is minimized with a fixed density $W(R_B)$ at the external boundary which then yields a certain current density $\underline{j}$ whereas M is minimized for a given flux through the external surface which in turn yields a defect density $W(\underline{R})$. Because (7.19) requires that there are no sources inside V and thus the total flux through any surface enclosing the sink is constant, the effective rate constant is determined by the density and current density of the mobile defects far away from the sink. For spherically symmetric interaction potentials which vanish at infinity the relation between $W(\underline{R})$ and $\underline{j}(\underline{R})$, (7.17), becomes exactly equal to the potential free diffusion equation at large distance. Since for spherical symmetry the current density is totally fixed by the condition $\partial_R j_R = 0$ and the boundary condition at the external surface, the lower limit, (7.28), yields the exact result with an ansatz neglecting the potential as will be shown in Sect. 7.2.1. Of course, the potential enters into the calculation of the rate constant via $\tilde{D}(\underline{R})$.

7.1.4 Variational Principle with Production

If we use the functional L, (7.6), for nonzero production, as e.g., given by the boundary conditions (5.6,8), the minimum value for the exact solution of (7.2) is related to the density $W(\underline{R})$ at the production sites

$$L_{ex} = -\frac{1}{2} \int_V d\underline{R}\, \Pi(\underline{R})\, W(\underline{R}) \tag{7.29}$$

and not to the rate constant directly. However, using a constant production rate in a finite volume as given by the boundary condition (5.8), one can relate L_{ex} to the rate constant. For constant $\Pi(\underline{R}) = p$ we obtain

$$L_{ex} = -\frac{1}{2}\, p \int_V d\underline{R}\, W(\underline{R}) = -\frac{1}{2}\, p\, V\, \overline{W} \tag{7.30}$$

with the average density $\overline{W}$. As in Sect. 5.4 we define the rate constant by the ratio of the total flux , J , of defects into the sink to the average defect density $\overline{W}$:

$$K_{eff} = \frac{J}{\overline{W}} . \tag{7.31}$$

Since for the boundary condition (5.8) the flux J is equal to the total number of defects produced per unit time

$$J = pV \tag{7.32}$$

we obtain the relation

$$L_{ex} = -\frac{1}{2}\, (pV)^2 \frac{1}{K_{eff}} < 0 . \tag{7.33}$$

Because L_{ex} is the minimum of the functional L, (7.6), a lower limit is obtained for the rate constant from

$$K_{eff} = \frac{1}{2}\, (pV)^2\, [-L_{ex}]^{-1} \geq \frac{1}{2}\, (pV)^2\, [-L\{W\}]^{-1} \tag{7.34}$$

for any trial function $W(\underline{R})$ which satisfies the boundary conditions (5.3,8).

7.2 Effective Rate Constant for Spherical Sinks

As before, we shall confine our treatment to the independent sink approximation. To calculate the rate constant exactly we have to solve the stationary diffusion equation (7.9) with the boundary conditions (7.5,11) and then calculate the flux of defects into the sink

$$J = 4\pi D_o R_a^2 \int \frac{d\Omega}{4\pi} \exp[-\beta E_s(\underline{R})]\partial_R W(\underline{R})\Big|_{R_a} = K_{eff}\cdot W(R_B) . \tag{7.35}$$

For spherical potential this can be done exactly because then (7.9) reduces to an ordinary differential equation. Other potentials will be treated with a variational principle.

7.2.1 Temperature Dependence

For power potentials of the form (7.1), $E_s(\underline{R}) = -\alpha\left(f(\Omega)/R^n\right)$, we can deduce the temperature dependence of the rate constant in the limit of strong potentials without actually solving the diffusion equation (7.6).

We introduce a dimensionless variable $\underline{x}$ by

$$\underline{R} = \lambda_n \underline{x} \; ; \quad \lambda_n = (\alpha\beta)^{1/n} \; . \tag{7.36}$$

Then the effective rate constant, defined in (7.35), reads

$$K_{eff} = 4\pi D_o \lambda_n C_n(x_a) \tag{7.37}$$

with

$$C_n(x_a) = x_a^2 \int \frac{d\Omega}{4\pi} \exp[U_s(\underline{x})]\, \partial_x \left.\frac{W(\underline{x})}{W(R_B)}\right|_{x=x_a} \tag{7.38}$$

$$U_s(\underline{x}) = -\beta E_s(\underline{R}) = \frac{f(\Omega)}{x^n} \; ; \quad x_a = R_a/\lambda_n \; . \tag{7.39}$$

The integral $C_n(x_a)$ contains the temperature only implicity via x_a, because the equation which determines $W(\underline{x})$ does not contain the temperature:

$$\partial_{\underline{x}}\left(\exp[U_s(\underline{x})]\, \partial_{\underline{x}} W(\underline{x})\right) = 0 \; . \tag{7.40}$$

This equation can be obtained from the functional

$$\tilde{L}\{W(\underline{x})\} = \frac{2}{4\pi D_o \lambda_n} L\{W(\underline{R})\} = \int_V d\underline{x} \frac{1}{4\pi} \exp[U_s(\underline{x})]\, [\partial_{\underline{x}} W(\underline{x})]^2 \; . \tag{7.41}$$

Thus for strong potentials ($x_a = R_a/\lambda_n \ll 1$) the effective rate constant is given by

$$K_{eff} = 4\pi D_o \lambda_n C_n(0) \; . \tag{7.42}$$

where the detailed shape of the potential only enters into the factor $C_n(x_a = 0)$. By comparison with (5.15) we define an effective reaction radius

$$R_{eff} = \lambda_n C_n(0) = \left(\frac{\alpha}{kT}\right)^{1/n} C_n(0) \; . \tag{7.43}$$

For arbitrary strength of the potential, i.e., arbitrary values of x_a we have to calculate the full function $C_n(x_a)$, (7.38).

7.2.2 Spherically Symmetric Potentials

For spherically symmetric potentials, i.e.,

$$E_s(\underline{R}) = E_s(R) \; , \tag{7.44}$$

the diffusion equation (7.9) reduces to the ordinary differential equation

$$\frac{1}{R^2}\frac{d}{dR}\left[R^2 \exp[-\beta E_s(R)]\frac{d}{dR} W(R)\right] = 0 . \tag{7.45}$$

The solution satisfying the boundary conditions (7.5,11) is

$$\frac{W(R)}{W_\infty} = \int_{R_a}^{R} \frac{dy}{y^2} \exp[\beta E_s(y)] \left(\int_{R_a}^{\infty} \frac{dy}{y^2} \exp[\beta E_s(y)]\right)^{-1} . \tag{7.46}$$

The effective reaction radius is given by

$$R_{eff} = \left(\int_{R_a}^{\infty} \frac{dy}{y^2} \exp[\beta E_s(y)]\right)^{-1} . \tag{7.47}$$

For power potentials, $E_s(R) = \pm\, \alpha/R^n$; $U_s = \mp\, 1/x^n$ (upper sign for repulsive, lower sign for attractive potentials) we obtain

$$R_{eff} = \lambda_n C_n(x_a) \;; \quad C_n(x_a) = n \left(\int_0^{x_a^{-n}} dt\, t^{\frac{1}{n}-1} e^{\pm t} \right)^{-1} \quad \begin{matrix} + \text{ repulsive} \\ - \text{ attractive} . \end{matrix} \tag{7.48}$$

These functions are related to the incomplete Γ-function /7.11/

In Fig. 7.1 we have plotted $C_n(x_a)$ as a function of $x_a = R_a(kT/\alpha)^{1/n}$ for attractive and repulsive potentials with n = 1 (Coulomb) and n = 3. Small x_a correspond to

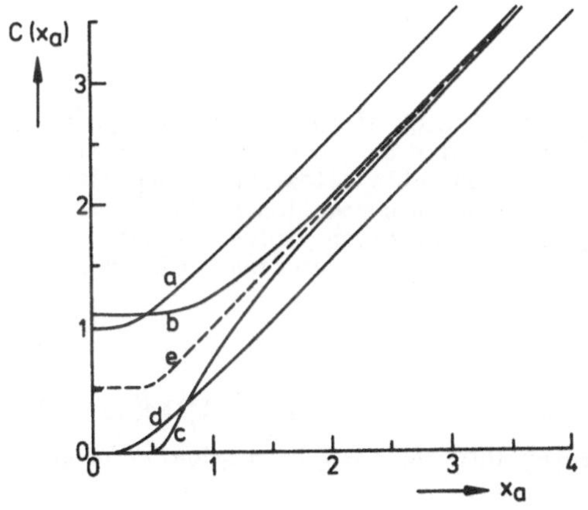

Fig. 7.1. $C_n(x_a) = R_{eff}/\lambda_n$ as a function of $x_a = R_a/\lambda_n$ with $\lambda_n = (\alpha/kT)^{1/n}$ for different interaction potentials $E_s(\underline{R})$ according to (7.48) (curves a - d) and (7.70) (curve e)

curve	a	b	c	d	e
potential	$-\alpha/R$	$-\alpha/R^3$	α/R^3	α/R	$-\alpha f(\Omega)/R^3$

strong potentials; then R_{eff} is essentially given by (7.43). Large x_a correspond to weak potential and the linear behaviour of $C_n(x_a)$ means that $R_{eff} \cong R_a$ in this case.

This is also seen from an expansion of $C_n(x_a)$ for small and large x_a /7.11/. We find for attractive potentials

$$R_{eff} = \begin{cases} R_a\left(1 + \frac{1}{n+1}\,x_a^{-n} + \dots\right) & x_a \gg 1 \\ \frac{\lambda_n}{\Gamma(1+\frac{1}{n})}\left(1 + \frac{x_a^{n-1}}{\Gamma(\frac{1}{n})}\exp(-x_a^{-n}) + \dots\right) & x_a \ll 1 \end{cases} \tag{7.49}$$

and for repulsive potentials

$$R_{eff} = \begin{cases} R_a\left(1 - \frac{1}{n+1}\,x_a^{-n} + \dots\right) & x_a \gg 1 \\ \lambda_n\, n\, x_a^{-n+1}\exp(-x_a^{-n}) + \dots & x_a \ll 1\,. \end{cases} \tag{7.50}$$

This is the expected result. For weak potentials ($x_a \gg 1$) the potential modifies the naked reaction radius very little whereas for strong potentials ($x_a \ll 1$) R_{eff} is entirely determined by the potential.

For strong attractive potentials R_{eff} is given by $\lambda_n/\Gamma(1+\frac{1}{n})$ which except for the factor $\Gamma(1+\frac{1}{n}) \cong 1$ is obtained from the condition

$$E_s(\infty) - E_s(R_{eff}) = \frac{\alpha}{R_{eff}^n} = kT\,. \tag{7.51}$$

This means that R_{eff} is essentially the distance from the sink where the potential energy gained in the saddle point configuration equals the thermal energy kT (see Fig. 7.2).

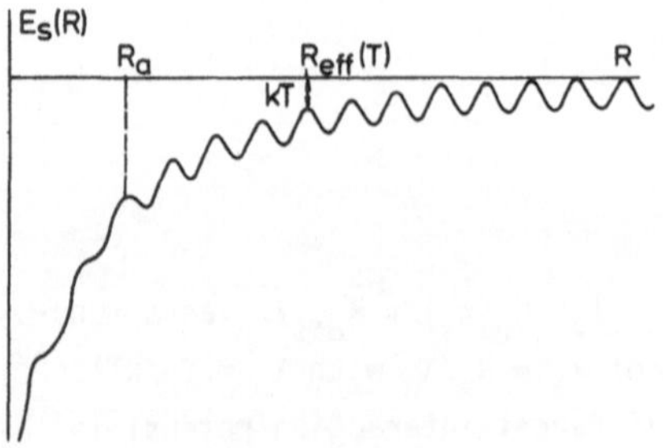

Fig. 7.2. Sketch of the interaction potential of a mobile particle with a sink at R = 0; the unstable sites are incorporated in the microscopic reaction radius R_a, the long-range interaction yields a temperature-dependent effective reaction radius $R_{eff}(T)$

For strong repulsive potentials R_{eff} goes to zero like $\exp[-\beta E_s(R_a)]$ because the barrier $E_s(R_a)$ which has to be overcome to enter into the reaction region is large against kT.

7.2.3 Test of Variational Principles

Although the rate constant can be calculated exactly in the case of spherical potentials it is interesting to test the success of the variational procedures with simple trial functions.

Using

$$W(R) = \begin{cases} 1 - \frac{R_o}{R} & \text{for } R_o \geq R_a \\ 1 - \frac{R_a}{R} & \text{for } R_a \geq R_o \end{cases} \qquad (7.52)$$

in the functional L, (7.7), with a free parameter R_o which is determined from $\partial L/\partial R_o = 0$ we obtain an upper limit for the reaction radius. In the case of an attractive Coulomb potential $E_s(R) = -\alpha/R$ this becomes

$$C_1(x_a) \leq \begin{cases} x_a\left(\exp(x_a^{-1}) - 1\right) ; & x_a = \frac{R_a}{\alpha\beta} \geq \frac{R_o^{min}}{\alpha\beta} = 0.628 \\ 1.544 ; & x_a \leq 0.628 \; . \end{cases} \qquad (7.53)$$

Using

$$j_R(R) = \frac{J_\infty}{R^2} \qquad (7.54)$$

in the functional M, (7.19), yields

$$C_1(x_a) \geq \left(1 - \exp(-x_a^{-1})\right)^{-1} \; . \qquad (7.55)$$

It turns out that the lower limit yields the exact result as can easily be seen from (7.48) which for n = 1 is an elementary integral. For strong potentials the upper limit is off by 50% whereas for weak potentials it shows the right behaviour.

One can show that the trial current density, (7.54), yields the exact result for all spherical potentials. The reason is that for spherical symmetry the condition $\partial_{\underline{R}}\underline{j} = 0$ and the boundary condition at the external boundary determine the current density uniquely. Unfortunately the lower limit does not yield good results for angular dependent power potentials with attractive and repulsive directions as we shall see in the next section.

7.2.4 Nonspherical Potentials

Realistic interaction potentials are angular dependent as we have seen in Sect. 3. The elastic interaction of point defects has attractive and repulsive directions with equal weight ($\int d\Omega\, f(\Omega) = 0$). We have shown that for strong power potentials the temperature dependence is determined by the R-dependence of the potential alone and that only the absolute value of R_{eff} depends on the detailed shape of the potential. From the results of the preceding section we would expect that in the strong potential limit the repulsive directions of the potential do not contribute at all to the current of defects into the sink whereas the attractive directions are fully operative. Thus we expect a reduction of the effective reaction radius by roughly a factor of 2 compared to a purely attractive potential.

To obtain more quantitative results we make use of the variational principles. If we insert a spherically symmetric trial function

$$W^{(1)}(\underline{x}) = W_\infty\, n(x) \tag{7.56}$$

into the functional $\tilde{L}\{W(\underline{x})\}$, (7.41), the only angular dependent function is $\exp[U_s(\underline{x})]$. Integrating over angles defines an effective spherical potential

$$g(x) = \int \frac{d\Omega}{4\pi} \exp[U_s(\underline{x})] = \langle \exp[U_s(\underline{x})] \rangle_\Omega \tag{7.57}$$

which enters the differential equation for $n(x)$

$$\frac{1}{x^2}\frac{d}{dx}\left(x^2\, g(x)\, \frac{d}{dx} n(x)\right) = 0\ . \tag{7.58}$$

The solution satisfying the boundary conditions (7.5,11) is

$$n(x) = \int_{x_a}^{x} \frac{dy}{y^2}\, g^{-1}(y) \left(\int_{x_a}^{\infty} \frac{dy}{y^2}\, g^{-1}(y) \right)^{-1} \tag{7.59}$$

and the upper limit for the reaction radius is given by

$$R_{eff} \leq R_{eff}^{(1)} = \lambda_n \left(\int_{x_a}^{\infty} \frac{dy}{y^2}\, g^{-1}(y) \right)^{-1} = \lambda_n\, C_n^{(1)}(x_a)\ . \tag{7.60}$$

To obtain numerical results one has to calculate the angular average, (7.57), which can be done approximately with the following arguments. For small x the attractive parts of the potential $[U_s(x) > 0]$ determine the angular average and $g(x) \cong \exp(x^{-n})$. These x values do not contribute much to the integral in (7.60) since $g^{-1}(x) \cong \exp(-x^{-n})$ is very small for small x. For large x however the potential is weak and one can expand the exponential in (7.57). For potentials with $\int d\Omega\, f(\Omega) = 0$ one obtains approximately

$$\begin{aligned} g(x) &= \left\langle 1 + U_s(\underline{x}) + \frac{1}{2} U_s^2(\underline{x}) + \dots \right\rangle_\Omega \cong 1 + \frac{1}{2}\left\langle U_s^2(\underline{x}) \right\rangle_\Omega \\ &\cong \exp\left(\frac{1}{2}\left\langle U_s^2(\underline{x}) \right\rangle_\Omega\right) = \exp(a/x^{2n})\ ; \quad a = \frac{1}{2}\left\langle f^2(\Omega) \right\rangle_\Omega\ . \end{aligned} \tag{7.61}$$

This shows that the effective potential determining $C_n^{(1)}(x_a)$ in this approximation is attractive with strength a and power 2n. For the case of the elastic dipole interaction, (3.15), we obtain a = 0.0105 and 2n = 6. We have plotted $C_n^{(1)}(x_a)$ for this case in Fig. 7.1 (curve e, dotted line). In the strong potential limit we obtain

$$C_3^{(1)}(x_a = 0) = 0.5367\ . \tag{7.62}$$

This result is only approximate; e.g., if we use $1 + \frac{1}{2}\langle U_s^2 \rangle_\Omega$ for g(x) instead of $\exp\left(\frac{1}{2}\langle U_s^2 \rangle\right)$, the result is $C_3^{(1)} = 0.4755$.

Unfortunately, an accurate lower limit for potentials with attractive and repulsive parts which diverge at the origin cannot be obtained with a spherically symmetric ansatz for the current density $\underline{j}(\underline{R})$. The only possible symmetrical ansatz compatible with $\partial_{\underline{R}}\underline{j} = 0$ is (7.54). Inserting this into $M\{\underline{j}\}$, (7.18), yields

$$M\{\underline{j}\} = 4\pi D_o J_\infty^2 \frac{1}{2} \int_{R_a}^{\infty} \frac{dR}{R^2} \int \frac{d\Omega}{4\pi} \exp[\beta E_s(\underline{R})] \ . \tag{7.63}$$

Here the angular average is mainly determined by the repulsive directions [$E_s(\underline{R}) > 0$] and small values of R contribute mostly to the integral. In fact for power potentials of the form (7.1) $M\{\underline{j}\}$ diverges in the strong potential limit and the lower bound for K_{eff} becomes zero. There seems to be no simple way of extracting a meaningful lower bound for strong potentials of this form. Evidently, one has to use angular dependent trial functions which take care of the fact that the current density goes to zero for small R in the repulsive regions. It is very hard to find such an ansatz compatible with $\partial_{\underline{R}}\underline{j}(\underline{R}) = 0$.

We rather use a different ansatz for $W(\underline{R})$ to see whether this yields a significant change of the upper limit for R_{eff}. Inserting

$$W^{(2)}(\underline{x}) = \exp[-\frac{1}{2} U_s(\underline{x})] \ z(x) \tag{7.64}$$

with a spherically symmetric function z(x) into the functional $\tilde{L}$, (7.41), and carrying out the angular integration yields the following differential equation for z(x)

$$\frac{1}{x} \frac{d^2}{dx^2} [x \cdot z(x)] + z(x) \ h(x) = 0 \tag{7.65}$$

which has the form of the stationary Schrödinger equation with the "effective potential" h(x) given by

$$h(x) = \left\langle \frac{1}{2} \Delta U_s(\underline{x}) + \frac{1}{4} [\partial_{\underline{x}} U_s(\underline{x})]^2 \right\rangle_\Omega \ . \tag{7.66}$$

For interaction potentials of the form of (7.1), i.e., $U_s(\underline{x}) = f(\Omega)/x^n$ we obtain

$$h(x) = \frac{a_n}{x^{2+n}} + \frac{b_n}{x^{2(n+1)}}$$

where a_n and b_n depend on the form of $f(\Omega)$. We shall consider only potentials where $f(\Omega)$ is a linear combination of spherical harmonics with $\int d\Omega \ f(\Omega) = 0$ as is, e.g., the case for the elastic dipole interaction of point defects. Then $a_n = 0$ and

$$h(x) = \frac{b_n}{x^{2(n+1)}} \ . \tag{7.67}$$

With this form of h(x) the solution of (7.65) satisfying the boundary conditions at R_a and at infinity is

$$z(x) = \Gamma(1 - \frac{1}{2n}) \left(\frac{t}{2}\right)^{1/2n} \left[I_{-1/2n}(t) - B(t_a) \ I_{1/2n}(t)\right] \tag{7.68}$$

where I_ν are the modified Bessel functions /7.11/,

$$t = \frac{b_n^{1/2}}{n\cdot x^n}; \qquad t_a = \frac{b_n^{1/2}}{n\cdot x_a^n} = \frac{b_n^{1/2}\,\beta_\alpha}{n\cdot R_a^n}; \qquad B(t_a) = I_{-1/2n}(t_a)\big/I_{1/2n}(t_a)\,. \tag{7.69}$$

Inserting this solution into the functional $\tilde{L}$, (7.41), (which can be transformed into a surface integral) the following upper limit for R_{eff} is obtained

$$\begin{aligned} R_{eff} &\leq \lambda_n\, C_n^{(2)}(x_a) \\ C_n^{(2)}(x_a) &= B(t_a)\, b_n^{1/2n}\,(2n)^{1-1/2n}\,\frac{\Gamma(1-\frac{1}{2n})}{\Gamma(\frac{1}{2n})}\,. \end{aligned} \tag{7.70}$$

For the elastic dipole interaction, (3.15), we obtain

$$b_3 = \frac{2^2\cdot 29}{3\cdot 5^2\cdot 7} = 0.221\,.$$

It turns out that for this interaction $C_3^{(2)}(x_a)$ and $C_3^{(1)}(x_a)$ coincide almost exactly for all values of x_a. For instance, in the strong potential limit

$$C_3^{(2)}(x_a = 0) = 0.5206 \tag{7.71}$$

which differs from $C_3^{(1)}(0)$, (7.62), by only 3%.

In view of the rather different angular dependence of the trial functions, (7.56, 64), this means that the upper limit obtained is close to the correct value of the effective rate constant. A numerical solution /7.12/ which takes into account non-spherical contributions to $z(\underline{x}) = W(\underline{x})\exp\left(\frac{1}{2}U_s(\underline{x})\right)$ agrees with our result for the strong potential limit, (7.71), up to the fourth digit.

[Due to a different definition of the potential strength and the angular function used in /7.12/, their b_n is larger by a factor 5^2 and their $C_3(x = 0)$ is thus larger by a factor of $5^{1/3} = 1.71$.]

7.3 Effective Rate Constant for Straight Dislocations

To include the effect of the elastic interaction in the calculation of the effective rate constant for the reaction of point defects with straight dislocations we can employ the methods discussed in the preceding sections. But as discussed earlier (in Sect. 5.1) we use an external boundary with a finite radius R_B. To obtain the independent sink approximation we can choose R_B so large that the interaction potential is negligible compared to kT at the external boundary. It turns out that the interaction can be incorporated into the reaction radius but since the rate constant depends on the logarithm of the reaction radius [see (5.21)] we cannot find the temperature dependence by simple scaling arguments as in the three-dimensional cases.

Of course, cylindrical symmetric potentials (in the plane perpendicular to the dislocation line, see Fig. 3.2) can be treated exactly. It turns out that also the interaction of a cubic defect with a straight edge dislocation, (3.26), can be solved exactly due to the special form of the potential /7.3,13/.

7.3.1 Cylindrically Symmetric Potentials

For cylindrically symmetric potentials $E_s(\underline{R}) = E_s(r)$ the stationary diffusion equation in two dimensions reads

$$\frac{1}{r}\,\partial_r\left(r\,\exp[-\beta E_s(r)]\,\partial_r W(r)\right) = 0 \tag{7.72}$$

which has the solution

$$W(r) = W(R_B)\int_{R_a}^{r}\frac{dy}{y}\exp[\beta E_s(y)]\left(\int_{R_a}^{R_B} dy\,\frac{1}{y}\exp[\beta E_s(y)]\right)^{-1} \tag{7.73}$$

satisfying the boundary conditions (7.5,12).

For the special case

$$E_s(r) = -\frac{B}{r} \tag{7.74}$$

which can be used as a rough approximation for the interaction of a noncubic defect with a screw dislocation /7.3/ where B is some angular average effective sink strength one obtains

$$W(r) = W(R_B)\,\frac{E_1\left(\frac{\beta B}{r}\right) - E_1\left(\frac{\beta B}{R_a}\right)}{E_1\left(\frac{\beta B}{R_B}\right) - E_1\left(\frac{\beta B}{R_a}\right)}\ . \tag{7.75}$$

Here

$$E_1(x) = \int_{x}^{\infty}\frac{dy}{y}\exp(-y) \tag{7.76}$$

is the exponential integral /7.11/.

The flux into the dislocation core (per unit length of dislocation line) is given by

$$J = 2\pi D_o R_a \left.\frac{dW(r)}{dr}\right|_{r=R_a} = K_{eff}\cdot W(R_B)\ . \tag{7.77}$$

Calculating K_{eff} with W(r) from (7.75) we obtain

$$K_{eff} = 2\pi D_o\,\exp\left(-\frac{\beta B}{R_a}\right)\left[E_1\left(\frac{\beta B}{R_B}\right) - E_1\left(\frac{\beta B}{R_a}\right)\right]^{-1} \tag{7.78}$$

which for the limit of strong interaction ($x_a = R_a/\beta B \ll 1$) and large cutoff radius

($x_B = R_B/\beta B \gg 1$) reduces to

$$K_{eff} = \frac{2\pi D_o}{\ln(R_B/\beta B e^\gamma)} \tag{7.78a}$$

with Euler's constant $\gamma = 0.5772$. Comparing this with the phenomenological results, (5.21), one can define an effective reaction radius

$$R_{eff} = \beta B e^\gamma \ . \tag{7.79}$$

As in the three-dimensional case R_{eff} is approximately the distance at which the interaction energy gained in the saddle point configuration equals kT.

7.3.2 Edge Dislocations

The interaction energy of a cubic defect with an edge dislocation in an isotropic material is given by (3.26), namely

$$E_s(\underline{R}) = -\frac{A \sin\theta}{r} \ . \tag{7.80}$$

With this potential the stationary diffusion equation (7.9) in two dimensions can be solved with the substitution

$$W(\underline{R}) = \exp\left(-\frac{A}{2}\frac{\sin\theta}{r}\right) z(r,\theta) \tag{7.81}$$

leading to the differential equation

$$\left[\frac{1}{r}\partial_r r \partial_r + \frac{1}{r^2}\partial_\theta^2 + \left(\frac{A\beta}{2r}\right)^2\right] z(r,\theta) = 0 \ . \tag{7.82}$$

Here the special properties of the potential, (7.80), become obvious: It is a harmonic function in two dimensions, i.e., $\Delta E_s = 0$, and the square of the gradient is independent of θ. Only due to these properties an exact solution for $z(r,\theta)$ can be found. The general form of the solution of (7.82) is

$$z(r,\theta) = \sum_n \exp(in\theta)\, u_n(r) \tag{7.83}$$

where the radial parts $u_n(r)$ are determined by Bessel's equation /7.11/

$$\left[t^2\partial_t^2 + t\partial_t - (t^2 + n^2)\right] u_n(t) = 0 \ ; \quad t = \frac{A\beta}{2r} \ . \tag{7.84}$$

For the cylindrically symmetric boundary conditions (7.5,11) only the term n = 0 remains and we obtain /7.13/

$$z(r,\theta) = W(R_B)\,\frac{K_o(t)\,I_o(t_a) - I_o(t)\,K_o(t_a)}{K_o(t_B)\,I_o(t_a) - I_o(t_B)\,K_o(t_a)} \tag{7.85}$$

with

$$t_a = \frac{\beta A}{2R_a}, \qquad t_B = \frac{\beta A}{2R_B}. \tag{7.85a}$$

The flux into the dislocation core (per unit length of the dislocation line) is given by

$$J = K_{eff} \cdot W(R_B) = 2\pi D_o R_a \int_0^{2\pi} \frac{d\theta}{2\pi} \exp(2t_a \sin\theta)\, \partial_r \left[\exp(-t \sin\theta)\, z(r,\theta)\right]_{R_a} . \tag{7.86}$$

The term coming from the differentiation of the exponential does not contribute to J, because $z(R_a,\theta) = 0$. With the definition of the modified Bessel function I_o and the Wronskian relation /7.11/

$$K_o'(t)\, I_o(t) - I_o'(t)\, K_o(t) = - t^{-1}$$

the remaining term can be written as

$$K_{eff} \doteq \frac{2\pi D_o}{K_o(t_B) - I_o(t_B)\, K_o(t_a)/I_o(t_a)} \tag{7.87}$$

which for the limit of strong potentials ($t_a = A\beta/2R_a \gg 1$) and large cutoff radius ($t_B = A\beta/2R_B \ll 1$) yields

$$K_{eff} \cong \frac{2\pi D_o}{\ln(4R_B/\beta A e^{\gamma})} . \tag{7.88}$$

Thus the effective reaction radius is given by

$$R_{eff} = \frac{1}{4}\, \beta A e^{\gamma} \tag{7.89}$$

showing a reduction by a factor of $\frac{1}{4}$ due to the angular dependence of the potential.

7.4 Discussion and Applications

We have shown in this section how a long-range interaction between sink and mobile defect can be incorporated into the calculation of the rate constant. An effective reaction radius can be defined which is determined by the interaction of the sink with the defect in the saddle point configuration.

The temperature dependence and dependence on the potential strength can be deduced rigorously for realistic angular dependent potentials of the form of (7.1) whereas the absolute value can be calculated only approximately using a variational approach. We have neglected the anisotropy of the diffusion tensor induced by the interaction (see Sect. 4) but this can also be incorporated into the variational approach /7.6/.

For potentials with attractive directions (as e.g., the elastic interaction) the effective reaction radius R_{eff} is approximately given by the distance at which the decrease of the potential energy in the saddle point configuration equals the thermal energy kT. This leads to a large increase of the effective rate constant especially at low temperature. For the extremely long-ranged interaction of point defects with straight dislocations ($\sim 1/r$) very large values for R_{eff} are obtained from the above criterion, e.g., for an interstitial with $\delta V \cong 1V_c$ the reaction radius with an edge dislocation in Cu at room temperature (300 K) is of the order of 100 lattice constants. For the dipole interaction of point defects reaction radii of the order of a few lattice constants are obtained, e.g., the effective recombination radius of a Frenkel pair in Cu is about 4 lattice constants at 50 K.

For a few cases we have demonstrated the application and usefulness of variational principles which yield upper and lower bounds for the rate constant. Work in this direction should be extended, in particular more significant lower bounds for the rate constant should be found in the case of potentials with attractive and repulsive directions.

One can apply the theory as presented to all reactions which lead to stable products, i.e., if the absorption boundary condition, (7.5), is applicable. However, one should carefully check from other evidence whether this is actually the case before using the theory to interpret experiments. This is illustrated by the following example: We have used the elastic dipole interaction of point defects, (3.15), whose strength depends on the product of the volume changes $\delta V^d \delta V^s$ of the defect d and sink s, respectively. The influence of the interaction on the rate constant does not depend on the sign of this product, i.e., whether both δV have the same sign or opposite signs. This just changes the crystallographic direction in which the potential is attractive or repulsive and this is not significant in our continuum approach. From this one would conclude that an interstitial ($\delta V^d > 0$) reacts in the same way with undersized ($\delta V^s < 0$) or oversized ($\delta V^s > 0$) impurities. However, the long-range elastic interaction does not allow any conclusion about the structure and binding energy of the complexes formed. This has to be determined by experiment or more microscopic calculations, e.g., by computer simulation. Such investigations /7.14/ have shown that undersized impurities in Al can form "mixed dumbbells" with self-interstitials. These dumbbells have approximately the structure of a <100>-dumbbell but consist of the impurity atom and a host lattice atom. They are quite stable against dissociation with a binding energy of about 1eV for $\delta V^s = -0.2V_c$. On the other hand, oversized impurities form rather loosely bound complexes with binding energies of about 0.2eV for $\delta V^s = +0.2V_c$. Whereas in the first case our theory should be applicable over a wide temperature range, in the latter case it only can be applied at low temperatures. Another complication can arise from the thermal population of less tightly bound configurations of a complex which we have assumed to decay into the most stable configuration. Indeed, experiments of WOLLENBERGER and his group /7.15,16/ on the temperature dependence of the trapping rate of interstitials at various impurities in Cu and Al are not satisfactorily explained by our theory and show a need for a more sophisticated, time-dependent treatment which takes into account the possibility of escape.

An example well suited for comparison with the presented theory is the recombination of vacancies and interstitials. Here the complications discussed above do not exist because the binding energy (which is the formation energy of a Frenkel pair) is extremely large, typically of the order of 5eV, and close Frenkel pairs are thermally unstable at the temperature of interstitial migration ($T > 40$ K in Cu). Here BECKER et al. /7.17/ found that the effective recombination radius is $\sim T^{-1/3}$ in the temperature range $50\,K < T < 200\,K$ which is in accord with the elastic interaction $\sim R^{-3}$.

The dependence of the rate constant on the strength of the potential yields $K_{eff} \sim (\delta V^s \delta V^d)^{1/3}$ for point defect sinks and $\sim \ln \delta V^d$ for dislocation sinks. This variation with volume change has been observed for the trapping of self-interstitials at various impurity atoms in Cu /7.18/. Since δV is larger for interstitials than for vacancies a preferential absorption of interstitials at dislocations is predicted /7.19,20/. There is some controversy as to the magnitude of this "bias" and to the correct application of rate constants in this context /7.4/. The questions to be answered are (i) Should a dislocation in a dense network be described as a loop, i.e., as a "large point defect", or as a straight dislocation? Clearly, the first description should be used if the average diffusion length is large against the loop diameter because then the point defect feels the field of the entire loop $\sim R^{-3}$. The second description must be used for short diffusion lengths because then the defects only feel the field of one part of the network which can be approximated by a straight dislocation. In view of the large effective reaction radii for straight dislocations the transition from one case to the other is not very clear cut. (ii) Is the reaction at the dislocation core diffusion limited or rate limited? We have assumed the first to be true, because we believe that any reaction following at the dislocation line is fast compared to the diffusion in the bulk. (iii) How does one account for the interaction in a dense dislocation network, i.e., for large sink densities. Because of the long-range of the interaction and the size of the effective reaction radius this question is relevant for realistic dislocation densities. However, no attempt has been made so far to answer it.

8. Summary and Outlook

We have discussed various aspects of the theory of diffusion and reactions of point defects in metals.

In Sect. 2 we have reviewed the properties of the Green's function for diffusion on a discrete lattice. We have briefly discussed experimental methods which have been applied to determine the geometry and kinetics of the microscopic jumps of defects.

In Sect. 3 the long-range elastic interaction of point defects with each other and with dislocations was discussed. We have shown that there are always attractive directions irrespective of the nature of the defects, i.e., whether the lattice expands or shrinks due to the defects. This indicates a general trend towards agglomeration of defects. However, the actual structure and stability of complexes cannot be inferred from this continuum point of view, but has to be determined by other methods, e.g., experimentally by diffuse scattering of X-rays /8.1/ or neutrons /8.2/, channeling studies /8.3/, electron microscopy /8.4/, and theoretically, e.g., by computer simulation /8.5/.

The influence of external and internal force fields on the diffusion of point defects was discussed in Sect. 4. It turns out that in general the diffusion tensor becomes locally anisotropic even for cubic defects in cubic materials because different jump directions become inequivalent. The changes of diffusion rates for stationary conditions are particularly sensitive to the interaction of the defect in the saddle point configuration. Generally, changes of diffusivities due to external fields are small, of the order of a few per cent /8.6/, but they can decisively determine the reaction rates of defects with sinks as is demonstrated in Sect. 7.

The usual description of defect reactions by rate constants using phenomenological reaction radii to describe the interaction and continuum diffusion to describe the motion of defects is discussed in Sect. 5 for various situations: (i) the reaction of isolated defect-sink pairs with strong spatial correlation, (ii) the reaction of homogeneously distributed defects with homogeneously distributed sinks for small sink concentrations, and (iii) the cell approximation as a model to take into account larger sink concentrations. The validity of this approach was checked in Sects. 6, 7 using a more detailed description for the motion of defects and for the interaction with sinks, respectively.

In Sect. 6 we have taken into account the discrete structure of the lattice and have described the motion of defects as a random walk. It turns out that the continuum approach is a very good approximation even if the reaction region contains just a few lattice sites and the diffusion lengths are comparable to the lattice constant.

One can define a reaction radius for each microscopic reaction region and explicit expressions are given in terms of the stationary Green's function for a random walk on a lattice. The reason for the applicability of the continuum approach is the surprisingly good agreement of the stationary Green's function for the lattice with its asymptotic expansion which describes continuum diffusion.

In Sect. 7 the influence of long-range interaction potentials between sinks and mobile defects on the rate constant has been discussed. Variational principles are formulated which yield upper and lower bounds for the rate constant. The long-range potential can be incorporated in an effective temperature-dependent reaction radius $R_{eff}(T)$. It turns out that R_{eff} is determined by the interaction of the defect in the saddle point configuration alone. For power potentials of the form of (7.1) the temperature dependence and the dependence of the potential strength can be deduced rigorously from scaling arguments for three-dimensional diffusion; it is determined by the radial dependence of the potential alone whereas the absolute value depends on the details of the angular dependence. For potentials which have attractive directions the effective reaction radius is approximately given by the distance at which the potential energy gained by the defect in the saddle point configuration equals the thermal energy kT (see Fig. 7.2).

We have adopted a certain model in our calculations, namely that all defects actually reach the most stable configuration and that dissociation of a complex is impossible once the reaction has taken place. The validity of this assumption has to be checked in each case by microscopic considerations, and in some instances the situation will turn out to be more complicated. For instance, SCHOBER /8.7/ has shown that there are a large number of metastable configurations of interstitial agglomerates with only moderately lower binding energy than the most stable ones. Due to the anisotropy of the defects there might be cases where the activation energy for the transformation into the most stable configuration is comparable to the dissociation energy of the metastable configuration. Then our assumption is not fulfilled and we have to consider the intermediate steps separately. BALLUFFI /8.8/ has discussed the process of incorporating an arriving defect into an existing agglomerate which involves several thermally activated steps. It is not clear at present whether a suitable potential can be found to describe the varying activation energies and whether the total process can be described by one effective rate constant. Most probably a nonmonotonous potential has to be used while we have considered monotonous potentials only.

The reactions of point defects with dislocations is an important mechanism for swelling and creep of metals under irradiation. The elastic interaction leads to a dependence of the rate constant on the volume change of the mobile defect proportional to $\ln \delta V$. This results in a preferential absorption of interstitials at dislocations whereas the vacancies form voids which lead to swelling. The creep phenomenon is explained by the fact that the reaction rate of point defects with dislocations depends on the orientation of the dislocations with respect to the external stress. An important contribution to the difference of the rate constants can result from the anisotropy of the diffusion tensor induced by stress fields /8.9/ which we have neglected in Sect. 7.

An important question not tackled so far is the incorporation of the long-ranged interaction into the determination of the sink density dependence of the rate constant. In view of the very large effective reaction radii obtained for dislocations this seems to be essential for the understanding of the clustering phenomena in heavily irradiated materials.

Acknowledgements

I would like to express my gratitude to the late Prof. G. Leibfried who initiated some of the work contained in this paper and furthered it by his continuous interest.

I am grateful to my colleagues, Drs. P.H. Dederichs, H.R. Schober and H. Trinkaus for their encouraging comments and their patience in discussing the many questions that came up during the preparation of the manuscript.

I would also like to thank Prof. U. Felderhof for critically reading the manuscript and for his suggestions.

References

1.1 Proc. of a Conf. on "Properties of Atomic Defects in Metals", Argonne, Ill.: Oct. 1976, J. Nucl. Mater. 69/70 (1978)
1.2 C.P. Flynn: *Point Defects and Diffusion* (Clarendon Press, Oxford 1972)
1.3 P.H. Dederichs, C. Lehmann, H.R. Schober, A. Scholz, R. Zeller: In Ref. 1.1
1.4 F.W. Young, Jr.: In Ref. 1.1
1.5 W.M. Franklin: In *Diffusion in Solids, Recent Developments*, ed. by A.S. Nowick, J.J. Burton (Academic Press, New York 1975) p.1
1.6 J.R. Manning: *Diffusion Kinetics for Atoms in Crystals* (van Nostrand, Princeton 1968)
1.7 N.L. Peterson: In Ref. 1.1
1.8 A.S. Nowick, B.S. Berry: *Anelastic Relaxation in Crystalline Solids* (Academic Press, New York 1972)
1.9 T. Springer: "Quasielastic Neutron Scattering for the Investigation of Diffusive Motions in Solids and Liquids" in Springer Tracts in Modern Physics, Vol. 64 (Springer, Berlin, Heidelberg, New York 1972)
1.10 D. Wolf: Phys. Rev. 10, 2710, 2724 (1974)
1.11 I.G. Mullen, R.C. Knauer: In *Mössbauer Effect Methology*, Vol. 5, ed. by I.J. Gruveman (Plenum Press, New York 1969) p.197
1.12 P.H. Dederichs, K. Schroeder: Phys. Rev. B 17, 2524 (1978)
1.13 H. Schmalzried: *Solid State Reactions* (Verlag Chemie, Weinheim / Bergstraße 1974)

1.14 W. Strieder, R. Aris: *"Variational Methods Applied to Problems of Diffusion and Reaction"*, in Springer Tracts in Natural Philosophy, Vol. 24 (Springer Berlin, Heidelberg, New York 1973)
1.15 M. Smoluchowski: Z. Phys. Chemie (Leipzig) 92, 129 (1917)
1.16 K. Schroeder, E. Eberlein: Z. Phys. B 22, 181 (1975)
1.17 K. Schroeder, K. Dettmann: Z. Phys. B 22, 343 (1975)
2.1 J.R. Rowe, J.J. Rush, L.A. de Graaf, G.A. Ferguson: Phys. Rev. Lett. 29, 1250 (1972)
2.2 R.A. Johnson: J. Phys. F 3, 295 (1973)
2.3 H.G. Haubold: in *Fundamental Aspects of Radiation Damage in Metals*, Vol. 1 (ERDA-CONF-75 1006, 1975) p.254
2.4 R.A. Johnson, E. Brown: Phys. Rev. 127, 446 (1962)
2.5 P.H. Dederichs, C. Lehmann, H.R. Schober, A. Scholz, R. Zeller: In Ref. 1.1
2.6 F.W. Young, Jr.: In Ref. 1.1
2.7 T. Geszti: Phys. Status Solidi 20, 165 (1967); 35, 441 (1969); see also Ref. 2.22
2.8 C.H. Bennett: In Ref. 1.5, p.74
2.9 J.R. Manning: *Diffusion Kinetics for Atoms in Crystals* (van Nostrand, Princeton 1968)
2.10 L.C. Feldmann, E.N. Kaufmann, J.M. Poate, W.M. Augustiniak: In *Ion Implantation in Semiconductors and Other Materials*, ed. by B.L. Crowder (Plenum Press, New York 1973) p.491
2.11 P. Wombacher: Jül-942-FF Report of the Nuclear Research Center Jülich, Germany, April 1973
2.12 H.D. Carstanjen: In *Ion Beam Surface Layer Analysis*, Vol. 2, ed. by O. Meyer, G. Linker, F. Käppler (Plenum Press, New York 1976)
2.13 V. Lottner, A. Heim, K.W. Kehr, T. Springer: Int. Symp. on Neutron Inelastic Scattering IAEA Vienna, Oct. 1977
2.14 H.R. Schober: J. Phys. F 7, 1127 (1977)
2.15 R.A. Johnson: Phys. Rev. 134, A 1329 (1964)
2.16 H.E. Schaefer, D. Butterweg, W. Dander: In Ref. 2.3, p.463
2.17 P. Ehrhart: In Ref. 2.3, p.302
2.18 A. Scholz, C. Lehmann: Phys. Rev. B 6, 813 (1972)
2.19 R. Zeller, P.H. Dederichs: Z. Phys. B 25, 139 (1976)
2.20 V. Spiric, L.E. Rehn, K.-H. Robrock, W. Schilling: Phys. Rev. B 15, 672 (1977)
2.21 C.P. Flynn: *Point Defects and Diffusion* (Clarendon Press, Oxford 1972)
2.22 W.M. Franklin: See Ref. 1.5
2.23 K. Kehr: JOL-1211, Report of the Nuclear Research Center, Jülich, Germany, June 1975
2.24 G.H. Vineyard: Phys. Chem. Sol. 3, 121 (1957)
2.25 see, e.g., C.P. Flynn, A.M. Stoneham: Phys. Rev. B 1, 3966 (1970)
2.26 W. Dietrich, I. Peschel: Z. Phys. B 27, 177 (1977)
2.27 M. Lax: Phys. Rev. 97, 629 (1955)
2.28 T.H. Berlin, M. Kac: Phys. Rev. 86, 821 (1952)
2.29 H.B. Callen: Phys. Rev. 130, 890 (1963)
2.30 N.W. Dalton, D.N. Wood: Proc. Phys. Soc. (London) 90, 459 (1967)
2.31 B.V. Thompson, D.A. Lavis: Proc. Phys. Soc. (London) 91, 645 (1967)
2.32 J. Callaway, R. Boyd: Phys. Rev. 134 A, 1655 (1964)
2.33 J. Callaway: J. Math. Phys. 5, 783 (1964)
2.34 G.F. Koster, J.C. Slater: Phys. Rev. 96, 1208 (1954)
2.35 E.W. Montroll, R.B. Potts: Phys. Rev. 100, 525 (1955)
2.36 G.L. Montet: Phys. Rev. B 7, 650 (1973)
2.37 G.N. Watson: Quart. J. Math. (Oxford) 10, 266 (1939)
2.38 G. Joyce: J. Math. Phys. 12, 1390 (1971)
2.39 M. Inonue: J. Math. Phys. 15, 704 (1974)
2.40 T. Morita, T. Horiguchi: J. Math. Phys. 12, 981 (1971)
2.41 G.S. Joyce: J. Phys. C 4, L53 (1971)
2.42 S. Katsura, S. Inawashiro: J. Math. Phys. 12, 1622 (1971)
2.43 T. Horiguchi: J. Math. Phys. 13, 1411 (1972)
2.44 G.L. Montet: J. Math. Phys. 14, 1022 (1973)
2.45 K. Schroeder: JOL-1083, Report of the Nuclear Research Center Jülich (Germany), July 1974, p.43

2.46 T. Morita, T. Horiguchi: J. Math. Phys. 13, 1243 (1972)
2.47 S. Katsura, T. Morita, S. Inawashiro, T. Horiguchi, Y. Abe: J. Math. Phys. 12, 892 (1971)
2.48 A. Maradudin, E. Montroll, G. Weiß, R. Herman, H. Milnes: Mem. Acad. Roy. Belgique, Classe de Science XIV, No. 1709 (1960)
2.49 H.R. Schober, M. Mostoller, P.H. Dederichs: Phys. Status Solidi (b) 64, 173 (1974)
2.50 W. Ludwig: "Recent Developments in Lattice Theory", in Springer Tracts in Modern Physics, Vol. 43 (Springer, Berlin, Heidelberg, New York 1967)
2.51 O. Bender, K. Schroeder: JÜL-1469, Report of the Nuclear Research Center, Jülich, Germany, December 1977; see also Phys. Rev. B 19, 3399 (1979)
2.52 P.H. Dederichs, R. Zeller: Phys. Rev. B 14, 2314 (1976)
2.53 N.L. Peterson: "Diffusion in Metals", in *Solid State Physics*, Vol. 22, ed. by H. Ehrenreich, F. Seitz, D. Turnbull (Academic Press, New York 1968) p.409
2.54 H. Mehrer: In Ref. 1.1
2.55 N.L. Peterson: "Isotope Effects in Diffusion", in Ref. 1.5, p.115
2.56 N.L. Peterson: Solid State Commun. (to be published)
2.57 A.S. Nowick, B.S. Berry: *Anelastic Relaxation in Crystalline Solids* (Academic Press, New York 1972)
2.58 J.L. Snoek: Physica 6, 591 (1939)
2.59 M. Weller: Dissertation, University of Stuttgart, Germany (1972)
2.60 J. Völkl: Ber. Bunsenges. Phys. Chem. 76, 797 (1972)
2.61 G. Schaumann, J. Völkl, G. Alefeld: Phys. Rev. Lett. 21, 891 (1968)
2.62 H. Kronmüller: In *Vacancies and Interstitials in Metals*, ed. by A. Seeger, W. Schilling, D. Schumacher, J. Diehl (North Holland, Amsterdam 1970) p.667
2.63 T.Springer: See Ref. 1.9
2.64 K. Sköld, G. Nelin: J. Phys. Chem. Sol. 28, 2369 (1967)
2.65 N. Stump, W. Gissler, R. Rubin: Phys. Status Solidi (b) 54, 295 (1972)
2.66 J.M. Rowe, K. Sköld, H.E. Flotow, J.J. Rush: J. Phys. Chem. Sol. 32, 41 (1971)
2.67 J.M. Rowe, J.J. Rush, H.G. Smith, M. Mostoller, H.E. Flotow: Phys. Rev. Lett. 33, 1297 (1974)
2.68 G. Alefeld, J. Völkl (eds.): *Hydrogen in Metals I*, Topics in Applied Physics, Vol. 28 (Springer, Berlin, Heidelberg, New York 1978)
2.69 M. Ait-Salem: JÜL-1322, Report of the Nuclear Research Center Jülich, Germany July 1976
2.70 K.S. Singwi, A. Sjölander: Phys. Rev. 120, 1093 (1960)
2.71 M.C. Dibar-Ure, P.A. Flinn: Phys. Rev. B 15, 1261 (1977); see also D. Wolf: Appl. Phys. Lett. 30, 617 (1977)
2.72 S.J. Lewis, P.A. Flinn: Appl. Phys. Lett. 15, 331 (1969)
2.73 R.C. Knauer, J.G. Mullen: Phys. Rev. 174, 711 (1968)
2.74 R.C. Knauer, J.G. Mullen: Appl. Phys. Lett. 13, 150 (1968)
2.75 Ch. Janot: J. de Phys. 37, 253 (1976)
2.76 D. Wolf: Phys. Rev. 10, 2710, 2724 (1974)
2.77 D. Wolf: Z. Naturforsch. A 26, 1816 (1971)
2.78 J. Völkl, G. Alefeld: "Hydrogen Diffusion in Metals", in Ref. 1.5, p.231
2.79 R. Messer, F. Noack: Appl. Phys. 6, 79 (1975)
3.1 V.K. Tewary: Adv. Phys. 22, 757 (1973)
3.2 H. Kanzaki: J. Phys. Chem. Sol. 2, 24 (1957)
3.3 P.H. Dederichs, G. Leibfried: Phys. Rev. 188, 1175 (1969)
3.4 P.T. Heald: In *"Vacancies '76"*, Proc. of a Conf. on Point Defect Behaviour and Diffusional Processes, Bristol, 13-16 Sept. 1976 (The Metals Society, London 1977) p.11
3.5 H.R. Schober, M. Mostoller, P.H. Dederichs: Phys. Status Solidi (b) 64, 173 (1974)
3.6 K.M. Miller, P.T. Heald: Phys. Status Solidi (b) 78, 341 (1976)
3.7 P.N. Kenney, A.J. Trott, P.T. Heald: J. Phys. F 3, 513 (1973)
3.8 J.D. Eshelby: The Continuum Theory of Lattice Defects in *Solid State Physics*, Vol. 3, ed. by F. Seitz, D. Turnbull (Academic Press New York 1956) p.79
3.9 H. Trinkaus: Proc. Conf. on "Fundamental Aspects of Radiation Damage in Metals", Gatlinburg, Tennessee, Oct. 6-10, 1975 USERDA-CONF 75001-P1, p. 254

3.10 P.H. Dederichs, J. Pollmann: Z. Phys. 255, 315 (1972); JÜL-836-FF, Report of the Nuclear Research Center Jülich, Germany, 1972
3.11 R.A. Masumura, G. Sines: J. Appl. Phys. 41, 3930 (1970)
3.12 R. Siems: Phys. Status Solidi 30, 645 (1968); JÜL-545-FN, Report of the Nuclear Research Center Jülich, Germany 1968
3.13 J.R. Hardy, R. Bullough: Philos. Mag. 15, 237 (1967)
3.14 N. Breuer, P.H. Dederichs, C. Lehmann, G. Leibfried: In Ref. 3.9, p.227
3.15 L.E. Rehn, K.H. Robrock, V. Spiric: In Ref. 3.9, p.234
3.16 P.T. Heald, R. Bullough: In Ref. 3.4, p.134
3.17 P.B. Hirsch: In Ref. 3.4, p.95
3.18 H. Trinkaus: Phys. Status Solidi (b), 54, 902 (1972)
3.19 M. Peach, J.S. Keohler: Phys. Rev. 80, 436 (1950)
3.20 G. Leibfried: Z. Phys. 135, 23 (1953)
3.21 P.H. Dederichs: Private communication
3.22 J.P. Hirth, J. Lothe: *Theory of Dislocations* (McGraw Hill, New York 1968)
3.23 J.D. Eshelby, W.T. Read, W. Shockley: Acta Met. 1, 251 (1953)
3.24 R. Bullough, R.C. Newman: Rep. Prog. Phys. 33, 101 (1970)
3.25 see Ref. 2.2
3.26 see Ref. 2.14
3.27 see Ref. 1.3
3.28 K. Schroeder: In Ref. 3.4, p.51
3.29 see Ref. 2.18
3.30 G. Duesing, W. Sassin, W. Schilling, H. Hemmerich: Cryst. Lattice Defects 1, 55 (1969)
3.31 W. Schilling: In Ref. 1.1
3.32 H. Wollenberger: In Ref. 2.3, p.582
3.33 B.L. Eyre, M.H. Loretto, R.E. Smallman: In Ref. 3.4, p.63
3.34 N. March: In Ref. 1.1
3.35 R. Jank: Dissertation Universität Innsbruck (1975); M. Doyama, K. Kuribayashi, S. Nanao, S. Tanigawa: Appl. Phys. 4, 153 (1974)
3.36 B. Lengeler, S. Mantl, W. Triftshäuser: J. Phys. F, Metal Physics 8, 1691 (1978)
4.1 P.H. Dederichs, K. Schroeder: Phys. Rev. B 17, 2524 (1978)
4.2 see Ref. 1.2, p.411
4.3 J.S. Koehler: Phys. Rev. 181, 1015 (1969)
4.4 H. Kronmüller, W. Frank, W. Hornung: Phys. Status Solidi (b) 46, 165 (1971)
4.5 E.J. Savino: Philos. Mag. 36, 323 (1977)
4.6 K.W. Ingle and A.G. Crocker: In Ref. 1.1
4.7 see Ref. 2.45, Appendix A
4.8 see Ref. 2.45, p.33
4.9 see Refs. 2.2 and 2.14
4.10 K. Forsch, J. Hemmerich, H. Knöll, G. Lucki: Phys. Status Solidi (a) 23, 223 (1974)
4.11 V. Spiric, L.E. Rehn, K.H. Robrock, W. Schilling: Phys. Rev. B 15, 672 (1977)
4.12 H. Johnson: Scripta Met. 4, 771 (1970); Ref. 1.2
4.13 H. Trinkaus: Phys. Status Solidi (b) 51, 307 (1972)
4.14 P. Ehrhart, H.G. Haubold, W. Schilling: In *Festkörperprobleme* (Advances in Solid State Physics) Vol. 14, ed. by H.J. Queisser, Vieweg Verlag, Braunschweig 1974) p.87
4.15 see Ref. 2.57
5.1 T.R. Waite: Phys. Rev. 107, 463 (1957)
5.2 F.S. Ham: J. Phys. Chem. Sol. 6, 335 (1958); see also Ref. 1.2, p.477ff
5.3 see Refs. 1.16 and 2.45
5.4 see Refs. 1.17 and 2.45
5.5 H.S. Carslaw, J.C. Jaeger: *Conduction of Heat* (Clarendon Press, Oxford 1959)
5.6 see Ref. 1.2, p.482
5.7 J.W. Corbett, R.B. Smith, R.M. Walker: Phys. Rev. 114, 1452, 1460 (1959); see also K. Schroeder: Rad. Effects 17, 103 (1973)
5.8 see, e.g., Ref. 1.13
5.9 W. Strieder, R. Aris: "Variational Methods Applied to Problems of Diffusion and Reactions", in Springer Tracts in Natural Philosophy, Vol. 24 (Springer, Berlin, Heidelberg, New York 1973)

5.10 B.U. Felderhof, J.M. Deutch: J. Chem. Phys. 64, 4551 (1976)
5.11 A.D. Brailsford: J. Nucl. Mater 60, 257 (1976)
5.12 D. Peak, K. Pearlman, P.J. Wantuck: J. Chem. Phys. 65, 5538 (1976)
5.13 M. Smoluchowski: Z. Phys. Chem. (Leipzig) 92, 129 (1917)
5.14 I.G. Margvelashvili, Z.K. Seralidze: Sov. Phys. Solid State 15, 1774 (1974); see also P.T. Heald: Philos. Mag. 31, 551 (1975)
5.15 A.D. Brailsford, R. Bullough, M.R. Hayns: J. Nucl. Mater. 60, 246 (1976)
6.1 see, e.g., W. Schilling, G. Burger, K. Isebeck, H. Wenzl: In *Vacancies and Interstitials in Metals*, ed. by A. Seeger, W. Schilling, D. Schumacher, J. Diehl (Noth-Holland, Amsterdam 1969) p.255 ff
6.2 R.C. Fletscher, W.L. Brown: Phys. Rev. 92, 585 (1953)
6.3 H. Mehrer: Rep. JÜL-CONF-2 (Vol. I, II), p.643; Papers presented at the International Conference on Vacancies and Interstitials in Metals, Jülich, 23-28. Sept. 1968
6.4 D. Peak, P. Brosius, Y.H, Lee, M.St. Peters, H.L. Frisch, J.W. Corbett: Rad. Effects 15, 61 (1972)
6.5 K. Schroeder, E. Eberlein: Z. Phys. B 22, 181 (1975)
6.6 Ref. 2.45, p.28 ff
6.7 E. Eberlein: Diplomarbeit Aachen (1974)
6.8 Ref. 2.45, p.66 ff
7.1 H.R. Schober: J. Phys. F 7, 1127 (1977)
7.2 P. Debye: Trans. Electrochem. Soc. 82, 265 (1942)
7.3 F.S. Ham: J. Appl. Phys. 30, 915 (1959)
7.4 A.D. Brailsford and R. Bullough: In Ref. 3.4, p.108
7.5 P.T. Heald, R. Bullough: In Ref. 3.4, p.134
7.6 K. Schroeder, K. Dettmann: Z. Phys. B 22, 343 (1975)
7.7 P.H. Dederichs, K. Schroeder: Phys. Rev. B 17, 2524 (1978)
7.8 W. Strieder, R. Aris: See Ref. 5.9
7.9 A.M. Arthurs: *Complementary Variational Principles* (Claredon Press, Oxford 1970)
7.10 see Ref. 5.2
7.11 M. Abramowitz, I.A. Stegun (eds.): *Handbook of Mathematical Functions* (Ntl. Bureau of Standards, New York 1968)
7.12 M. Profant, H. Walther: JÜL-1027-MA, Report of the Nuclear Research Center Jülich, Germany 1973; M. Profant, H. Wollenberger: Phys. Status Solidi (b) 71, 515 (1975)
7.13 I.G. Margvelashvili, S.K. Seralidze: Sov. Phys. Solid State 15, 1774 (1974)
7.14 see Ref. 1.3
7.15 F. Dworschak, R. Lennartz, H. Wollenberger: J. Phys. F 5, 400 (1975)
7.16 F. Dworschak, Th. Monsau, H. Wollenberger: J. Phys. F 6, 2207 (1976)
7.17 D.E. Becker, F. Dworschak, H. Wollenberger: Phys. Status Solidi (b) 54, 455 (1972)
7.18 A. Kraut, F. Dworschak, H. Wollenberger: Phys. Status Solidi (b) 44, 805 (1971)
7.19 P.T. Heald: Philos. Mag. 31, 551 (1975)
7.20 W.G. Wolfer, M. Ashkin: J. Appl. Phys. 47, 791 (1976)
8.1 P. Ehrhart, H.G. Haubold, W. Schilling: See Ref. 4.14
8.2 W. Schmatz: In *Treatise on Mat. Science and Technology*, Vol. 2 (Academic Press, New York 1973) p. 105
8.3 see, e.g., Ref. 2.12
8.4 B. Eyre, M.H. Loretto, R.E. Smallmann: In Ref. 3.4, p.63
8.5 see Refs. 2.2 and 2.14
8.6 see Ref. 1.12
8.7 see Ref. 2.14
8.8 R.W. Balluffi: In Ref. 2.3, Vol. 2, 852
8.9 E.J. Savino: Philos. Mag. 36, 323 (1977)

G. Leibfried, N. Breuer

Point Defects in Metals I

Introduction to the Theory

1978. 138 figures, 22 tables. XIV, 342 pages
(Springer Tracts in Modern Physics, Volume 81)
ISBN 3-540-08375-8

Contents: Introduction and survey. – Harmonic approximation and linear response (Green's function) of an arbitrary system. – Lattice theory. – Continuum theory. Transition from lattice to continuum theory. – Statics and dynamics of simple point defects. – Scattering of neutrons and X-rays by crystals. – Probability, distributions and statistics. – Properties of crystals with defects in small concentration. – Appendix.

This is an introduction to the basic theoretical concepts and methods for the description of point defects in metals. Particular emphasis is placed on the influence of point defects (in small concentration) on the mechanical properties of metals (elastic data, phonon dispersion, static displacements due to defects, elastic interaction between defects, defect vibrations).

The dynamics of the lattice and the continuum are reviewed in comprehensive chapters on Harmonic Approximation and Linear Response, Lattice Theory, Continuum Theory and in one on the Transition from Lattice to Continuum. Phenomenological force-constant models including indirect (electronic) interactions as well as the basic concept of Green's functions are discussed in great detail. For elasticity theory, a new notation has been chosen, which is particularly convenient for cubic symmetry. The statics and dynamics of single point defects are explained, employing the simplest models available. The theory of the interaction with external strain fields and of X-ray and neutron scattering are presented as the background for the experimental investigation of point defects in metals. A short chapter on statistics highlights several statistical problems of defect physics which are important for the interpretation of experimental results and needed in the final chapter on the Properties of Crystals with Defects in Small Concentration.

Particular emphasis is put on simplicity; therefore, many figures and many simple examples have been included, and the most important mathematical techniques are explained in 16 appendices.

Springer-Verlag
Berlin
Heidelberg
New York